Innovative Approaches to the On-Site Assessment and Remediation of Contaminated Sites

NATO Science Series

A Series presenting the results of scientific meetings supported under the NATO Science Programme.

The Series is published by IOS Press, Amsterdam, and Kluwer Academic Publishers in conjunction with the NATO Scientific Affairs Division

Sub-Series

I. Life and Behavioural Sciences	IOS Press
II. Mathematics, Physics and Chemistry	Kluwer Academic Publishers
III. Computer and Systems Science	IOS Press
IV. Earth and Environmental Sciences	Kluwer Academic Publishers
V. Science and Technology Policy	IOS Press

The NATO Science Series continues the series of books published formerly as the NATO ASI Series.

The NATO Science Programme offers support for collaboration in civil science between scientists of countries of the Euro-Atlantic Partnership Council. The types of scientific meeting generally supported are "Advanced Study Institutes" and "Advanced Research Workshops", although other types of meeting are supported from time to time. The NATO Science Series collects together the results of these meetings. The meetings are co-organized bij scientists from NATO countries and scientists from NATO's Partner countries – countries of the CIS and Central and Eastern Europe.

Advanced Study Institutes are high-level tutorial courses offering in-depth study of latest advances in a field.
Advanced Research Workshops are expert meetings aimed at critical assessment of a field, and identification of directions for future action.

As a consequence of the restructuring of the NATO Science Programme in 1999, the NATO Science Series has been re-organised and there are currently five sub-series as noted above. Please consult the following web sites for information on previous volumes published in the Series, as well as details of earlier sub-series.

http://www.nato.int/science
http://www.wkap.nl
http://www.iospress.nl
http://www.wtv-books.de/nato-pco.htm

Series IV: Earth and Environmental Sciences – Vol. 15

Innovative Approaches to the On-Site Assessment and Remediation of Contaminated Sites

edited by

Danny Reible
Hazardous Substance Research Center S/SW,
Lousiana State University, Baton Rouge, LA, U.S.A.

and

Katerina Demnerova
Institute of Chemical Technology,
Department of Biochemistry and Microbiology,
Prague, Czech Republic

Kluwer Academic Publishers

Dordrecht / Boston / London

Published in cooperation with NATO Scientific Affairs Division

Proceedings of the NATO Advanced Study Institute on
Innovative Approaches to the On-Site Assessment and Remediation of Contaminated Sites
Prague, Czech Republic
24 May–2 June 2001

A C.I.P. Catalogue record for this book is available from the Library of Congress.

ISBN 1-4020-0956-9 (HB)
ISBN 1-4020-0957-7 (PB)

Published by Kluwer Academic Publishers,
P.O. Box 17, 3300 AA Dordrecht, The Netherlands.

Sold and distributed in North, Central and South America
by Kluwer Academic Publishers,
101 Philip Drive, Norwell, MA 02061, U.S.A.

In all other countries, sold and distributed
by Kluwer Academic Publishers,
P.O. Box 322, 3300 AH Dordrecht, The Netherlands.

Printed on acid-free paper

Printed in the Netherlands.

TABLE OF CONTENTS

Acknowledgement

The Editors and Participants would like to express their appreciation to the North Atlantic Treaty Organization, the Assistant Secretary General for Scientific and Environmental Affairs, and the Advisory Panel on Environmental and Earth Science & Technology for support of the Advanced Study Institute under Grant EST.ASI.976725.

The facilities and support of the host institution, the Institute of Chemical Technology in Prague, is also gratefully acknowledged.

Finally, the Advanced Study Institute would not have been possible without the support and assistance of Nedra Korevec of Louisiana State University and Pavel Jenc and Irena Krumlova of the Institute of Chemical Technology.

NATO ADVANCED STUDY INSTITUTE – PRAGUE, CZECH REPUBLIC 2001

CO-DIRECTORS

Katerina Demnerova
Professor
Dep. of Biochemistry and Microbiology
Institute of Chemical Technology
Technická 3
166 28 Prague 6
CZECH REPUBLIC
Phone: 420 2 2435 5172
Fax: 420 2 245 3075
Email: demnerok@vscht.cz

Danny Reible
Chevron Professor of Chem. Egr. Director, Hazardous Substance Research Center S/SW
3221 CEBA Bldg.
Louisiana State University
Baton Rouge, LA 70803
UNITED STATES
Phone: 225-578-6770
Fax: 225-578-5043
Email: reible@che.lsu.edu

PARTICIPANTS

Peter Adriaens
Associate Professor
Associate Director, IESET
Environmental & Water Resources Engineering
University of Michigan
180 EWRE Bldg. 1351 Beal St.
Ann Arbor, MI 48109-2125
UNITED STATES
Phone: 734-763-1464
Fax: 734-763-2275
Email: Adriaens@engin.umich.edu

Tanil Akyuz
Cekmece Nuclear Research And Training Center
Cnaem, P.K. 1, Havalimani,
34831 Istanbul
TURKEY
Phone: + 90 212 5484050 /ext. 2815
Fax: + 90 212 5482230
Email: sakyuz@istanbul.edu.tr

Doina Alexandra Botzan
National Research and Development
Institute of Environment
Splaiul Independentei nr. 294, C.P.78-316
Bucharest 77703
ROMANIA
Phone: +40 (1) 2215770
Email: doina.botzan@usa.net

Mehmet Ali Yukselen
Marmara University
Environmental Engineering Department
81040 Göztepe - Istanbul
TURKEY
Phone: +90 216 348 0292 Ext. 269
Fax: +90 216 348 0293
Email: yukselen@eng.marmara.edu.tr

Pedro Alvarez
5317 Seamans Center
The University of Iowa
Dept. of Civil and Env. Engr.
Iowa City, IA 52242

UNITED STATES
Phone: 319-335-5065
Fax: 319-335-5660
Email: pedro-alvarez@uiowa.edu

Sukru Aslan
Envrionmental Engineering Dept
at Dokuz Eylul University,
Kaynaklar Kampusu
Buca, Izmir,35160
TURKEY
Phone: 0 232 453 11 43/1089-1188
Fax: 0 232 453 11 53
Email: sukru.aslan@deu.edu tr

Constantin-Horia Barbu
University of Sibiu
7-9 Ioan Ratiu St.
2400 SIBIU,
ROMANIA
Phone: 0040-69-211338
Fax: 0040-69-216490
Email: horia@verena.ro

Mutlu Berik
Electrical Power Resources Survey And Development Administration
Bahadyrlar Sok. No:19/19
Kurtulu, 06600 Ankara
TURKEY
Phone: +90 312 2955292
Fax: +90 312 2955005
Email: mberik@eie.gov.tr

Lorenzo Bertin
DICASM, Engineering,
University of Bologna
via Marche, 2
40139 Bologna
ITALY
Phone: (39)051-209-3217
Fax: (39)051-209-3218
Email: bel2565@iperbole.bologna.it

Claudia Bock
Manager of organic analysis
Zentrales Analytisches Labor

LG 5, Keller, Theodor-Neubauer Str. 6
03044 Cottbus
GERMANY
Phone: 0355/69-4182
Fax:
Email: bock@zal.tu-cottbus.de

Balazs Bodo
Oerloe u.30. 1.em/4
TERRA ENVIRONMENTAL TECHNOLOGIES
Terra Kornyezettechnika es Terinformatika
1031 Budapest,
HUNGARY
Phone: +36 (-1) -250-6703
Fax: +36 (-1) -242-1544
Email: info@terra-technologies.com

Corneliu Bogatu
Institute for Industrial Ecology
(ECOIND)
P.O.Box 254 Of.1 P-ta Victoriei
no.2 et.2 1900 Timisoara
ROMANIA
Phone: +40 (56) 220369
Fax: +40 (56) 220369
Email: gpetca@rectorat.utt.ro

Alexander Boronin
Institute for Biochemistry & Physiology of Micro-organisms of Russian
Academy of Sciences
Pushchino, 142290
RUSSIA
Phone: (7 095) 956 33 70
Fax: (7 095) 956 33 70
Email: vera@ibpm.serpukhov.su

Radek Borovka
Dep. of Biochemistry and Microbiology
Institute of Chemical Technology
Technická 3
166 28 Prague 6
CZECH REPUBLIC
Email: rborovka@cygnus.cz

Rebecca C. Daprato
Rice University

6100 Main St. MS-317
Houston, TX 77005
UNITED STATES
Phone: +1 (713) 348-8093
Fax: +1 (713) 348-5203
Email: daprato@rice.edu

Monica Carnevali
Department of Biology University of Bologna
42, Irnerio
40126 Bologna
ITALY
Phone: +390512091286
Fax: +39051242576
Email: monica@alma.unibo.it

Lucia Cavalca
DISTAM-MAAE,
Università degli Studi di Milano
Via Celoria 2
20133 Milano
ITALY
Phone: +39 02 23955825
Fax: +39 02 70630829
Email: lucia.cavalca@unimi.it

Lida Chroma
Dep. of Biochemistry and Microbiology
Institute of Chemical Technology
Technická 3
166 28 Prague 6
CZECH REPUBLIC
Phone: +420 24353021
Fax: +420 2435 3075
Email: ludmila.chroma@vscht.cz

Natalia Chubar
Department for Sorption and Fine Inorganic Synthesis,
Institute for Sorption and Endoecology Problems,
National Academy of Science of Ukraine,
General Naumov str.13, 03680,
Kyiv-164, 03680,
UKRAINE
Phone: +380 (+38) 044 - 452 79 07; (452 93 28)
Fax: +380 (+38) 044 452 93 27
Email: chubar@ispe.kiev.ua

Francisco Cota Rodrigues
University of the Azores Islands
Departement of Agricultural Sciences
9700 Angra do Heroísmo
Azores
PORTUGAL
Phone: 351 295 204560
Fax:
Email: cota@angra.uac.pt

Michelle Crimi
Colorado School of Mines,
Division of Environmental Science and Engineering
1500 Illinois Street
Golden, CO 80401
UNITED STATES
Phone: 303-384-2219
FAX: 303-279-0129
Email: mcrimi@mines.edu

Nedra Davis Korevec
LWRRI/ Hazardous Substance Research Center S/SW
Louisiana State University
3221 CEBA Bldg.
Baton Rouge, LA 70803
UNITED STATES
Phone: (225) 578-6027
Fax: (225) 578-5043
Email: nkorevec@lwrri.lsu.edu

Patrizia Di Gennaro
University of Milano
Dept. Genetics and Biology of Microrganisms
Via Celoria,
26-20133- Milano
ITALY
Phone: 039/02/26605229
Fax: 039/02/2664551
Email: Patrizia.DiGennaro@unimi.it

Ludo Diels
Environmental Technology
Vlaams Institute voor Technologisch Onderzoek (VITO)
Boeretang 200
BE-2400 Mol

BELGIUM
Phone: 321-433-5100
Fax: 321-458-0500
Email: ludo.deils@vito.be

David Ellis
Dupont Corporation
Barley Mills Plaza
Bldg. 27
Room 2234
Wilmington, DE 19880-0027
UNITED STATES
Phone: +1 (302) 892-7445
Fax:
Email: David.E.Ellis@USA.dupont.com

Nuray Erdogan
Mining Engineering Department
Middle East Technical University
ODTU Maden Muh. Bol.
06531 Ankara,
TURKEY
Phone: +90-312-210 2656
Fax: +90-312-210 1265
Email: nerdogan@metu.edu.tr

Marco Estrela
Centro de Tecnologias Ambientais
Environmental Technologies Centre
ISQ - Instituto de Soldadura e Qualidade
Av. Prof. Dr. Cavaco Silva, N 33,
Talaíde, TagusPark
EC Oeiras - 2781-951 Oeiras –
PORTUGAL
Phone: 351-21-422 90 05
Fax. 351-21-422 81 04
E-mail: maestrela@isq.pt
Zdenek Filip
Umweltbundesamt / Federal Environmental Agency
Dienstgebäude Langen / Langen Building
Paul-Ehrlich-Strasse 29
63225 Langen
GERMANY
Phone: +49 6103-704-160
Fax: +49 6103-704-147
Email: zdenek.filip@uba.de

Fabio Fava
DICASM, Faculty of Engineering,
University of Bologna
Viale Risorgimento 2,
I-40136, Bologna,
ITALY
Phone: +39 051 209-3212/209-3217
Fax: +39 051 209-3218.
Email: fabio.fava@mail.ing.unibo.it

Katerina Francova
Dep. of Biochemistry and Microbiology
Institute of Chemical Technology
Technická 3
166 28 Prague 6
CZECH REPUBLIC
Email: katerina.francova@vscht.cz

Jonathon Fruchter
Battelle,
Pacific National Laboratory
P.O. Box 999 / MS K6-96
Richland, WA 99352
UNITED STATES
Phone: 509-376-3937
Fax: 509-372-1704
Email: john.fruchter@pnl.gov

Sergey Galchenko
Moscow State University
Department of Chemistry, Division of Chemical Technology
119899 Moscow
RUSSIA
Phone: +7 (-095) -939-1279
Fax: +7 (-095) -939-2158

Georgiy Geletukha
Scientific Engineering Centre "Biomass"
P.O. Box 964,
Kiev-67, 03067,
UKRAINE
Phone: +380 (44) 446 9462 , 441 7378
Fax: +380 (44) 484 8151
Email: geletukha@biomass.kiev.ua

Jaan Habicht
University of Tartu
Mureli 16
Tartu 50405
ESTONIA
Phone: +372 (56) 560 392
Fax: +372 (7) 420 444
Email: h@elmar.ee

Joe Hughes
Associate Professor | Department Chair
Department of Environmental Science and Engineering
Rice University MS-317
6100 Main
Houston, TX 77005-1892
UNITED STATES
Phone: 713-348-5903
Fax:
Email: hughes@owlnet.rice.edu

Gytautas Ignatavicius
Environmental Protection Agency
Vilnius Gediminas Technical University
Sauletekio al 11-2306
2040 Vinius
LITHUANIA
Phone: +370 (2) 765298
Fax: +370 (2) 700497
Email: gytisi@takas.lt

Tissa Illangasekare
Colorado School of Mines
Dept. of Civil Engineering
1500 Illinois St.
Golden, CO 8041-1887
UNITED STATES
Phone: (303) 384-2126
Fax: (303) 273-3413
Email: tillanga@mines.edu

Eduard-Emil Iojoiu
Technical University "Gh.Asachi"
Faculty of Industrial Chemistry
General Chemistry Department
Bvd. D.Mangeron Nr. 71

6600 Iasi
ROMANIA
Phone: +40 (32) 278683 / int. 2251
Fax: +40 (-32) -271311
Email: eiojoiu@ch.tuiasi.ro

Eva Jindrova
Dep. of Biochemistry and Microbiology
Institute of Chemical Technology
Technická 3
166 28 Prague 6
CZECH REPUBLIC
Phone: +420 2 475 2752
Fax: +420 (2) 472 2257
Email: eva.jindrova@vscht.cz

Jim Jordahl
CH2M HILL
825 NE Multnomah
Suite 1300
Portland, OR 97232
UNITED STATES
Phone: 503-235-5022 ext. 4353
Fax: 503-736-2022
Email: jjordahl@ch2m.com

Craig Just
The University of Iowa
Department of Civil and Environmental Engineering
120 ERF
Iowa City, IA 52242
UNITED STATES
Phone: (319) 335-5051
Fax:
Email: craig-just@uiowa.edu

Ugur Karaaslan
Middle East Technical University
Dept. of Petroleum & Natural Gas Eng.
06531 Cankaya
Ankara, TURKEY
Phone: 90 312 2104886
Fax: 90 312 2101271
Email: ugurk@metu.edu.tr

Sevgi Kocaoba

Yildiz Technical University.
Faculty of Art and Science.
Department of Chemistry.
Davutpasa Caddesi.
No:127 34210
Esenler- Istanbul-
TURKEY
Phone:
Fax:
Email: kocaoba@yildiz.edu.tr

Gennady Kochetov
National University of Construction and Architecture
31 Povitroflotskii prospekt
Kiev 03037
UKRAINE
Phone: +380 (-44) -2415530
Email: v_zaitsev99@yahoo.com

Alexandre Kotchetkov
TYPHOON
Institute of Experimental Meteorology
82 Lenin Avenue
Obninsk 2490
Kaluga Region
RUSSIA
Phone: +7 (08439) 71996
Fax: +7 (08439) 40910
Email: kochet@obninsk.com

Petra Kucerova
Dep. of Biochemistry and Microbiology
Institute of Chemical Technology
Technická 3
166 28 Prague 6
CZECH REPUBLIC
Email:

Rafal Kucharski
Institute for Ecology of Industrial Areas
40-832 KATOWICE
Kossutha 6 str.
POLAND
Phone: +48 32 254 00 29
Fax: +48 32 254 17 17

Email: sas@ietu.katowice.pl

Tamara Kukharchyk
Institute for Problems of Natural Resources Use & Ecology
10 Staroborysovski tr.
220114 Minsk,
BELARUS
Phone: +375 (17) 264 23 12
Fax: +375 (17) 264 24 13
Email: kakareka@ns.ecology.ac.by

Mihaela Lazarescu
National Research and Engineering
Institute for Environment
Spl. Independentei 294, sector 6
Bucharest 78
ROMANIA
Phone: +40 (1) 221 09 90
Fax: +40 (1) 221 92 04
Email: mihaela.lazarescu@usa.net

Thomas Macek
Inst. of Org. Chemistry & Biochemistry
Academy of Sciences of the Czech ublic
Flemingovo n. 2
166 10 Prague,
CZECH REPUBLIC
Phone: +420-2-20183340
Fax: +420-2-24310090
Email: tom.macek@uochb.cas.cz

Jim Mack
New Jersey Institute of Technology
Otto H. York Building/CEES
138 Warren St.
Newark, NJ 07102
UNITED STATES
Phone:
Fax:
Email: mack@njit.edu

Martina Mackova
Dep. of Biochemistry and Microbiology
Institute of Chemical Technology
Technická 3
166 28 Prague 6

CZECH REPUBLIC
Email: mackovam@vscht.cz

Lili Macovei
Department of Plant Physiology of
University Of Agricultural Science
And Veterinary Medicine
Str. Manastur Nr.3-5,
3400 Cluj-Napoca,
ROMANIA
Phone: +40 (-64) -196384
Fax: +40 (-64) -193792
Email: macoveiusamv@netscape.net

Dave Major
Geosyntec,
160 Research Lane, Suite 206
Geulph, ON N1G 5B2
CANADA
Phone: 519-822-3150
Fax: 519-822-3151
Email: torontoinfo@geosyntec.com

Dmytro Malytskyj
Carpathian Branch of Subbotin Institute of Geophysics
Department of Seismotectonic Researches
3-b,Naukova st.
79053 Lviv
UKRAINE
Phone: +380 (322) -649777
Fax: +380 (322) -648563
Email: dmytro_malytskyj@mail.ru

Dale Manty
US EPA HSRC Program
1200 Pennsylvania Ave.
Washington, DC 20460
UNITED STATES
Phone: 202-564-6922
Fax: 202-565-2448
Em:MANTY.DALE@epamail.epa.gov

Perry McCarty
Director, Western Region Hazardous Substance Research Center
Stanford University
Dept. of Civil & Environmental Engineering

Stanford, CA 94305-4020
UNITED STATES
Phone: 650-723-4131
Fax: 650-725-9474
Email: mccarty@ce.stanford.edu

Steven McCutcheon
USEPA National Exposure Research Laboratory/ORD
Ecosystems Research Division
960 College Station Road
Athens, GA 30605-2720
UNITED STATES
Phone: 706-355-8235
Fax: 706-355-8202
Email: Mccutcheon.Steven@epa.gov

Madeleine McMullen
IC Consultants Ltd.
47 Prince's Gate, Exhibition Road
London SW7 2QA
UNITED KINGDOM
Phone: 020 7594 6565
Fax: 020 7594 6570
Email: m.mcmullen@ic.ac.uk

Ana Melato
Instituto de Ciência Aplicada
Edifício ICAT
Campus da Faculdade de CiênciasLisboa
Estremadura
1749-016 PORTUGAL
Phone: 00351 217500006
Fax: 00351 217500172
Email: Ana.Melato@icat.fc.ul.pt

Ivana Melenova
Dep. of Biochemistry and Microbiology
Institute of Chemical Technology
Technická 3
166 28 Prague 6
CZECH REPUBLIC
Email: ivana.melenova@vscht.cz

Elena Moreno Barbero
Colorado School of Mines
Environmental Science and Engineering Division

Coolbaugh Hall
1500 Illinois St.
Golden, CO 80401
UNITED STATES
Phone: 303-384-2237
Fax:
Email: emorenob@mines.edu

Stephanescu Mugur Cristian
Institue of Biology of Romanian Academy,
Spl. Independentei 296,
Sector 6, CP 56-53
Bucharest 79651
ROMANIA
Email: mstef@ibol.ro

Vera Munteanu
National Institute of Ecology
3 Academiei Str.
MD-2028 Chisinau
REPUBLIC OF MOLDOVA
Phone: +373 (-2) 739614
Fax: +373 (-2) 739954
Email: munteanu@citc.asm

Lee Newman
University of South Carolina
Norman J. Arnold School of Public Health
Columbia, SC 29208
UNITED STATES
Phone: 803-777-4795
Fax: 803-777-3391
Email: lnewman@sph.sc.edu

zgur Nilgun Akkus
Yildiz Technical University
Faculty of Arts and Sciences
Department of Chemistry
Davutpasa Cad. No:127, 34210 Esenler, Istanbul
TURKEY
Tel: +90 212 449 16 41
Fax: +90 212 259 57 92
Email: ozakkus@hotmail.com

Hana Novakova
Dep. of Biochemistry and Microbiology

Institute of Chemical Technology
Technická 3
166 28 Prague 6
CZECH REPUBLIC
Phone: +420 (2) 2435 3021
Email: novakovh@seznam.cz

Mikhail Novitsky
TYPHOON
Institute of Experimental Meteorology
82 Lenin Avenue
Obninsk 2490
Kaluga Region
RUSSIA
Phone: +7 (08439) 71941
Fax: +7 (08439) 71941
Email: novitsky@typhoon.obninsk.org

Yuksel Orhan
Ondokuz Mayis University
Environmental Engineering Department
TR-55139 Samsun
TURKEY
Phone: 90-362 457 60 00/2824
Fax: 90-362 457 60 35
Email: yukselo@omu.edu.tr

Igar P. Samsonenka
Institute for Problems of Natural
Resources Use and Ecology of National
Academy of Sciences of Belarus
Staroborisovsky trakt 10
220114 Minsk,
BELARUS
Fax: +375 (0172) -642312
Email: samson@ns.ecology.ac.by

Tatyana Parshikova
Kiev University
Department of Plant Physiology and Ecology
Vladimirskaja st. 64
01017, Kiev,
UKRAINE
Phone: +380 (-44) 234-6870
Email: parshik@ukrpack.net

Jarmila Pazlarova
Dep. of Biochemistry and Microbiology
Institute of Chemical Technology
Technická 3
166 28 Prague 6
CZECH REPUBLIC
Email: pazlaroj@vscht.cz

Gary Pope
Dept. of Petroleum & Geosystems Engineering
University of Texas at Austin
Austin, TX
UNITED STATES
Phone:
Fax:
Email: gpope@mail.utexas.edu

Valentin Popov
TYPHOON
Institute of Experimental Meteorology
82 Lenin Avenue
Obninsk 2490
Kaluga Region
RUSSIA
Phone: +7 (08439) 71914
Fax: +7 (08439) 40910
Email: vpopov2@obninsk.com

Marek Prouza
Dep. of Biochemistry and Microbiology
Institute of Chemical Technology
Technická 3
166 28 Prague 6
CZECH REPUBLIC
Phone: +420 (2) 475 2752
Fax: +420 (2) /4722257
Email: marek.prouza@vscht.cz

Radu Rautiu
IC Consultants Ltd.
47 Prince's Gate, Exhibition Road
SW7 2QA London
UNITED KINGDOM
Phone: 44(0) 20 7594
Fax: 44(0) 20 7594 6570

Email: r.radu@ic.ac.uk

Martin Reinhard
Department of Civil and Environmental Engineering
Stanford University
Stanford, CA 94305-4020
UNITED STATES
Phone: 650-723-0308
Fax:
Email: reinhard@ce.stanford.edu

Elena Rovdan
Institute for Problems of Natural Resources Use and Ecology,
National Academy of Sciences of Belarus
10, Staroborisovsky trakt,
220114 Minsk,
BELARUS
Phone: +375 (172) 64 32 82
Fax: +375 (172) 64 24 13
Email: elenarov@yahoo.com

Julia Russkova
Business Center PARUS, PARSONS Delaware Inc.
23 1st Tverskaya-Yamskaya st.,
125047 Moscow,
RUSSIA
Phone: +7 (095) 755-8313
Fax: +7 (095) 135-0420
Email: Julia.Russkova@rcwdp.ru

Zehra S. Can
Marmara University, Faculty of Engineering,
Environmental Eng. Dept.
Marmara Universitesi,
Mühendislik Fak. Çevre Müh. Böl.
81040 Göztepe / Istanbul
TURKEY
Tel: +90 (216) 348-0292 / 250
Fax: +90 (216) 348-0293
Email: zcan@eng.marmara.edu.tr

Aleksandra SAS-NOWOSIELSK
Institute for Ecology of Industrial Areas
40-832 KATOWICE
Kossutha 6 str.

POLAND
Phone: +48 (32) 254 00 29
Fax: +48 (32) 254 17 17

Jerry Schnoor
Dept. of Civil & Environmental Engineering
116 Engineering Research Facilities
330 South Madison St.
University of Iowa
Iowa City, IA 52242
UNITED STATES
Phone: 319-335-5649
Fax: 319-335-5585
Email: jschnoor@cgrer.uiowa.edu

Kenichi Soga
University of Cambridge
Engineering Department
Trumpington Street
Cambridge
CB2 1PZ
UNITED KINGDOM
Phone: +44 (1223) 332713
Fax: +44 (1223) 339713
Email: ks@eng.cam.ac.uk

Jim Spain
AFRL/MLQL
139 Barnes Dr.
Tyndall AFB, FL 32403-5323
UNITED STATES
Phone: 850-283-6058
Fax: 850-283-6090
Email: jim.spain@tyndall.af.mil

Lori Swadly
Colorado School of Mines,
Division of Environmental Science and Engineering
1500 Illinois Street
Golden, CO 80401
UNITED STATES
Phone:
Fax: (303) 273-3629
Email: lswadley@mines.edu

Pavel Szerbin

Div. of Environmental Radiohygiene
Natl. Res. Inst. for Radiobiology and Radiohygiene (NRIRR)
Anna u. 5,
Budapest, 1221
HUNGARY
Phone:+36-1-229 1968
Email: pavel@hp.osski.hu

Boytcho Tenev
Institute for Land Reclamation
136 Tzar Boris III blvd.
Sofia 1618
BULGARIA
Phone: + 359 2 551145
Fax: + 359 2 955 52 36 - fax
Email: tenevs@techno-link.com

Senja Totevova
Dep. of Biochemistry and Microbiology
Institute of Chemical Technology
Technická 3
166 28 Prague 6
CZECH REPUBLIC
Email: totevova@biomed.cas.cz

Stella Triantafyllou
National Technical University of Athens
Dept of Mining/Metallurgical Engr.
Laboratory of Metallurgy
Heroon Polytechniou 9
157 80 Zografou Campus
Athens, GREECE
Phone: +30 1 7722177
Fax: +30 1 3314328
Email: triantaf@metal.ntua.gr

Victor Tsvetkov
Engineering Radioecology and Radiochemical Technology Department,
St. Petersburg Institute of Technology
Moskovsky pr., 26
St. Petersburg 198013
RUSSIA
Phone: +7 (812) 315 10 36
Fax: +7 (812) 315 10 36

Email: tvi@tu.spb.ru

Richard Tykva
Institute of Organic Chemistry
Academy of Sciences of the Czech Rep.
CZ-16610 Prague 6
Flemingovo 2
CZECH REPUBLIC
Phone: +420 (-2) 20183269
Fax: +420 (-2) 3111714
Email: tykva@uochb.cas.cz

Ülkü Ündeðer
Hacettepe University
Faculty of Pharmacy
Dept. of Toxicology
Sihhiye Ankara 06100
TURKEY
Tel: + +90 312 309 29 58
Fax: + +90 312 311 47 77
Email: uundeger@hacettepe.edu.tr

Fusun Uysal
Trakya University Corlu Engineering Faculty
Environmental Engineering Dept. PK 355
59860, Corlu
TURKEY
Email: ffusun@mail.koc.net

Andon Vassilev
Agricultural University of Plovdiv
Mendeleev 12
Plovdiv 4000
BULGARIA
Phone: +359 (32) 861395
Fax: +359 (32) 861395
Email: vassilev@au-plovdiv.bg

Mustafa Versan Kok
Dept. of Petroleum and Natural Gas Eng.
Middle East Technical University
06531, Ankara-
TURKEY
Tel: +90-312-210 4891
Fax: +90-312-210 1271
Email: kok@metu.edu.tr

Konstantin Volchek
Environmental Technology Program
SAIC Canada
3439 River Road
Ottawa, Ontario K1A 0H3
CANADA
Phone: (613) 990-7147
Fax: (613) 991-1673
E-mail: konstantin.volchek@saic.com

Herb Ward
Director, Energy & Environmental Systems Institute
6100 Main Street, Mail Stop 316
Rice University
Houston, TX 77005-1892
UNITED STATES
Phone: 713-348-4086
Fax: 713-348-5948
Email: wardch@rice.edu

Walt Weber
GLMAC
University of Michigan
181 EWRE Building
Ann Arbor, MI 48109-2125
UNITED STATES
Phone: 734-763-2274
Fax: 734-936-4391
Email: wjwjr@engin.umich.edu

Peter Werner
Technische Universität Dresden
Department Of Water Sciences
Mommsenstraße 13,
01062 Dresden
GERMANY
Phone: 0351) 463 3382
Fax: (0351) 463 7108/7288
Email: pwerner@Rcs1.urz.tu-dresden.de

INTRODUCTION

DANNY D. REIBLE[1] AND KATERINA DEMNEROVA[2]
[1]*Hazardous Substance Research Center/South and Southwest, Louisiana State University, Baton Rouge, LA 70803*
[2]*Department of Biochemistry and Microbiology, Institute of Chemical Technology, Prague, Czech Republic*

On May 24, 2001, a total of 102 students and lecturers participated in an Advanced Study Institute (ASI) sponsored by the North Atlantic Treaty Organization (NATO) under our direction. The Institute was focused on in situ and onsite management of contaminated sites. The objective of the Institute was to balance state of the art science with techniques for field application of a variety of technologies for in situ assessment and remediation of contaminated sites. Many of the lecturers were drawn from the ranks of the Hazardous Substance Research Centers, multi-university consortia that have been funded by the US Environmental Protection Agency to conduct research and technology transfer designed to promote risk-based management and control of hazardous substances for the nation. The Centers have made special contributions to the areas of in situ and onsite assessment and remediation of contaminated sites. Such approaches have the potential for being significantly less expensive than other assessment and remediation approaches while maintaining accuracy and effectiveness. Cost-effective remedial and management approaches that are also effective in minimizing exposure and risk to human health and the environment are a critical need throughout the world but particularly in Eastern Europe and the former Soviet Union where resources that can be devoted to environmental cleanup are especially limited.

The ASI began with a series of presentations identifying both the types of contaminants commonly encountered and the means currently being used to manage these contaminants. This was followed by an intensive series of workshops on field analytical techniques. Discussions and practice focused on field portable gas chromatograph-mass spectrometry, field enzyme test kits and portable x-ray fluorescence equipment. This equipment has revolutionized the speed and accuracy of field characterization and assessment.

The core of the ASI, however, was focused on the techniques to manage contaminants in situ. Exposure and risk to contaminants is largely defined by the fate and transport processes that they exhibit. Developing issues on bioavailability and the unique transport and exposure processes associated with contaminated sediments and nonaqueous phase liquid subsurface contamination were identified. The role that these processes may play in the future management of such sites were discussed. The balance of the ASI addressed the state-of-the-art of in-situ management of contaminants by microbial biotransformation, phytoremediation, and other technologies. Much attention was concentrated on chlorinated solvents which are important groundwater and soil contaminants. The application of fundamental knowledge on microbial biotransformation to the assessment of natural attenuation in soils and groundwater was also emphasized. Discussions in this area were focused by a series of workshops examining natural attenuation on specific contaminants and sites.

Although much of the focus was on natural processes and how they might be enhanced for the management of contaminants at a site, there was also recognition that a portion of sites may require more invasive approaches. These approaches are especially useful in management of source areas that are contributing to the migration of contaminants in groundwater and other media. Among the more invasive approaches discussed were solvent extraction of contaminants, direct bioremediation of PCBs, source areas and munitions, and chemical oxidation.

The ASI led to an exchange of information on these important topics among scientists and engineers from North America and European countries. The Institute attracted a good balance of participants. Approximately 1/3 of the students were currently pursuing advanced degrees at various institutions in the NATO and NATO partner countries. Approximately 1/3 of the students were recent graduates or early to mid career technical practitioners responsible for managing contaminated sites. The remaining 1/3 represented a wide range of backgrounds from senior academic or government regulator to scientists and engineers from consulting firms or private industry. These groups all represent individuals that are most likely to put the lessons learned in the Institute directly into practice.

All of the participants realize, however, that the lessons of the Institute could be helpful to many individuals that were not able to participate directly. It is toward those individuals that this volume is directed. This volume provides a summary of the key technical information discussed in the Institute. It is hoped that the recognition of the range of contaminants and sites with which we must deal and an improved understanding of in situ management approaches and their effectiveness will be the result, leading to an enhanced ability to respond to these sites both within Europe and North America. In both the developed and the developing world there is a clear need to pursue remedial strategies that are both effective and less costly than previous approaches. The Editors sincerely hope that publication of the lectures from the ASI will assist in the dissemination of such approaches.

USING A FIELD ANALYTICAL APPROACH TO ACCELERATE SITE ASSESSMENTS

A. KINNEY, J. MACK AND G MCKENNA
Northeast Hazardous Substance Research Center
New Jersey Institute of Technology

1. History or Environmental Management

1.1 DEVELOPMENT OF NEED FOR SITE MEASUREMENT

Prior to 1950 the environment was viewed as an untamed beast that required development so that "civilized" people could inhabit it. Analytical methods were limited and rarely if ever applied to the environment for measurement of contamination. Worker safety was determined in mining by the canary approach – if the canary died when sent into the mine then it was considered unsafe for workers.

It wasn't until the 1960's and the publication of "Silent Spring" that an activist movement to preserve and "save" the environment began the first efforts at trying to stop the decimation of our environment. Between the 1960's and the 1970's in the United States efforts at "beatifying" the area where we lived took form. A campaign of "Give a Hoot Don't Pollute" was launched to increase awareness of trash disposal and the basic visual pollution. Pollution prevention at the industrial waste level had not yet been recognized or action started.

In the 1970's the "Love Canal" story hit which literally scared our country into acting at the federal level to pass laws that would protect and clean-up the environment and stop industrial dumping. The story of Love Canal brought to light many important issues related to industrial dumping and waste disposal. The actions taken by the companies that used the canal as a disposal ground were not irresponsible since current environmental agencies actually licensed and permitted this form of disposal. The impact on the earth that this careless practice, many of which were licensed by various environmental bodies at the state level, was not recognized as destructive to future generations and the Earth.

Between the 1970's and the year 2000 focus was directed at remedial activities to "clean-up" past contamination. This activity required very large and expansive site investigation efforts requiring voluminous amounts of samples and analysis, carried out for long periods of time. The data used was no longer a single one, like before where you were generally interested in compliance against single health or environmental standards. Site investigations were used to answer many different questions about the source, pathway and receptors that were affected by the contamination. Many trips back and forth to the site were required to collect samples. Investigation stretched out over years and costs were very high. This required a great deal of resources, rules, regulations and overall framework to accomplish. At the same time great advances in measurement technology started to come about as a result of the microchip and better electronic circuitry. Technology is able to perform at lower detection limits, higher resolution and data storage capacity exceeds even the wildest fantasies of laws instituted in the 1970's. The bricolage of developments that have taken place over the past ten years have made on-site, *in situ* analysis of various contaminants a reality.

D. Reible and K. Demnerova (eds.), Innovative Approaches to the On-Site Assessment and Remediation of Contaminated Sites, 1–30.

1.2 TREND FORWARD USE OF FIELD ANALYTICAL METHODS IN SITE ASSESSMENT

As we begin the 21st century, challenges for environmental measurement will be focused on sustainable development. This new focus is galvanic to the entire environmental preservation and business community. The objective is to allow growth at the economic level and continue to protect the environment from harm. Improvements in technology have allowed for continuous monitoring and system-shutdowns when compliance is exceeded.

Site characterization, remedial action and site monitoring all require analytical methodologies to determine the extent of contamination. The best approach to addressing this need are field-based methods that allow the user the ability to take measurements on-site. Field analytical methods (FAMs) are instruments and techniques that can be used to evaluate on-site the nature and extent of contamination. There are numerous instruments, ranging from basic probes and meters to sophisticated field portable gas chromatography/mass spectrometers. The value of the information provided from these methods is dependent upon the goals of the project and the skill of the user. Recently, EPA has linked the use of FAMs with two other elements of the site assessment process, systematic planning and dynamic workplans, to create a new approach that will vastly improve upon the existing or traditional approach, reduce site characterization costs and optimize the overall process, from site characterization to remediation implementation. This new approach includes the complete process of equipment, personnel, application methods and data interpretation techniques that are used to perform an assessment of the nature and extent of environmental contamination at a site. Currently EPA is holding internet-based training sessions for federal and state project managers with the intent to modernize the environmental business with regard to collecting and interpreting environmental data.

One principal motivating factor is the need to quickly and inexpensively characterize the thousands of Brownfield sites that exist throughout the United States. EPA and others have recognized that the traditional remedial investigation (RI) process, with its multiple field events and repetitive sampling, is too cumbersome to react to the Brownfield redevelopment need. Brownfields is an effort to recycle and reuse our land: land that is idle, abandoned or underutilized due to the presence of contamination. Additionally, many Brownfield programs are being financed by public money, forcing regulators, planners and politicians to be aware of the cost of investigations and the value of the information obtain. Thus the key players in Brownfield programs, the owners, seller, buyers, lenders, investors, local government officials, planning boards and regulators are all looking for ways to achieve adequate characterization without excessive cost. The solution is a new approach to site characterization, which produces greater amounts of information with an experienced and qualified field team.

The overall approach that EPA is promoting is called the Triad Approach and is intended to optimize the site assessment process while substantially reducing uncertainty in environmental decision making. The Triad consists of Systematic Planning, Dynamic Workplans and Field Analytical Measurement Technology (i.e. Field Analytical Methods or FAMs). According to EPA, all phases of the remedial process can benefit from the adoption of the triad approach. This includes site investigations, remedial action and long-term monitoring of remedial action performance.

2. Value of Field Methods in Site Assessment

Field methods provide the user increased efficiency in sample analysis and site assessment process. The information obtained can be processed on-site and guide the field engineer in sample collection that fully and completely delineates the contaminant area. Conventional methods which take samples at pre-defined areas and send these to an off-site lab requires additional field mobilizations to obtain additional information about the site.

Real-time information obtained from FAMs can be used to define contaminant boundaries and direct sampling in real-time versus conventional methods which require re-mobilization of the field team. This is a far more efficient approach, but the only way it can work is through the use of realizable, accurate FAMs. Technology has evolved to the point where these methods are available cost effectively.

Many of the assessment technologies are non-intrusive and require little to no sampling. This enables a "real-time" look at the contamination situation without disruption to the contaminant environment. In summary FAMs provide:

- Increased efficiency
- Real-time information for on-site decision making
- Cost effective monitoring of remedial actions
- Non-intrusive characterization & identification
- Decreased cost
- Flexibility
- Better representation of contaminant distribution
- Reduced/focused remedial efforts(i.e. accurate hot spot delineation)

Field analytical methods have been applied recently to many sites across the U.S. as well as in other parts of the world. The advantages offered by them are:

- Increased efficiency
- Decreased cost
- Flexible
- Comprehensive
- Analytically equivalent

Unfortunately, historically field methods have been viewed primarily as "screening" techniques. Thus they were considered to provide data of such inferior quality that they could be used for only gross understanding of site conditions and not appropriate for detailed, regulatory based site characterization. However, FAM technology has advanced tremendously, to the point where today many FAMs are based upon definitive analytical methods which have been subject to rigorous performance analysis. Thus the term "field screening" is misleading and should not be used when referring to FAMs.

Thus, the terminology best applied is effective data, which refers to the integration of sampling and analysis uncertainty management as a comprehensive system, rather than strict reliance of analytical method as the only basis for acceptable data quality. For example, FAMs will allow higher sampling density because of reduced costs, while the information uncertainty provide by one sampling point maybe slightly less than a fixed based analytical method, the vastly increased size of the data set due to higher sample density provided for greater information on the distribution of site contaminants.

Field methods create obvious convenience over off-site analysis but require the a more experienced field team. Many FAMs are designed to make use easy and are field rugged. However, knowledge of the interferences that can exist when analyzing samples on-site are is needed and thorough training of this is required.

Field analysis allows for real-time decision-making reducing the number of field mobilizations required to characterize a site. This methodology not only applies to characterization but also remediation and monitoring activities. When sufficient data quality elements are employed the results can reduce time and costs of these activities by orders of magnitude. For example, $3,000 will analyze approximately 20 samples off-site for VOC's using GC. This sum would analyze 150 to 250 samples on-site, including a field chemist, in a week. This savings does not include the savings of being able to delineate the site more accurately, reducing the number of field trips, decreasing the re-mobilization costs. Even if analysis costs were similar, the savings experienced through contractor and equipment re-mobilization is more than enough to justify its use.

Field methods performance have progressed substantially over the past ten years. Advances in technology experienced in personal computing have been utilized in the on-site measurement arena. Analytical technologies can in many cases give quantitative results on-site and sometimes even *in situ*. The result gives the site characterization, remediation and monitoring activities a new set of tools, which require training and technical rules, data quality objectives and field quality control. Understanding how FAMs performance are affected by matrix interferences and the skill of the field user is also important. Once these issues have been completely addressed, data generated from these techniques can expedite the clean-up and characterization process saving thousands of dollars in the process.

The limitations faced when using FAMs need to be considered.

1) Government approval - Obtaining government approval (in the U.S.) that has been difficult to achieve. Many states and federal agencies have initiated a large training campaign for teaching their employees how to use these methods on-site.
2) Experienced field personnel - on-site measurement requires experienced scientists in the field rather than collection technicians. The lab has to come to the field requiring testing and evaluation expertise.
3) Higher start-up costs - The start-up costs with FAMs can be higher than a characterization that uses conventional fixed based laboratory approach since you are sometimes required to buy equipment upfront, that will eventually be used over and over again. This cost can be amortized over many site investigations but requires initial upfront higher costs.
4) Site history - site knowledge is also required to best utilize the information collected and assessed with FAMs. Understanding the factors that impact the technologies used can limit there use also. For example, if you are on a site where there is high water content in the soil this will affect the results when using a technology like XRF, which requires less than 20% water saturation to give meaningful information.

3. Overview of Common and Emerging Field Analytical Methods

There are numerous FAMs available that can be implemented to reduce cost, gain higher confidence of site contamination and reduce time. To choose a FAM you need to consider the media that the contaminant is housed in, the type of contaminant, and the level of contamination. Each of these elements are essential to actually measuring the contamination on-site. Additionally, the field engineer needs training and instruction on field instrumentation calibration, QC protocol, calibration and measurement technique. What was once simply a matter of taking a sample, recording the location and ensuring that the chain-of-custody is complete now involves increased capabilities understanding and making field decisions. Using FAMs requires the following elements be address to obtain data of significant quality:

- Compatibility with site contaminants
- Knowledge of sample matrix
- Performance QA/QC requirements
- Operator skill level
- Regulatory action levels

3.1 INSTRUMENTS AND METHOD INSTRUCTION

Field instrumentation requires the user to understand how field conditions impact measurements as well as how to operate the instrument and perform the method reproducibly and correctly. Instruction on proper instrument calibration procedures, routine calibration checks and chemical and physical interferences that affect the results by the field method are examples of activities for ensuring usable data.

Field instruments should have set training protocol, which should be obtained by all users of the equipment. This should include starting up the instrument, waiting appropriate times for instrument stability, checking calibrations if necessary, positioning of the sample, understanding the effects of field factors such as heat, humidity, moisture and matrix on the instrument, and overall operator use. Each field person using a test kit or field method should be trained in the method use and understand factors that can affect the results prior to operator use.

3.2 COMMON TYPES OF FIELD ANALYTICAL METHODS

There are numerous FAMs available for the environmental scientist. The various method types typically focus on contaminants and media. There does not currently exist a technology that can measure every contaminant in all medias, though it is hoped that this will exist in the future. The categories of field technology include both:

- Inorganics
- Organics

The challenge for field methods is to overcome the "contaminated" environment influences while still providing for easy "turn-key" extraction and testing approaches. The following discussion focused on the three of the most commonly used FAMs:

- X-Ray Fluorescence – for measurement of inorganics in soil, sediment and filters
- Field Portable Gas Chromatography/Mass Spectrometry – for measurement of volatile and semi-volatiles in air, water and soil.
- Immunoassay test kits – for measurement of certain organic compounds in soil and water

3.2.1 *X-Ray Fluorescence (XRF)*

Description of Technology

XRF uses x-ray energy to cause fluorescence at the atomic level. This fluorescence occurs by emission of x-rays from a radioactive source to the sample. The x-ray ejects an electron from the atomic sub-shells within the atom. Each atom requires specific x-ray energies to release this energy and therefore emits fluorescent energy that is unique to the element under analysis. The fluorescence occurs in the form of an x-ray emission. This emission is sensed by a detector which records the emitted x-ray according to their relative frequencies. In XRF the higher the number of emitted x-rays the higher the concentration of the element in the sample.

When analyzing a soil sample with XRF you can take samples in several ways. There are four testing modes for soil analysis direct, 1) in situ (point and shoot), 2) bagged, 3) sieved, 4) pulverized then sieved samples. Each of these types of analysis have their own error associated with it. To perform an analysis of soil via XRF you must account for the soil matrix effects, debris, positioning of the instrument over the sample, moisture content in the sample, operator differences and the particle size of the soil.

When performing in situ analysis the XRF is placed on a bare spot on the ground. The area is clear of debris and analyzed for approximately 30 to 60 source seconds. This information has been shown in other studies to give correlations to laboratory analysis of approximately 80%.

The bagged sample analysis method consists of typically collecting material from the top ½ inch of surface soil, though this can vary depending on the sampling protocol. Then the sample is placed in a bag and analyzed directly. The sample can then be sent directly to the laboratory for confirmatory analysis. The correlation of this sampling mode to laboratory is approximately 85%.

Grinding samples requires taking a sample and placing it in a grinder or using a motar and pestle to grind up the soil particle to a very fine consistency. Once ground the sample is placed in a sample cup and testing stand and a reading is taken for 60 source seconds. This type of analysis gives correlations of about 95%.

The grinding and sieving procedure expands upon the grinding mode mentioned above by taking the ground sample and placing it through a series of sieves. The finest mesh used is a 0.6 micron. The finely sieved particles are then placed in a soil cup and measured in the soil stand for 60 source seconds. The lab to field correlations for this sample preparation mode is between 95% to 99%.

Training Required for XRF Use

The training required for using an XRF is radiation safety and instrument use. This typically requires a day of training and is offered as part of the purchase for an XRF. All states and even countries require that radioactive sources be licensed. This process requires simply filling out paper work for some states or paying fees and applying for others. Each state has specific rules for this and users should contact the instrument manufacturer to determine what regulations they fall under.

Applying XRF to the Field

XRF can be used to delineate a site very quickly as it is small and very portable and can provide *in situ* measurements. Sites that have unknown contamination boundaries or that have unknown inorganic contamination history would gain advantage to using this technology. The analysis does not cost extra as the cost for the instrument is paid up-front therefore each analysis simply decreases the cost/sample ratio.

3.2.2 *Description of Technology Application*

Immunoassay technical description

Immunoassay methodology uses the ability of antibodies to selectively bind to the target analytes in a sample matrix. The primary mechanism for environmental detection of contaminants with immunoassay is ELISA (enzyme-linked immunosorbent assays. In this methodology the target analyte is coupled with an enzyme conjugate which acts as a tracer. The labeled enzyme can be detected later by a color change and the analyte is proportional to the change in color.

ELISA is the preferred method for immunoassay chemistry because it can be optimized for speed, sensitivity, and selectivity. The operation of the test requires that the antibodies be coated onto a small test tube or microwell or they can be attached to magnetic particles, or latex particles. This creates a limited number of binding sites. A sample containing both the analyte and the analyte conjugate are added to the container and allowed to competitively bind. The analyte conjugate is of known quantity and it competes with the analyte for the limited binding sites. Therefore, if you have higher amounts of analyte in your solution than the enzyme conjugate then you will a higher amount bound to the limited antibody sites. An enzyme substrate and a chromogen are added to the test tube to cause the color formation. After a certain amount of time a solution is added to stop the reaction. The amount of bound enzyme conjugate determines the amount of color. Therefore the more color the less analyte in your sample.

To obtain a qualitative result the user would need only to observe the color density. For a quantitative result a differential photometer is required to measure the absorbance of light through the substance. In this case a calibration curve of absorbance versus concentration is generated and the concentration can be calculated.

The types of compounds that are measured by immunoassay are gasoline, diesel fuel, jet fuels, BTEX, and PAHs. Additionally immunoassays can measure pesticides such as triazine herbicides, 2,4-dichlorophenoxyacetic acid, organophosphates, cyclodienes, carbamates, and dichlorodiphenyl trichloroethane (DDT). Of these pesticides some kits can detect only one while others do entire classes.

Training required prior to use

The field engineer will need minimal training on this technology to be able to apply this in the field. The primary consideration when doing the ELISA analysis on-site is to ensure that each step is followed and that timing of the development phases is followed to the exact time. The user will need to understand some graphing techniques to obtain final answers as well. Generally the entire process is very simple and straight forward, requiring less than a half day to master for field use.

3.2.3 *Portable Gas Chromatography/Mass Spectrometry (GC/MS)*

Description of the technology application

Gas chromatography has been used for years to analyze contaminants in environmental media. Recently, portable GC systems have been available to the field engineer to rapidly assess various environmental media on-site in minutes. This technology is relatively inexpensive, approximately \$12,000 to \$25,000 for a complete instrument. This enables quick detection of volatile and semivolatile contaminants that can spread quickly if present in or near the water table. The application of bringing mass spectrometry (MS) – typically a large laboratory based instrument – into the field is a unique and challenging task. Technology has advanced to enable GC/MS technology to be brought into the field thus providing the "ultimate" in site assessment technology on-site. The components of a GC/MS system are:

a. injection port
b. column
c. integrator or data acquisition system
d. MS detectors

Separation occurs by adding a carrier gas and introducing the sample through an injection port. The sample is volatilized and swept into the column where separation of a mixture of analytes occur. The columns can be either open tubular or packed. The material in an open column is silica based material coating the inside of the column.

Separation occurs by the analytes having a different affinity for stationary phase. A compound that "likes" being in the stationary phase more than another will be retained in the column longer. The mobile phase – the inert gas – is simply a carrier by which the analytes can enter and be carried down the column while interacting with the stationary phase along the way.

Controlling the speed and process of separation has been the primary challenge of chromatographers over the past two decades. Separations can be controlled in several ways by varying the:

1. stationary phase
2. carrier gas and flow rate
3. length of the column
4. diameter of the column

The stationary phase can be changed to cause less or more separation to occur. The flow rate of the carrier gas can improve or lengthen the time the analyte spends in the stationary phase and cause a more efficient sweep of the injection port. The length of the column can be changed to speed up or slow down the time spent in the column. By shortening the column you will decrease analysis time but may eventually degrade the resolution of the separation if you are not careful. For environmental applications 30 to 60 meter columns are typical.

The diameter of the column can be increased or decreased. The smaller the diameter the finer the resolution and the greater the sensitivity but the volumes will be decreased as well. Typical diameters are about 1 millimeter with volume sizes of 1 to 2 microliters.

3.3 ADDITIONAL FIELD ANALYTICAL METHODS

The previous sections have discussed these commonly used FAMs that have a history of site application and have been subject to rigorous performance and verification testing. This history of application and testing has resulted in well-established operations procedures that provide consistent results when compared to fixed based laboratory methods. Additionally, these three basic FAMs can provide field analysis for all the commonly found compounds at contaminated sites, which are heavy metals in soil, volatile organic compounds in soil, air and water and heavy organic compounds, such as PAHs, fuel oils, pesticides and PCBs in soil and water. However, the science of FAMs is rapidly developing and there are other methods that have either evolved from "test kits" or use innovative measurement methods that are available for specific compounds or matrixs. This section provides brief description of several of these FAMs.

3.3.1 *Colorimetric indicators*

Colorimetric indicators are another source of hand-held measurement technologies that gives on-site information, which is considered qualitative. These methods and technologies measure various organic compounds in air soil gas and water. There are several to choose from and each is typically either specific to a given class of compounds or to an entire group.

3.3.2 *Fluorescence*

Fluorescence analyzers for organic compounds can be extremely sensitive and versatile for field analysis. This technology uses ultra-violet light to excite the compound and cause fluorescence. This provides a very selective method of identifying compounds that fluoresce at different energy levels. This is used in conjuction with down hole technology to get "on-the-fly" measurements. This can be used to determine areas of contamination in subsurfaces and also can give quantitative data from on-site analysis. The amount of sample preparation performed will determine the level of sensitivity that one will get from this technique.

3.3.3 *Infrared*

Infrared spectroscopy is used for organic compound identification and gives a "fingerprint" of the compounds identity. This has been primarily a laboratory based method of analysis but has recently developed into field measurement techniques. Some of these include Long Range Gas Monitors, Open Path Infrared Flammable Gas Detectors, and Infrared Ambient Air Monitors, and Long Path Open Path Fourier Transform Infrared (FTIR) systems. The primary application of this technology to date has been in fugitive industrial emissions, industrial health, and safety monitoring, and indoor air assessments. They have recently been made more rugged and can be used outside. The detection limits for this technology can be as low as low ppb for many compounds.

3.3.4 *Chemical*

Chemical sensors for a variety of organic compounds are defined as a transducer which gives information about the compound's composition. They are designed for mass, electrochemical, optical, or thermal sensing. There are currently three sensors being used in field analysis: 1) surface acoustic wave (SAWs), 2) fiber optic chemical sensors and 3) biosensors. The SAW detectors are primarily used to detect gases and VOCs, both halogenated and nonhalogenated, in the vapor phase. Fiber optic chemical sensors were developed to measure volatile petroleum constituents in soil gas, water and air. They can achieve detection limits as low as 1 ppm in water. Finally biosensors are chemical sensors that which use a biological mechanism to determine the analyte.

3.3.5 *Anode Stripping Voltammetry*

Anodic stripping voltammetry (ASV) is an older method that has been strictly used in the laboratory. The adaptation of this method for use in the field has occurred recently, within the last 6 years, with success. ASV can measure metals in water or a buffered solution to the part per trillion level. In the field it has been shown to measure consistently down to the ppb level. This technique can measure metals in soil, dust, paint and water with some sample preparation for all media. Applications of ASV have been the identity of metals in surface waters, acid mine drainage studies, laboratory research, and monitoring worker safety.

3.3.6 *Water Quality Measurement Techniques*

Water quality measurement techniques are vital for maintaining environmental compliance. Typical tests include temperature, pH, conductivity, turbidity, redox potential, dissolved oxygen and sometimes BOD. Many of the techniques that are commercially available are downhole sensors or probes that allow the field engineer immediate and *in situ* readings of water quality. These can be left to read continuously or at predetermined intervals and the information can be transmitted via satellite to an off site data collection systems to ensure compliance at all times. Other technologies are hand-held and require very little technical expertise to use. Most kits or probes use colorimetric indicator tubes to determine some of the above parameters. The costs are relatively low and disposal is very simple.

3.3.7 *Emerging Techniques*

The portable datalogging spectrophotometer is a portable unit weighing about 4.4 pounds, which can measure up to 1000 samples prior to downloading. It analyzes water samples for a variety of contaminants, which absorb light in the 400 to 900 nm wavelength range.

Laser-induced breakdown spectrophotometer (LIBS) is a portable technology for the detection of metal contaminants on surfaces. This can also be used for *in situ* detection of metals in the subsurface when coupled with a cone penetrometer. There is some adaptation of this technology occurring to look at metals on different types of matrixes.

Field portable ion chromatography may change the way metals and anions, such as nitrate, chloride, and sulfates, are measured in the field. This instrument has been adapted for field use and can give ppb detection of these contaminants in less than 5 minutes. The average operating range is 10 to 40 ppb.

Using on-site technologies requires up-front consideration to effectively use and obtain their true benefit. The user must know the types of contaminants that might be present on the site prior to going to the site for assessment. FAMs can provide ultimate benefit if by delineating the contamination quickly – with as little as one field trip, determining the actual site risk, and implementing corrective actions immediately to ensure the minimization of spread of contamination.

4. Strategic Use Of FAMs In Site Investigations

4.1 INTRODUCTION TO DYNAMIC WORK PLANS

Because of the advances in field analytical method technology, the opportunity is now available to implement site investigations in a more efficient and cost effective manner. Since FAMs allow field investigation personnel to obtain information about the nature and extent of site contaminants in a "real-time" manner, it is now possible to integrate this data into the daily decision making process required in site investigations. This allows personnel to constantly update the site database and make informed decisions about new sampling locations as the investigation unfolds. This type of approach is called a dynamic site investigation and the mechanism for implementing and guiding the work is a dynamic work plan.

Dynamic investigations are intended to combine the site investigation (SI) and remedial investigation (RI) components into one mobilization. In order to accomplish this, the investigation team needs to use field-based analytical methods to the maximum degree possible to evaluate sampling requirements on a daily basis. The program will then be performed in a dynamic mode such that data will be interpreted and mapped daily and judgmental sampling will be used to adapt the program on a daily basis in order to achieve cost-effective and high accuracy investigations. Important components that must be agreed upon before fieldwork begins are the basic sampling decision rules, the data requirements (including Field Analytical Methods (FAMs)) and the communication strategy.

Decision rules are guidance that the field personal use to modify sampling strategies and make on-site field decisions. They are provided to assist the stakeholders in assessing the results of sample analysis and determining the need for additional sampling. Data requirements establish the application of FAMs and identify the techniques that will be used to assure that analytical results reflect the site conditions. The communication strategy describes how information will be distributed to stakeholders, and personnel responsible for field decision-making and the methods to assure that critical decisions are made in a timely manner.

4.2 PURPOSE OF A SITE INVESTIGATION

Before discussing the components of a dynamic work plan and the dynamic site investigation process, it is important to understand the overall purpose of a site investigation. In general, site investigations are intended to accomplish the following objectives:

a. Identify the contaminants of concern
b. Delineate the horizontal and vertical extent of the contaminants of concern
c. Identify the risk to human health and the environment associated with the contaminants
d. Determine appropriate remedial action requirements

FAMs allow achievement of these goals to be cheaper, faster and with better representation of the site.

4.3 CREATING A DYNAMIC WORKPLAN

In the traditional RI approach, major decisions concerning the direction of the fieldwork are made after the data has been collected, analyzed and the field team is off-site. This always results in multiple mobilizations, increased costs and inflexible use of personnel. A significant reason for the "step like" approach is the traditional RI work plan. This is usually a document that carefully specifies sample locations, analytical methods and data quality requirements, but in so doing removes the ability of field personnel to react to site conditions. Thus, it is a "cookbook" approach, usually written is such detail that junior level and entry-level personnel are sent to implement it. In contrast, the dynamic workplan seeks to provide a framework within which to conduct the investigation, but leaves much of the detailed decision making to the field personnel. The focus of a dynamic work plan is what decisions will be made, how will these decisions be made, what instruments will be used to collect data for these decisions and who will make the decisions. The principal components of a dynamic work plan are:

- Identification of the Technical Team
- Description of the Conceptual Site Model (CSM)
- Description of the Decision Making Framework and Decision Logic
- Identification of Field Analytical Methods (FAMs) to be Used
- Discussion of Program Objectives and End Points
- Data Management Plan

The creation of a dynamic work plan consists of two parts, planning the investigation and preparing the work plan

4.3.1 *Planning the Investigation*

The planning step sets cleanup goals, data quality objectives, field lines of communication, site closure requirements, team members and best technology are all discussed by all individuals who will be involved in the investigation (who have a stake in the outcome) and agreed upon prior to any fieldwork. This planning is then used throughout the program. The components of systematic planning are:

Areas of Concern (AoCs)

Areas of concern are locations at the site that could possibly be the sources of contamination. They may be underground tanks, chemical storage areas, spills/stains of unknown chemicals, landfills, lagoons/sludge deposits, industrial process areas, and discharge points. Sources of information that will assist in locating these areas are old aerial photographs, compliance records, interviews with current and former employees, manufacturing processes/disposal records and fire insurance maps. Also, a visit to the local regulatory office is important. Tables 1 and 2 provide information on common AOCs and typical of contaminants found at hazardous waste sites.

TABLE 1. Common Sources of Hazardous Substances and/or Areas of Environmental Concern

Area of Concern	**Items to Investigate**
Storage and staging	Drum storage pads/loading docks, dumpsters, rail sidings, lagoons and ponds
Drainage systems	Floor drains, storm sewers, leach fields, dry wells, sumps
Bulk Storage	Underground tanks, piping, above ground tanks
Buildings and facilities	Electrical transformers, insulation and building materials with asbestos, floors, walls, trenches, drains, sumps, boilers, vents, laboratories, chemical storage areas
Areas of discharges	Stressed vegetation, drain pipe discharges, lagoons beneath pipes and sewers, compressor blowdown
Discharge from off-site	Drainage or flooding, surface spills, groundwater flow on and off site
Historic Fill	Coal and coal ash residues, slag (metal waste), dredged material

TABLE 2. Primary Contaminants of Concern

Top 20 Contaminants of Concern	
Trichloroethylene	Chrominum
Dibenzo(a,h)anthracene	Dieldrin
Hexachlorobutadiene	DDE
Arsenic	Lead
Mercury	Vinyl chloride
Benzene	Polychlorinated biphenyls (PCBs)
Cadmium	Benzo(a)pyrene
Polycyclic aromatic hydrocarbons (PAHs)	Benzo(b)fluoroanthene
Chloroform	DDT

Development of a Conceptual Site Model (CSM)

This is a critical component of the dynamic approach and a basis for field decision-making. The CSM contains the best information available before the start of fieldwork. It is created from any information that can be developed from background searches, site histories, aerial photograph review, etc. and is an assemblage of roads, buildings, hydrogeology, geology, hydrology, critical receptors, transportation pathways, contaminant concentrations and any other pertinent information. The CSM is the basis to develop a consensus among the stakeholders regarding site components and more importantly, is dynamic in nature, changing to reflect the increased site knowledge gained from FAMs. It provides the basis to integrate new information into the program and guides the field decision logic. Figure 1 provides an example of a typical CSM.

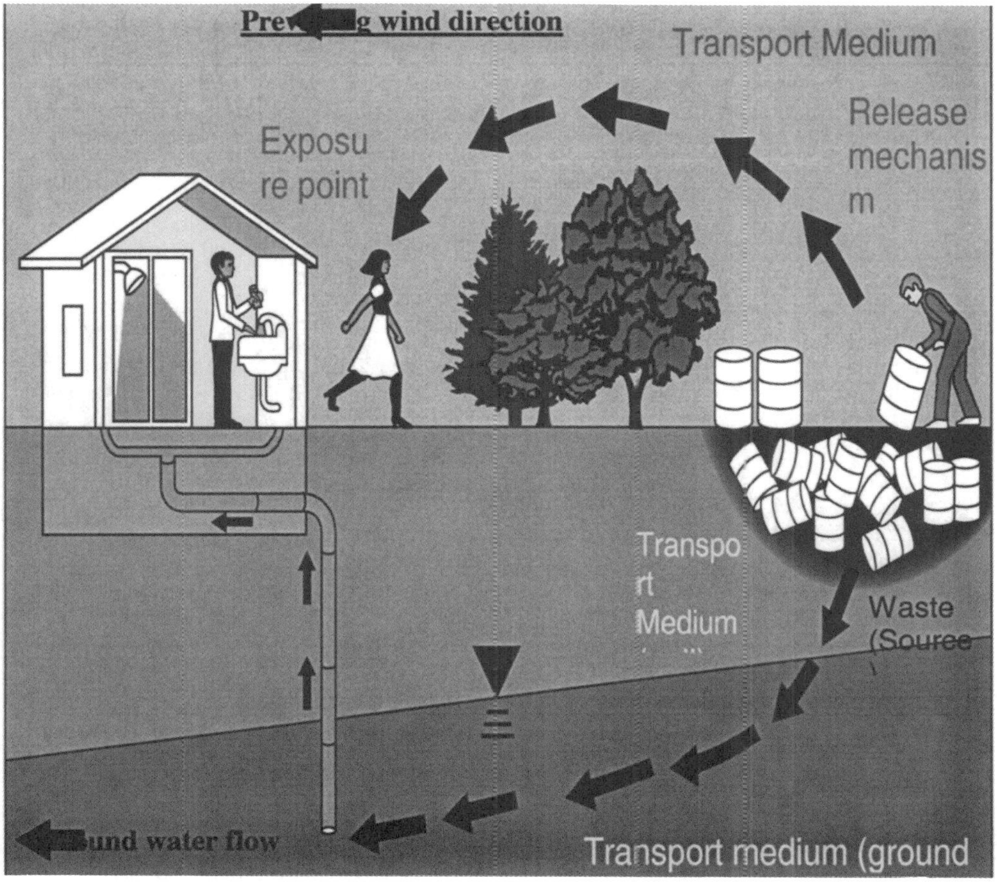

Figure 1 Depiction of a Conceptual Site Model

Technical Team
A fundamental criteria for a successful dynamic investigation program is field decision making. This mandates that the technical field team be composed of experience-qualified personnel to interpret the data as it becomes available. It is especially important that personnel involved with the planning be members of the field team so that there is a continuity of knowledge throughout the process.

Identify the Communication Strategy
This defines the lines of communication needed for field decisions, designates who will interact with regulators and overseers, and establishes methods to assure optimum transfer of data such as software visualization, graphical displays and database management.

Identification of Contaminants of Concern (COCs)
In addition to identifying potential areas of concern, waste type and medium of interest, this also includes considerations of the effects of matrix interference on FAMs, fate and transport potential, exposure scenarios and any general chemistry parameters needed such as pH, DO, Eh, etc.

Data Quality Objectives
This involves the information value of the data set. It includes a statement of the problem, identification of the measurements to be made, action levels for field decisions, acceptable limits on measurement and decision errors, allowable modification to measurement technologies to account for site-specific interference's and end points to the characterization process. Embodied in the data quality objectives is the concept of effective data. This is defined as data of known quality that are useable (i.e. effective) to directly support primary project decisions. Many data points of "lower quality" data can provide higher information value than fewer data points of "higher quality". The former will produce an information set that reflects the true site conditions. An illustration of this concept is contained in Figure 2. Note that fewer higher quality data points are less likely to find a "hot spot" than many slightly lower quality data points.

The following are the major stages of the data quality objectives in the planning process

- State the problem to be addressed
- Identify the decision(s) to be made
- Identify all the inputs to the decision(s)
- Narrow the boundaries of the study
- Develop a decision rule(s)
- Develop uncertainty constraints
- Optimize the design for obtaining the data

Establish Cleanup Goal/Action Levels
As part of the strategic planning, this component includes defining regulatory requirements and the appropriate regulatory agency that implements them, establishing background or anthropogenic levels for organic/inorganic COCs, determining preliminary risk-based cleanup goals and establishing cleanup goals in concert with the regulators. Additionally, this includes
identification of the data needs to support decisions, defining the boundaries of the study area and determining the limits of sample collection.

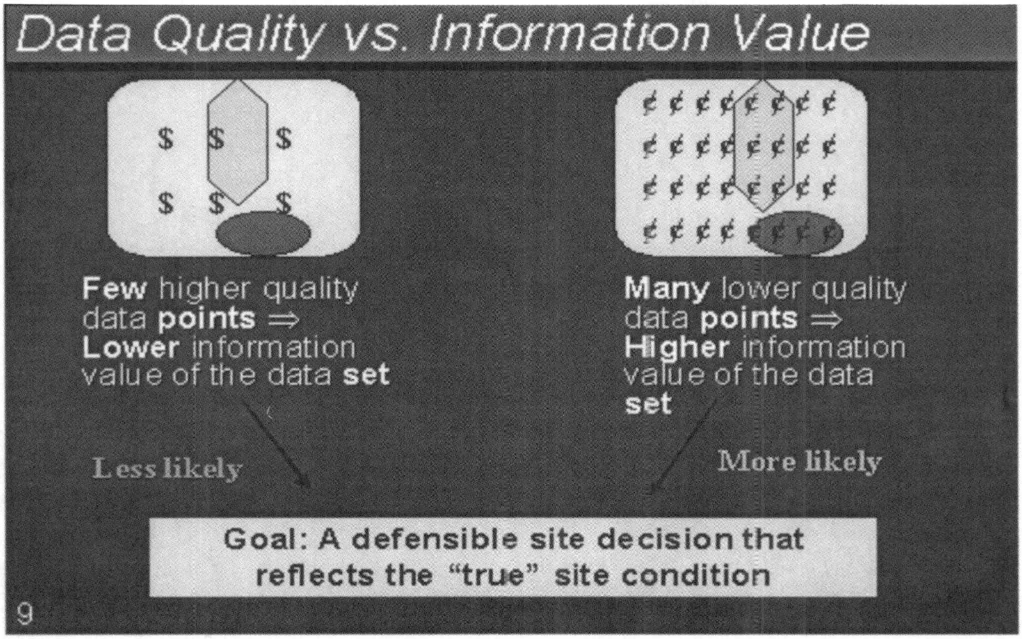

Figure 2 Data Quality versus Information Value

The Planning is not completed until all stakeholders are in agreement. Thus, this aspect of an ESC program may go longer than a traditional RI because the planning objectives are more rigorous than the traditional RI process, but results in an alignment of stakeholder understanding. However, this is well worth the effort because once established, the out come of this planning is carried throughout the program, from preparation of the Dynamic Work plan to the field program

4.3.2 *Components Of A Dynamic Workplan*

Creating a Conceptual Site Model (CSM)

The work plan should describe the CSM, including maps, cross sections, three-dimensional images and GIS layouts. The detail of the CSM descriptions is related to the site complexity and magnitude of the ESC program. The work plan should describe the method for integrating new information into the CSM and provides for daily revisions of CSM. GIS can be incorporated into the budget and used as a powerful technique for daily presentation of field data as it is collected and helps in refining the CSM.

Choosing a Technical Team

The technical team needs to be identified in the work plan and will possess the expertise needed for the site. Typically this includes a sampling team, analytical team data management staff and field team leader (FTL). The FTL has final decision making authority and responsibility for the ESC program. Ideally, the field technical team should be primarily composed of the same personnel who were involved with the Planning. The communication strategy should be specified and mechanisms for regular contact with federal and state regulators are necessary. The field team must be experienced scientists, technicians and other professionals.

Decision Making Framework

This involves both the decision logic and decision rules. A decision logic tree should be developed and described in the work plan. An example is shown in Figure 1. It outlines the basic steps that should be followed in the field decision making process. It provides for decision points that determine the next action to be taken, depending upon a field observation or analytical result. This establishes the overall framework for the program and the following provides more specific rules depending upon certain conditions.

This section of the work plan should include

- Statement of the Problem
- Identification of the decisions to be made that will address each problem
- Identify the information needed for the decision
- Develop decision rules
- Specify acceptable limits on decisions
- Optimize the CSM

The decision rules are guidelines that integrate decision outputs from previous steps into an "if…then…" type statement that defines the conditions that would cause personnel to choose alternative actions. The process requires constant comparison of data acquired to the CSM and adjustments if necessary. This section also needs to discuss data quality objectives, particularly as it relates to decision-making. This would involve a discussion of data requirements needed to support decisions, strengths and limitations of field analytical methods, site specific conditions that could influence the accuracy of a field analytical method and what is the intended use of the data (i.e. remediation design, risk assessment, site closure)

Decision Rules

Field decision rules are used on a daily basis to guide a dynamic investigation. The dynamic investigation mode combines the SI and RI into one phase. Under this mode the results from the previous days sample analysis are used to make next-day decisions about additional sampling (called judgemental sampling), with the goal of completing the delineation of an impacted area in one field mobilization. Thus, in the dynamic workplan mode, data is assessed daily and decisions made about the delineation process.

In order to guide the field team during the implementation of the sampling and analysis, decision rules are developed. These are "if…then" type statements that define the conditions that would cause the decision maker(s) to choose alternate actions and/or take different directions to solve problems. All sites require an initial evaluation to determine the type of environment they exist in. For example, an urban environment with a long history of industrialization will require substantial flexibility in the program to allow for field adjustments in response to unexpected conditions and judgmental sampling to accomplish the delineation. Examples of decision rules are provided in Table 3.

The methodology outlined in Figure 5 embodies the concept of effective data. Effective data depends heavily upon the interaction between sampling design, analytical design and the intended use of the data. Thus, "screening level" and "definitive level" data are integrated to produce the data for making well-defined project level decisions. Screening level data is normally defined as data produced by field instrumentation that does not receive rigorous quality assurance normally associated with "approved" laboratory based analytical methods. Definitive data is considered data that has been subject to data validation analysis and is usually considered a precise measurement of the quantity of a contaminated present in a particular media. Both types of data are necessary to achieve a complete delineation of an impacted area. The goal of a dynamic approach to site characterization is to combine these two types of data in the most efficient and cost effective manner.

TABLE 3. Examples of Decision Rules

Activity	**Decision Rules**
Initial sampling at an AOC	1. If results of FAMs indicate concentrations below action level, then collect laboratory sample for verification 2. If results of FAMs indicate concentrations above the action level, then begin "step out" delineation sampling
Step out sampling	1. If concentrations at first "step out" sampling location are greater than action level, then collect next sample similar distance. 2. If concentrations at first "step out" are less then action limit, then step inward one half distance and re-sample 3. If sample is 25% less then action level, then collect sample for laboratory verification of delineation.
Depth sampling	1. If contamination of concern (COC) exceeding action level has a water solubility greater than 100 mg/L at 20 C, then collect depth soil sample directly above saturated zone. 2. If COC has a water solubility less then 100 mg/L at 20 C, then collect soil samples at two foot intervals until concentration are below action levels.

The flow chart (contained in Figure 3) identifies points in the data assessment process where different levels of analytical method definitiveness are applied. The assumption that the use of definitive "approved" analytical methods is the only way to produce decision quality data ignores a fundamental aspect of site characterization. This type of data is a result of a chain of events that include sample collection, sample strategy, preservation, transportation and analytical methodology. In many cases a more accurate picture of the site can be gained when numerous samples are analyzed with a less precise methods (i.e. screening level), even if the analytical method is less accurate. Considering this concept, the flow chart attempts to blend the use of FAMs for maximum sample analysis with more acceptable "approved methods" for QA/QC. The actual AOC specific degree of blending is at the discretion of the field team manager.

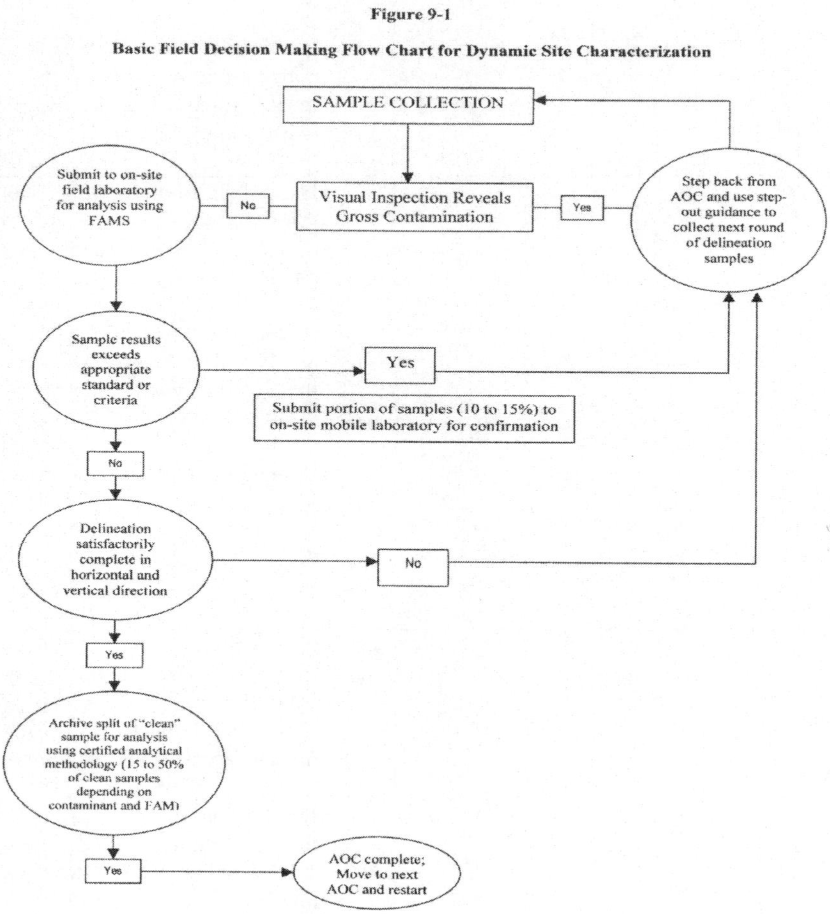

Figure 3- Field Implementation of Data Feedback Loop

Figure 2 illustrates the concept between the information value of a data set and the quality of individual datum. Assume that the data points depicted in Scenario B were generated using a "screening method", which may have less precision and potentially more interferences than the more expensive method shown in Scenario A. However, the Scenario B method is less expensive and can be used to generate results within hours of sample collection. Since Scenario B can provide more density and the real time results can be used to direct the sampling during the investigation, then the analytical method in Scenario B can produce a more accurate representation of the distribution of site contamination and hence have a higher information value. If the analytical method used in Scenario B is occasionally field checked by an approved methodology (either in a mobile or fixed based laboratory), then the team has confidence in the data quality as well as higher density for more accurate delineation. Thus Figures 2 and 3, coupled with decision rules form the core of the dynamic approach to site investigations. FAMs, which produce more data in a real time setting can provide higher density to sampling patterns. Decision rules guide the use of the FAMs as the sampling program expands and progress. More precision is introduced into the field program as needed, for example in situations where action level concentrations are approached and verification samples are collected to prove that an impacted area has been delineated. Higher density of sampling, using FAMs, provides a much greater assurance to the field team that the imported area will be properly delineated and the appropriate remedial approach selected.

Program Objectives and Endpoints

One major concern with dynamic field programs is how to define endpoints to complete the field program. The objectives of the program should be clearly established in the workplan such that end points are identified. For example, the endpoint maybe a "hot spot" delineation or another endpoint could be plume delineation. The workplan should discuss the level of data sufficient to define the hot spots, and including density and spatial distribution. Higher density, lower quality data points will more likely define a hot spot than few high quality data points.

Data Management Plan

Since the success of a dynamic investigation program using FAMs is based upon the teams ability to integrate a variety of field information on a daily basis, data management is critical. Protocols should be established for logging, analysis, data reduction and site mapping. The work plan should define responsibilities for data collection, assimilation and presentation. A computer database should be used as much as possible. The data can be downloaded to a computer containing site visualization software for conceptual model update and review.

A system should be in place to compile data and assemble information that is collected on site following specific protocol. This is especially important with field-based measurements since data can be generated quickly, continuously and not necessarily in a prescribed sequence. The cataloging and compiling of this information in a systematic process is imperative to the proper interpretation of the data and ultimately success of the program.

Once data has been collected interpretation of the data is necessary. Typically this process takes into account the following elements:

1. Statistics
2. Findings
3. Comparison of findings
4. Standards, guidelines or criteria

The statistics applied to evaluate the site will be dependent on the amount of samples and the field methods used during the analysis process. Findings from the site characterization and comparison of those findings to laboratory confirmation is essential to validate the data quality and ensure data confidence. Using this information to identify the receptor exposure and provide site findings is the final step in managing and processing the data. The conclusions that are drawn from the process will assist the site engineers in determining site contamination, remedial action steps, monitoring frequency, and site clean-up effectiveness.

Choosing the Correct Field Analytical Method (FAMs)
The work plan should identify the FAMs that will be used, and provide descriptions of their standard operating procedures (SOPs). The SOPs include identification of operator expertise required, QC procedures, data reporting format and other information pertinent to a particular field method generation of useable data. The following elements must be addressed when choosing the appropriate FAM:

TABLE 4. Elements to be Considered when Choosing a FAM

FAM Elements
Sample preparation,
Instrument operation
Method ranges
Potential interference's
Calibration procedures
Reliability of the method and procedures
Operator training requirements
Availability
Sample through-put
Detection limits versus action levels
History of use

Sample collection and preparation are fundamental components of obtaining high quality data in the field. The sagacious approach to using field-based methods will consider every element listed above. If samples are not digested sufficiently or prepared to give the instrument the ability to 'see' the analytes of interested then the method will be feckless in giving meaningful information about a site's contaminant concentration.

Users must be taught proper instrument operation to ensure quality data is generated. This component of FAMS is significantly different than the conventional way of performing site investigations. The user is the field engineer and not a laboratory chemist. Therefore, it is important that all field personnel that will be in the sampling and testing process receive proper training.

The instrument operation can vary given different environmental conditions. For example, in extremely cold conditions the instrument's LCD screens can freeze requiring the user to transport samples into cars or vans on-site for analysis. Understanding the effects of temperature, humidity and contaminate concentrations will enable data of known quality to be collected. This can also relate to interferences by other contaminants or by the substrate measured. If shielding is occurring due to matrix effects or the contaminant concentration is higher than the instrument's linear dynamic range then the results will be meaningless. To effectively address this problem the user must be trained on the instruments limits of detection and range of concentrations as well as any interferences that may effect the results.

Creating a routine check of the instrument performance and looking at user variations will also show problems that may not otherwise be detected that could affect the data quality. This will identify if the instrument is functioning correctly, if users have a specific bias and if the checks are bracketed through-out a given number of measurements then instrument drift can be identified and corrected for on those samples without loss of significant information.

5. Quality Assurance/Quality Control For Field Analytical Methods During Site Assessment

5.1 INTRODUCTION

Barriers have existed against the acceptance of using FAMs in site assessment often because the methodologies were perceived to lack rigorous QA/QC. The methods were believed to be only screening in nature and not as technically sound as the traditional, fixed-laboratory methods. In fact, field analytical approaches to site assessment may not only be faster and cheaper in providing site decision information, but also provide data of higher decision-making quality. This is provided as long as a carefully designed and implemented QA/QC practice is adopted.

5.2 MAJOR ELEMENTS OF A QA/QC FOR USING FAMS

Essential to this is that the site assessment be undertaken with: (1) carefully planned and well defined measurement objectives, (2) undertaken with trained, qualified people, (3) employ well developed and proven measurement methods, (4) utilize sound QA/QC practice which provides a known state of control around the measurements as well as definition of achieved data quality (5) a complete documentation of the sampling/analytical process and (6) a final review of the data. Careful planning of the whole QA/QC process must be conducted and managed in a way that can address real-time changes because of the often challenging and unforeseen situations which arise while carrying out a dynamic sampling/analytical activity.

5.2.1 *Planning and Statement of Objectives*

USEPA has spent considerable effort in developing planning guidance, known as Data Quality Objectives (DQO's) which can be applicable to site assessment planning.(USEPA "Guidance for Data Quality Objective Process," QA/G-4(USEPA, 2000c). This process includes seven steps.

- State the problem
- Identify the decision
- Identify inputs to the decision
- Define study boundary areas
- Develop a decision rule
- Specify decision error limits
- Optimize design for obtaining data

There are many potential questions to be answered when carrying out site assessments about the source of contamination, the pathways and the receptors, as well as how to manage the remedial process. It is very important for quality information and sound decision making that the specific objectives for a field assessment are clearly stated and well understood by all involved. Often the site conceptual model (CSM), discussed previously, is used as an important tool to understand, state and communicate these objectives. Development and statement of the objectives includes use of all available background information, an understanding of the decision that has to be made regarding the site, an understanding of available resources and constraints and an idea from the decision makers how much risk they are willing to take with data quality when making the site related decision. For example, a decision might be based on how far the hot zone of contamination extends. Or it might be to decide if a site is cleaned up to standards. Both of these can use site assessment field data and it is important to know how good the data needs to be; how much the decision makers are concerned about the consequence of a wrong decision. Once the objectives have been clearly worked out, the technical design of the sampling and analysis is undertaken. For dynamic studies this includes selection of Field Analytical Methodologies, sampling strategies, decision rules and all necessary quality controls to assure that the data meets the quality requirements for the decision at the defined risk levels.

5.2.2 *Use of Qualified, Trained Personnel*

Dynamic investigations based by using FAMs will generally save time, money and even provide better representative data on conditions that they are undertaken by well trained scientists. Since FAM approaches affect quick field decisions, they generally require on site a variety of multi-disciplined field personnel who have a higher level of professional training and experience. For site assessments this might include personnel experienced in geology, hydrology, soil science, statistics, chemistry, biology, field sampling and quality assurance/ quality control practices. In the United States, there has been some concern by regulatory officials about the inconsistency in personnel requirements between data generated in the laboratory versus in the field. While there has been much activity in building environmental certification programs for personnel in off-site facilities engaged in environmental testing, there has been little in the way of certification for field data. Regardless of how this is resolved, one thing is clear - equal or even higher quality standards of practice need to be adopted for field analytical data.

5.2.3 *Identifying Appropriate FAMs*

Basically, the same selection rationale is used for FAMs selection as is used for selecting methods for fixed based laboratories. However, selecting FAMs also requires two additional considerations. The first is that trade-offs might be allowed in measurement precision and accuracy, if use of the FAM provides many more data points resulting in increased ability to represent actual site conditions. The second is the need to select methods and associated technologies that will stand up and be practical under field conditions.

The first step in the selection process includes identifying all candidate methods and technologies that might be applicable. These comparisons are obviously made in terms of measurement time, sample flow through time, project cost and performance. Ideas for candidate FAM's might be obtained from manufacturers' advertising, recommendations from other professionals, or various organizations using web sites (see Appendix A).

After candidates are identified, a careful review must be made of their field performance for selectivity and sensitivity. This review needs to be done with an understanding of project measurement objectives determined previously in the planning process. In considering the selection, one should be open to considering any possible trade-offs of FAM performance. For instance by combining two FAMs together, both with lesser selectivity power, might produce more certainty about contaminant identity.

Review should be made of all manufacturers claims and documentation of their FAM's performance. This should be supplemented with looking at independent evaluation information. These are usually available in written methods or in reports published on web sites. Several sources of information are the EPA Clu-In site, SW-846 drafts, Environmental Technology Verification (ETV) reports, the Federal Remediation Technologies Roundtable, and locations on the USEPA website such as REACH IT and FATE (see Appendix A). Sometimes it will be necessary to conduct field evaluation on samples that are representative of the soil matrix of interest. This will include comparing the candidate FAM's performance to known values or those obtained from reference laboratory methods.

In addition to meeting performance requirements, FAM's must be able to be logistically applicable for use at the site. This includes consideration of a number of factors including ruggedness to hold up under field and transport conditions, weight and size for easy handling, usability, licensing requirements, utility requirements, security, and training needs. Also, a review should be made of any issues pertaining to safety or disposal of testing materials.

5.2.4 *QA/QC practice*

Quality Assurance (QA)

Quality Assurance is a system of management activities designed to ensure that a service or product meets the required levels of quality. As we have said, the QA process starts at the early planning stages and continues to manage the entire information collection to assure that it will provide needed quality for the specified decision. Classical QA activities include quality assurance plans, standard operating procedures and QA audits.

Quality Assurance Plans (QAP's) link between the data quality objectives (DQO's) in the decision making process (DQO's), and the actual field implementation. They provide the optimized designs for the data collection projects. These address the personnel and equipment requirements, sampling design and decision rules, methodology that will be used, calibration procedures, quality control, data handling, interpretation and documentation.

Standard Operating Procedures (SOP's) are cook-book like versions of carrying out the various site assessment tasks such as sampling, field measurement and handling of samples and data handling. These are generally referenced in the QAPP's and maintained at the site with the field team. They assure all important details are followed and errors are prevented.

Quality Assurance audits are carried out by individuals who are independent of the routine sampling and analysis. They include Technical System Audits which are reviews of the sampling and analysis procedures, and Performance Audits which introduce check samples either blind or known into the measurement process to check the proficiency of the field measurement personnel

Quality Control (QC)

Quality Control consists of designing controls around the entire sampling/measurement system to assure that undesired error (determinate error) error in the process is identified and corrected and that anticipated error (indeterminate error) is measured and known when using the data for decision making. There are many different quality controls used when carrying out field analytical approaches to site assessment. Basically, they fall into five broad categories: calibration standards, blanks, spikes, replicates and split samples with off-site fixed laboratories.

Calibration standards are integral to analytics and they are used as part of most methods to determine the instrument or kits response against a known quantity. These materials or devices are also used to periodically check and monitor the FAM's response. Field instruments move around and are subject to losing calibration and they must be checked initially and at frequent intervals throughout the day. Field kits need to be checked for stability of packaged reagents which can deteriorate over time.

Blank samples are introduced into the process at various times to detect and monitor contamination and to establish a zero baseline condition. They can be used during sample collection, preparation, both in the field and laboratory analysis. They can help identify undesired contamination getting into the process and include trip blanks, field blanks, sample storage blanks, equipment rinse blanks, reagent blanks and method blanks.

Spiked samples are those to which a known concentration of target analyte is added to an environmental sample or blank before the analysis in order to see if the analyst and/or method can recover that same amount back. These are useful in monitoring method interference and performance as well as personnel proficiency. They include performance evaluation samples, laboratory control samples, internal standards, surrogates and matrix spikes.

Replicate samples are introduced in the process to understand and monitor precision of the measurement. This is usually accomplished by introducing two or more samples and measuring the dispersion as relative percent difference or relative difference.

Splitting samples with off-site laboratories is often used with FAM's. A comparison of the FAM results is made with the off-site laboratory, which uses the reference method. This is sometimes done on a pilot scale before the actual field activity begins in order to assess the performance of the FAM under the condition of the actual site. The use of this cross-comparison is often done in four ways during the actual carrying out of the field work portion of the site assessment. First, a higher frequency of sample splitting with the fixed laboratory occurs during the early phase of field analysis. This tracks the FAM performance closely. As confidence is gained there is a dropping off the amount of splitting. The second way is to send samples to the off-site laboratory at an even rate, for example, 10% of FAM's are checked. The third way is to send only samples that are very close to an action level or standard where a critical decision needs to be made. The fourth way is to send only samples for cross check that are taken at a final decision point, for example, post-cleanup samples to show that remediation is complete and that no further action needs to be taken at that site (see Figure 6).

5.2.5 *Documentation*

Even if highly skilled workers generate excellent quality field data during a site assessment, it is of little or no value if it can't be supported and defended at a later date. Because of the dynamic nature of FAM approaches, it becomes even more imperative to maintain a rigorous documentation process. This includes documentation of plans, FAM selection pilot studies, detection limit studies, QC information, sampling documentation, chain of custody records, FAM SOPs and field sampling/analytical logbooks. The documentation should be able to reveal the who did what, when and how. It should also clearly show all changes and corrections with the supporting rationale why these were implemented.

5.2.6 *Final Review of the data*

This data review consists of data verification, data validation and data quality assessment. When FAM approaches are being used these reviews are often carried out in the field while the sampling and analysis is being conducted.

Data verification generally is aimed at making sure the data that is reported was the actual data that was generated. Reviews are made for any transcription errors from manual records and logbooks that have been entered into electronic databases. Data validation is the review to make sure that the data is the quality needed for the site assessment project. It is made by an independent person who looks at various records including chain of custody, shipping reports, holding times, calibrations, chromatograms, spectra, spike and replicate QC data, and corrective action reports. Data quality assessment is the final step in the review process and these reviews are used to determine if the data set, as a whole, is appropriate for the decision to be reached.

6. Summary

Technology advances during the 1990's have vastly improved the performance of field analytical methods and instruments, to the point where they are capable of having a significant impact on the way in which environmental site investigations are performed. Traditional site investigations, which rely on fixed base analytical laboratories to perform sample analysis using ridged standard methods, are being supplemented by dynamic site investigations, which use FAMs to analyze samples on site, thus generating real time data for in field decision making. Studies by the Department of the Defense and the USEPA have shown that dynamic site investigations using FAMs are more efficient, saving time and money while providing more accurate identification and delineation of site contaminants.

There is a wide range of FAMs available, though only a few have been applied consistently enough to have an adequately documented performance history. Perhaps the most commonly used FAMs are X-Ray Fluorescence (XRF), field portable gas chromatography (GC) and GC/MS systems and immunoassay (IAA) test kits. XRF uses x-rays to cause florescence at the atomic level and is used to determine the presence and concentration of metals on soil. Field portable GC and GC/MS systems are used to measure the concentrations of volatile and semi-volatile compounds in soil water and air. With the advent of field portable MS systems, compound libraries can be taken into the field, enabling the rapid identification complex organic site contaminants. Immunoassay methodologies use the ability of antibodies to selectively bind to the target analytes in a sample matrix. IAA test kits are easy to use and are capable of detecting a wide range of organic contaminants, including gasoline, diesel fuel, jet fuel, BTEX, PAHs, PCBs, TPH and certain pesticides. In addition to these common FAMs, there are other methods that have evolved from "test kits" or use innovative developing technology to extract samples and detect specific compounds. However, it must always be remembered that FAMs need to be operated by trained personal, QA/QC procedures carefully followed and that the appropriate regulators contacted prior to their application. Additionally, it is recommended that that a short pilot test be performed before full-scale use to account for any unusual site matrix interferences.

Field analytical methods are most powerful when coupled with a dynamic site investigation. This is because FAMs provide information about the distribution of site contaminants in a "real time" mode, which in turn allows for daily decision-making about the progress of the investigation. Dynamic site investigations are designed to provide flexibility in selecting sample locations, leaving much of the detailed decision making to the field personnel. Thus, FAMs play a critical role in that they supply the needed information to the field team so they can make decisions on sample placement and adequacy.

Preparation for a dynamic site investigation consists of two steps, detailed planning and a dynamic work plan. The planning step includes identification of the areas and contaminants of concern, creation of a conceptual site model, selection of the skill sets and personnel that will make up the technical team, establishment of the communication strategy, identification of the data quality objectives, agreement on the cleanup goal/remedial action levels. The dynamic work plan consists of a description of the conceptual site model, identification of the technical team (particularly the team leader), a discussion of the decision-making framework (this involves both the decision logic and the decision rules), description of the decision rules, a data management plan and identification of the selected FAMs (SOPs, QA/QC procedures, operator training, etc). Overall the dynamic work plan seeks to provide a framework within which to conduct the investigation and provide guidance for field decision-making. The focus of a dynamic work plan is what decision will be made, how will these decisions be made, what instruments will be used to collect the information for the decisions and who will make the decisions.

Finally, it is important to emphasize the need for strong QA/QC when using FAMs. Barriers have existed against the use of FAMs often because these methods were perceives to lack rigorous QA/QC. Therefore, it is even more critical that QA/QC procedures be established and closely followed. Major elements of a QA/QC plan for FAMs include carefully planned and well defined measurement objectives, properly trained experienced personnel, well developed and proven measurement methods, definition of achieved data quality and complete documentation of the sampling/analytical process.

7. Resources For More Information About Innovative Site Characterization And Field Technologies

General Environmental Sites

· *Air Force Center for Environmental Excellence (AFCEE)*
Provides a complete range of environmental, architectural and landscape design, planning and construction management services and products
Contact: http://www.afcee.brooks.af.mi

· *AuditSource*
Environmental/risk assessment audits for industry
Contact: www.afcee/brooks.af.mil

· *American Society for Testing Materials (ASTM)*
A not-for-profit organization that provides a forum for producers, users, ultimate consumers, and those having a general interest (representatives of government and academia) to meet on common ground and write standards for materials, products, systems, and services.
Contact: http://www.astm.org/

· *clay.net –*
Work platform designed specifically for environmental consultants and remediation professionals
Contact: www.clay.net

· *Department of the Navy Environmental Programs*
The Department of the Navy's cleanup program identifies, studies, and cleans up past hazardous waste disposal sites on Navy and Marine Corps installations within the United States and its territories. The Navy began the Installation Restoration Program in the early 1980s based on the requirements found in the Comprehensive Environmental Response, Compensation, and Liability Act and the Resource Conservation and Recovery Act.
Contact: http://enviro.navy.mil/

· *Envirobiz-*
International Environmental Information Network
Contact: www.envirobiz.com

· *EPA Brownfields Initiative*

EPA's Brownfields Initiative will empower States, communities, and other stakeholders in economic development to work together in a timely manner to prevent, assess, safely clean up, and sustainable reuse brownfields

Contact: http://www.epa.gov/swerosps/bf/

· *EPA Government Information Locator Service (GILS)*

Locator service for all EPA government information

Contact: http://www.epa.gov/gils

· *Environmental Protection Agency (EPA) Home Page*

Provides information on EPA publications, offices, laboratories, initiatives, contracts and grants, and much more

Contact: http://www.epa.gov/

· *EPA National Library Network Program*

The mission of the EPA Library Network is to improve access to information for EPA decision making and environmental awareness.

Contact: www.epa.gov/natlibra/

EPA Office of Solid Waste and Emergency Response (OSWER)

Information about the hazardous waste program including identification, generation, management and disposal of hazardous wastes

Contact: http://www.epa.gov/swerrims/index.htm

· *Fedworld*

A one-stop location for locating and ordering government information.

Contact: www.fedworld.gov

· *Global Network of Environment and Technology (GNET)*

Brings together the information, resources, and people that shape the environmental technology marketplace

Contact: www.gnet.org

Groundwater.com

Groundwater.com will cover groundwater equipment, groundwater published papers, groundwater news items, groundwater magazines, groundwater consultants, groundwater software, a groundwater job search page, groundwater links, and groundwater publications, among others.

Contact: www.groundwater.com

· *Ground Water On-Line*

Ground Water On-Line® is a database containing more than 82,000 ground water literature citations with key words, abstracts, chemical compounds, biological factors, geographic locations, authors, titles, publication source names, and more. Each citation may contain up to 25 fields of information. Documents that are indexed include scientific, technical, and trade journals; newsletters; books; government documents; university reports; dissertations and theses; state publications; and proceedings of national and international conferences and symposia. The collection is the largest and most comprehensive of its kind in the world. NGWA members automatically receive free, unrestricted use of the on-line database, while nonmembers can subscribe for an access fee.

Contact: http://www.ngwa.org/gwonline/index.html

· *Hazardous Substance Research Centers (HSRC)*

The Hazardous Substance Research Centers (HSRC), a national organization, carries out an active program of basic and applied research, technology transfer, and training. Activities are conducted regionally by five multi-university centers, which focus on different aspects of hazardous substance management.

Contact: www.hsrc.org

· *Illinois Waste Management and Research Center Home Page*

The Waste Management and Research Center (WMRC) helps businesses and citizens reduce and manage solid and hazardous wastes released to air, water or land. WMRC makes its headquarters on the campus of the University of Illinois at Urbana-Champaign and is a division of the Office of Scientific Research and Analysis in the Illinois Department of Natural Resources. WMRC also operates offices in Springfield, IL, Oak Brook, IL, and Southern Illinois.
Contact: http://www.wmrc.uiuc.edu/

· *International Environmental Technology Centre*

Promoting Transfer of Environmentally Sound Technologies (ESTs)
Contact: http://www.unep.or.jp

· National Association of Environmental Professionals

The NAEP HomePage is now located (mostly1) at Verio.com; but our website architecture is as dispersed as our chapters, reflecting both the geographic and professional diversity of our membership. A list of the State, Regional, and Student Chapters is maintained by the Hawaii Chapter (HAEP) with links to all those chapters that have developed websites.
Contact: www.enfo.com/NAEP/

· *National Risk Management Research Lab (NRMRL)*

The National Risk Management Research Laboratory, (NRMRL) which is part of EPA's Office of Research and Development, conducts research into ways to prevent and reduce risks from pollution that threaten human health and the environment. The laboratory investigates methods and their cost-effectiveness for prevention and control of pollution to air, land, water, and subsurface resources; protection of water quality in public water systems; remediation of contaminated sites, sediments and ground water; prevention and control of indoor air pollution; and restoration of ecosystems.
Contact: http://www.epa.gov/ORD/NRMRL/index.html

· *National Technical Information Service (NTIS)*

The National Technical Information Service is the U.S. Government's central source for the distribution of scientific, technical, engineering, and related business information. This information is produced by or for the U.S. Government and complementary material from international sources. Nearly 3 million products are available from NTIS in a variety of formats, including microfiche, paper, diskette, audiovisual, CD-ROM, and online.
Contact: www.fedworld.gov/ntis/

· *National Technology Transfer Center (NTTC)*

The National Technology Transfer Center (NTTC) was established in 1989 by congress to provide American companies and individuals with access to federal R&D to better enable them to compete in the international marketplace. Staffed with technology transfer specialists, technical area experts and information management professionals, the Center provides technology transfer support via several different areas: access to $70 billion worth of research and development, 100,000 research professionals at over 700 federal laboratories and universities; technology assessment services; product testing and prototyping and professional testing and development. These services build upon each other to create NTTC's full service commercialization center.
Contact: www.nttc.edu

· *Navy Environmental Leadership Program (NELP)*

Contact: http://nelp.navy.mil/

· *Oak Ridge National Laboratory Environmental Sciences Division (ESD)*

The Environmental Sciences Division (ESD) of Oak Ridge National Laboratory (ORNL) is an interdisciplinary research and development organization with more than 30 years of achievement in local, national, and international environmental research. Scientists in ESD conduct research, develop technology, and perform analyses to understand and assess responses to global and regional change, environmental stress, and resource use. By expanding scientific knowledge and developing technological solutions, we strengthen the nation's leadership in addressing important environmental problems.

Contact: http://www.esd.ornl.gov/

· *Pennsylvania Department of Environmental Protection - Bureau of Land Recycling and Waste Management*

This program includes cooperative multi-site remediation agreements, interim response contracting at contaminated sites, and general technical assistance contracting (GTAC). Learn about the cooperative multi-site agreements with Penn Fuel Gas, the military services and BP Amoco.

Contact: http://www.dep.state.pa.us/dep/deputate/airwaste/wm/

· *Robert S. Kerr Environmental Research Lab - Ada OK*

The Subsurface Protection and Remediation Division of the National Risk Management Research Laboratory in Ada, Oklahoma, conducts EPA-investigator led laboratory and field research to provide the scientific basis to support the development of strategies and technologies to protect and restore ground and surface water quality within a watershed perspective. The Division's research programs include basic studies to enhance understanding of the physical, chemical, and biological processes that control the transport of mass and energy in surface and subsurface ecosystems through the movement of water; the impact of these processes on surface and subsurface ecosystems; and, the application of this process understanding to protect and restore water quality throughout a watershed.

Contact: http://www.epa.gov/ada/kerrlab.html

· *Technical Information Services (TIS)*

The ES&H Information Portal is dedicated to making current information available to the Environment, Safety and Health professionals of the Department of Energy.

Sponsored by the Office of Environmental, Safety and Health, the Portal fosters DOE's commitment to excellence by striving to provide the right information to the right people at the right time.

Contact: http://tis.eh.doe.gov/

· *U.S. Army Corps of Engineers Waterways Experiment Station*

The U.S. Army Engineer Waterways Experiment Station (WES) is the largest Civil Engineering and Environmental Quality Research and Development (R&D) complex in the United States. The strong relationship among the Department of Defense (DoD) Research, Development, Test and Evaluation (RDT&E), U.S. Army RDT&E, and Army Civil Works R&D programs creates a unique research environment that has spawned numerous research breakthroughs and transitions that have enhanced the national capability in meeting warfighting technological challenges. Each WES project is conducted in response to the requirements of Army Civil Works R&D (coastal engineering, flood control and navigation, materials and structures, and environmental quality) and military R&D (airfields and pavements, survivability and protective structures, sustainment engineering, environmental cleanup, computer aided design and drafting, high performance computing, and software engineering).

Contact: www.wes.army.mil

· *Centre for Research in Earth and Space Technology (CRESTech)*

In support of its role to create jobs and wealth in Ontario, and as part of its contribution to the Ontario Centres of Excellence program, CRESTech's mission is to:

· Stimulate and promote collaborative research, innovation and new enterprise through academic, industrial and government partnerships fostered through networking;

· Bridge the gap between pure science and the successful application of science, engineering and technology in profitable new ventures;

- Exploit technology transfer and commercialization endeavours;
- Provide access to technical infrastructure, services and support, assisting industry to achieve their new millenium goals;
- Incubate and foster entrepreneurial spirit while training highly qualified personnel,
- Market the credibility of graduate students and academic researchers to industry and the public sector.

Contact: www.crestech.ca

MOBILITY AND AVAILABILITY OF CONTAMINANTS

P. CORBISIER[1], L. DIELS[1], TISSA ILLANGASEKARE[2], DANNY REIBLE[3], MARTIN REINHARD[4], J. VANGRONSVELD[5]
[1]Vito, Vlaamse Instelling voor Techologisch onderzoek, Boereta Belgium, [2] Colorado School of Mines, [3] Louisiana State University, [4] Stanford University, [5]LUC, Limburgs Universitair Centrum, Diepenbeek, Belgium

The ability to accurately determine the effects of pollutants upon individual organisms or to local ecosystems is hampered by the uncertainty in the quantification of receptor exposure. Through laboratory studies and field observations, it has been demonstrated that the response of an ecosystem or any at-risk population is not controlled by the total concentration of a contaminant in the environment in which a receptor resides, but instead by only that fraction which is biologically available and which can be transported in a phase accessible by the organism. This is particularly true in cases where contaminants are associated with soils or sediments and complex physical and chemical interactions result in a poorly reversible "entrapment" of contaminant molecules. But it is also generally true of any contaminants that are not present in the air or water phases through which exposure typically occurs.

In order to set or to establish realistic endpoints for remediation processes, the physical-chemical and biological mechanisms that govern contaminant release, transport and fate in soils, sediments and associated fluid phases must be understood and quantified. In this chapter the processes that govern portioning between these various phases and the processes that lead to release and migration from soils, sediments and nonaqueous phases will be considered. Both metal and organic contaminants will be considered. A variety of physicochemical and biological processes influence speciation and partitioning in metals and a priori prediction in a particular soil or sediment is not generally possible. As a result, the discussion will focus on experimental assays for determination of partitioning and availability. For organic contaminants, physical properties such as partitioning characteristics can often be predicted and approaches for such predictions are discussed. The use of these partitioning characteristics to describe contaminant fate and transport behavior in both aqueous and nonaqueous phases is then discussed.

1. Metals Partitioning And Availability: Improvement Of The Biomet Kit For Toxicity And Metal Bioavailability Measurements In Soils

1.1. INTRODUCTION

The bioavailability of pollutants in soil has now been recognised as an important parameter when risk associated to a pollution needs to be taken into account [1,2]. Most metals are persistent in soils and change in chemical form can occur over time. Those changes can profoundly influence the solubility and bioavailability and hence the toxicity of the metal. Different analytical methods have been designed to evaluate those changes [3]. But the most trivial way to determine the availability of a metal for a biological entity is to measure the amount of metal, which has been accumulated upon exposition. Those approaches have been widely used for example by measuring the metal accumulation in plants [4], in oligochaete

D. Reible and K. Demnerova (eds.), Innovative Approaches to the On-Site Assessment and Remediation of Contaminated Sites, 31–65.

worms [5, 6] or even in rats [7]. Other approaches concern cytotoxicity or reproduction criteria measurements on soil invertebrates. Most biological methods are very relevant but time consuming and often require food to be added in case of the test system. On the other hand theoretical consideration have been followed and reliable mathematical models were also developed [8, 9]. Other systems for the estimation of free ions initially developed for waters samples [10] such as the diffusive gradients in thin films techniques have been later on adapted to soil solutions [11]. The prediction of soil trace metal bioavailability using extractions has been studied extensively [12].

To detect sub-lethal concentrations of metals in a soil organism, the induction of certain metal-induced genes or stress genes was also studied. During those last ten years, we have been following the expression of specific metal induced genes in a soil bacterium [13-16]. The expression of the genes was followed in situ and on-line by measuring the activity of *lux* reporter genes fused to metal-induced promoter sequences. Those bacterial strains and the all processes were optimised and developed as a kit later called BIOMET test [17, 18]. The main characteristics of those bacterial strains used are listed in table 1.

The BIOMET test has been compared to chemical and biological tests and appeared to be very relevant for the detection of bioavailable heavy metal pollution in soils or other matrices such as sediments, fly-ashes [19], and slags.

The bioluminescence signal is proportional to the amount of bioavailable metals present in the test as shown for example in Figure 1 for a Zn-sensor. However because of the nature of the reporter system used some limitations of the test were noticed. The principle of the BIOMET test is to measure a bioluminescence when the bacterial sensor is mixed with a soil slurry. All chemical parameters affecting the availability of the metal for the soil bacteria, such as soil pH or organic matter are integrated in the measurement. The sensor is indeed measuring the soil matrices effect on the availability of the specific metal.

TABLE 1. Properties of the most used metal induced BIOMET® sensors

Name	Bacterial Host	Metal ions	DL (μM)	DR (μM)	Reference
AE1239	*Ralstonia silverii DS185*	Cu^{2+}	2	2-40	[13]
AE1433	*Ralstonia metallidurans CH34*	Zn^{2+} Cd^{2+}	5 5	5-250 1-25	[13]
AE2440	*Ralstonia metallidurans CH34*	Cr^{6+}	1	1-40	[15]
AE1110	*Ralstonia metallidurans CH34*	Tl^{+}	1	1-50	[13]
AE2515	*Ralstonia metallidurans CH34*	Ni^{2+}	0.08	0.08-100	[17]
AE2450	*Ralstonia metallidurans CH34*	Pb^{2+}	0.5	0.5-18	[15]
AE2613	*Alcaligenes eutropha N9A*	Hg^{2+}	0.01	0.01-1	Not published

DL : Detection level with a signal/noise ratio of 2.
DR : Dynamic range of the sensor.

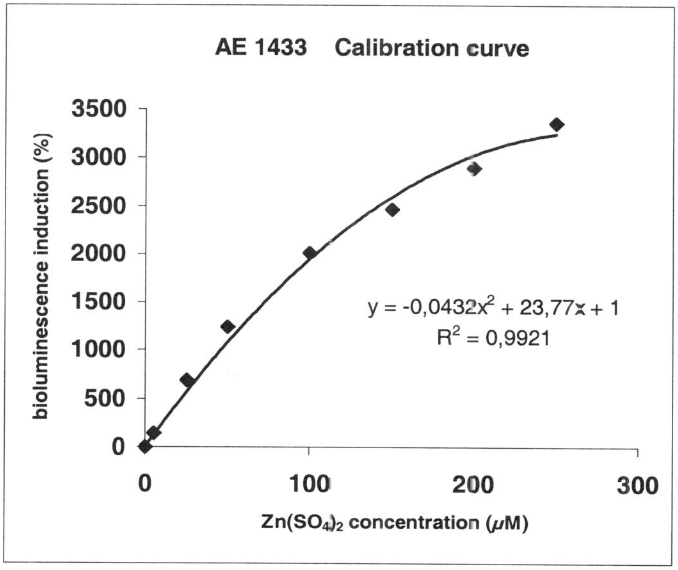

Figure 1: Calibration curve of strain AE1433 realised with $Zn(SO_4)_2$. This salt is considered has being 100% bioavailable for the strain.

However during the measurement, we need to be sure that no other interfering toxic components affecting the bioluminescence are present in the sample present. This bioluminescence can indeed be affected by other parameters, which are not related to the metal concentration. The presence of other organic pollutants can either inhibit or even simulate the luciferase genes. The same miscalculation will be done in samples containing bioavailable metals mixed with biocides or other traces of toxic heavy metals.

1.2. RESULTS AND DISCUSSION

Therefore we have decided to include a global toxicity measurement in the BIOMET assay. The hosts strains (*Ralstonia*) used for this toxicity measurement are the same as those used for the specific sensors. This means that all BIOMET strains have the same cellular background and therefore a very similar cellular metabolism. Both groups of strains use also the same reporter system, the l*uxCDABE* of *Vibrio fischeri*. This means that interfering chemicals affecting the luciferase production in the specific metal sensor will also affect the toxicity sensors. Two toxicity sensors were constructed by mating *E. coli* CM601 bearing the Tn*4431* on pUCD623 and two recipient *Ralstonia metallidurans* strains AE126 and AE104. The strain AE104 is the same as AE126 but has been cured from a megaplasmid pMOL30 bearing a number of metal-induced resistances [20]. The idea was to obtain two constitutive light emitting strains, one being metal-resistant (AE864) and one being metal-sensitive (AE2279) to differentiate in a test sample the nature of the toxicity. A sample that will reduce the bioluminescence level of AE2279 but not from AE864 will most probably contain toxic metals. A sample reducing the bioluminescence of both strains will be toxic and the toxicity will be related to the presence of organic pollutants or other biocides. A sample that will not affect the bioluminescence production of both strains will be considered as being not toxic. At least a sample that will increase the bioluminescence of one or both strains (false positive) will need to be withdrawn because some luciferase activating products (such as degraded by-product of DMSO) will also interfere with the other metal-specific sensor strains.

Two thousand trans-conjugants isolated colonies were obtained and tested for their bioluminescence phenotype. The best 2 clones were selected according to their ability to produce a strong and stable bioluminescent signal in absence of any exogenous inducers. Both selected strains AE864 and AE2279 were freeze-dried and dispensed in small glass vials under N_2 atmosphere. Cells were cooled and freeze-dried in a cryo- protector (1ml/0.1 g dry weight) to assure a survival of about 1 x 10^{-8} CFU/ml after resuspension.

The characteristics of both toxicity sensor strains are summarised in table 2 and their relative sensitivity was measured and compared in table 3 with each other and to published data from the Microtox^{TM} toxicity test using a marine bacteria *Vibrio fischeri* as cellular host. The Microtox EC_{50} is a acute measurement which needs to be measured after 30 minutes whereas the BIOMET toxicity can be either acute or a long term measurement.

TABLE 2. Properties of the BIOMET® toxicity sensors strains

Name	Bacterial host	Reporter system	Heavy Metal resistances
AE864	*Ralstonia metallidurans AE128*	*luxCDABE*	Cu^{2+}, Zn^{2+}, Cd^{2+}, Pb^{2+}, Hg^{2+} on pMOL30
AE2279	*Ralstonia metallidurans AE104*	*luxCDABE*	No plasmid-encoded resistances

TABLE 3: Sensitivity of the BIOMET® toxicity sensors strains to heavy metal ions and comparison with the Microtox™ system. The EC_{50} was measured continuously and is given here after 22 h of incubation.

Name	AE864 $EC_{50}(\mu M)$	AE2279 $EC_{50}(\mu M)$	Microtox (15°C, 30 min) $EC_{50}(\mu M)$ [20]
Zn (II)	>250	12	22
Cu (II)	60	22	2.05
Ni (II)	18	8	719
Pb (II)	>20	0.75	0.48
Cd (II)	>200	75	48.03
Co (II)	> 25	3	47.51
As (II)	> 3000	200	446
Hg (II)	1.2	0.03	0.15
Cr (VI)	28	17	307

It can be concluded that the metal sensitive strain of *Ralstonia metallidurans (AE2279)* has approximately the same sensitivity as the *Vibrio fischeri* strain for a number of metal ions or oxyanions (Hg^{2+},Ni^{2+}, Co^{2+}, Cr^{6+}, As^{5+}) for which the *Ralstonia* strain is more sensitive and therefore has a lower detection limit. The major advantage of the BIOMET toxicity strains is however to use a soil bacteria which is less sensitive to external factors and which do not need a standardised salt concentration to give reproducible results. The difference in metal sensitivity between AE864 and AE2279 was also evident which makes possible discrimination of the metal related toxicity if both strains are used to test a sample. The resistance genes present on the pMOL30 of the AE864 conferred, as expected, an increase resistance to a number of cations. The electroporation of pMOL28 into AE864 could increase this resistance to Ni^{2+}, Cr^{6+} and Co $^{2+}$ and improve the system.

Two typical sandy soils, one being highly polluted by heavy metals [Maatheide] and the other being taken from a non-polluted area [Teut] were tested with a Zn, Cd sensor (AE1433) and both toxicity-strains in order to illustrate the applicability of the method. The Maatheide soil was taken at a highly metal polluted acid sandy soil at the site of a former pyrometallurgical zinc smelter.

Both soils were chemically analysed and a number of soil parameters were determined. Total metal concentration, water extractable metal concentration as well as the exchangeable fraction [$Ca(NO_3)_2$ 0.1 M] and the carbonate fraction [$NH_2OH.HCl$ 1M] were determined for both soils (Table 4). Both soils had very similar soil pH and sand content.

TABLE 4: Chemical characterisation of the non-polluted (Teut) and polluted soil (Maatheide)

	Teut				**Maatheide**			
PH (H_20)	5.9				5.8			
PH (KCl)	4.3				4.6			
Conductivity (μS/cm)	68				192			
Sand (%)	92				90			
Cd (mg kg^{-1})	<3 [a]	n.d. [b]	0.2 [c]	n.m [d]	20 [a]	0.6. [b]	4.7 [c]	3.1 [d]
Cu (mg kg^{-1})	3 [a]	n.d [b]	1.7 [c]	n.m. [d]	829 [a]	0.9 [b]	12 [c]	207 [d]
Pb (mg kg^{-1})	12 [a]	n.d [b]	2.2 [c]	n.m. [d]	2362 [a]	1 [b]	65 [c]	963 [d]
Zn (mg kg^{-1})	13 [a]	n.d [b]	4.7 [c]	n.m. [d]	4798 [a]	96 [b]	416 [c]	1024 [d]
Fe (mg kg^{-1})	545 [a]	5.4 [b]	870 [c]	n.m. [d]	22827 [a]	2.8 [b]	n.d [c]	30 [d]
Mn (mg kg^{-1})	12 [a]	n.d [b]	4.8 [c]	n.m. [d]	577 [a]	8.6 [b]	30 [c]	22 [d]

a : total concentration; b: water extract; c: exchangeable fraction; d: carbonate fraction
n.d : not detectable; n.m : not measured.

Both polluted and non-polluted soils were tested with the new toxicity sensors and with the Zn- Cd sensor AE1433. A clear bioluminescence induction of 1240 % was obtained when the Zn-Cd sensor AE1433 was mixed with the highest tested Maatheide soil concentration (0.17 mg ml^{-1}) whereas no induction was detected with the control soil from Teut (Figure 2). These bioluminescence intensities detected by AE1433 correspond to a concentration of bioavailable Zn of 324,9 mg $kg^{-1}$for the Maatheide soil and 4,92 mg $kg^{-1}$for the Teut soil. Those concentrations are of course much lower than the total amount of Zn present in both samples, but higher than the concentration of Zn found in the water extract. The best approximation of the Zn bioavailable concentration (detected by *Ralstonia* soil bacteria) was given by the Zn concentration present in the exchangeable fraction extracted with 0.1 M of $Ca(NO_3)_2$. The sensor detected 324,9 mg kg^{-1} of Zn whereas 416 mg kg^{-1}were found in the exchangeable fraction. Only 96 mg Zn kg^{-1}were found in the water extract. Similar results were obtained before on soils isolated from the same area [22].

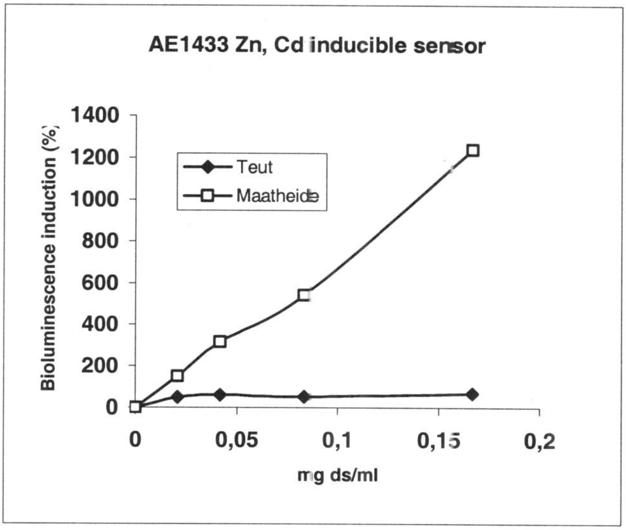

Figure 2: Bioluminescence induction of strain AE1433 in the presence of increasing concentration of both polluted (Maatheide) and not polluted soil (Teut).

The bioluminescence of the metal resistant toxicity sensor was not affected by the control soil (Figure 3). A small not relevant induction of 30 % was observed when using the AE2279 sensor. Because of this small artefact induction only induction of 200 % (signal/noise ration of 2) is considered as relevant when using the metal-induced sensors.

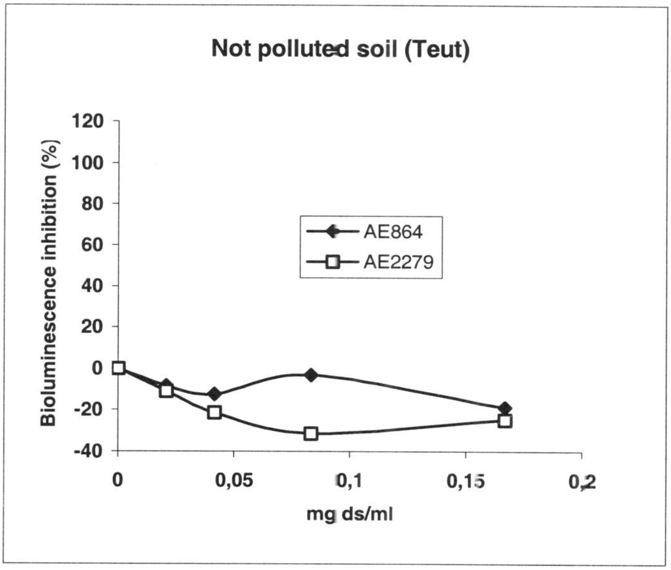

Figure 3: Bioluminescence inhibition of the toxicity strains AE864 and AE2279 in the presence of increasing concentration not polluted soil (Teut).

When the toxicity sensors where tested with the metal polluted soil from Maatheide a clear toxicity effect was measured with the metal sensitive strain (AE2279) whereas the metal resistant strain (AE864) was unaffected, confirming the metallic origin of the toxicity (Figure 4).

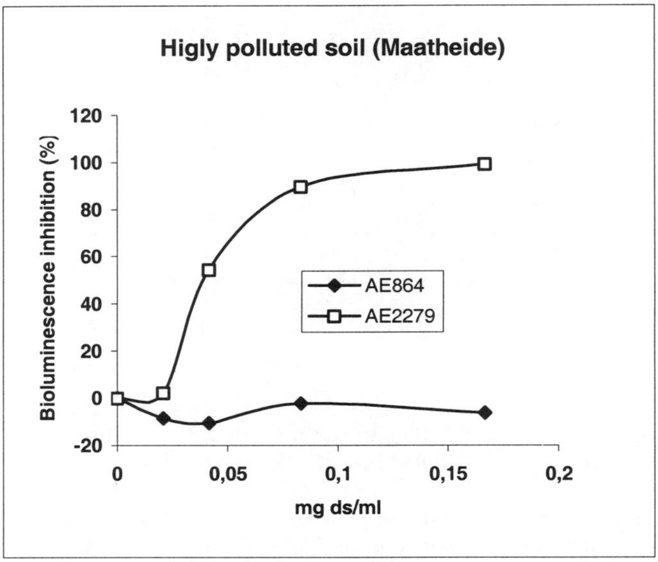

Figure 4. Bioluminescence inhibition of the toxicity strains AE864 en AE2279 in the presence of increasing concentrations of polluted soil (Maatheide).

1.3. CONCLUSIONS

From those preliminary tests it can be concluded that the combined use of heavy metal-induced bioluminescence emitters such as the BIOMET strains and the two toxicity strains having the same cellular background is a good tool to detect metal-related toxicity and metal availability in polluted soil samples. The bioluminescence measurements were performed in a microtiterplate format that allows the simultaneous detection of both parameters for 6 soil samples. False positive results as well as underestimation of the metal bioavailability because of interfering compounds were also eliminated with this combined use of bacterial sensors.

2. Organic Contaminant Physical Property Estimation

2.1. GENERAL APPROACH

Knowing the physicochemical properties of environmental contaminants is a prerequisite for predicting the fate of organic contaminants or new products in the environment, predicting the efficiency of treatment systems, and assessing ecotoxicological and human health risks. Unfortunately, such data are available only for a relatively small number of compounds, and if they are available, they apply only to standard conditions (ambient temperature and pressure). Obtaining experimental data is generally expensive. Furthermore, physicochemical measurements require compounds to be available in pure form, which is the case only in a limited number of cases. Even if reported data are available, they are often inconsistent and difficult to locate in the literature.

In the absence of measured data, estimating properties using estimation techniques is often the only option for assessing the fate of an environmental contaminant. The first step in this process involves selection of an appropriate estimation technique and data of compounds that are similar. Locating measured data and making estimates for compounds that have not been measured can be a time-consuming task that requires expertise and chemical insight. To facilitate this task, computer tools have been developed that combine property estimation with storage and retrieval of measured properties [23]. The property estimation techniques that are available are applicable for a limited number of compound classes and relatively simple compounds, i.e., compounds that contain few (no more than two to three) functional groups.

The properties relevant for environmental applications are summarized in Table 5 and include (1) inherent physicochemical properties, (2) properties that characterize compound behavior in chemical systems, and (3) behavior in natural systems. Inherent physicochemical properties only depend on the query chemical, and their measurement only involves the measurement of physical parameters. The second group of properties involves quantification of interactions with other chemicals. Included in this group are partitioning constants in two-phase systems, reaction rates in aqueous solutions, diffusion rates, transport in chromatographic columns, and reactivity with light. The third group includes behavior in soil and biodegradation.

TABLE 5. Properties of Environmental Relevance

Property	Symbol	Property	Symbol
Neat Phase Properties		**Partitioning Constants**	
Vapor Pressure	p_v	Air/Water	K_{AW}
Enthalpy of Vaporization	ΔH_{vap}	Octanol Water	K_{ow}
Boiling Point	T_b	Soil/Water (organic-carbon)	K_{oc}
Melting Point	T_m	**Chemical Properties**	
Surface Tension	σ	Aqueous Solubility	S_W
Density	ρ	Solubility in Organic Solvents	S
Refractory	N	Acid Dissociation Constants	
Viscosity	η	Photochemical Reactivity	

Estimation techniques exist for predicting compound behavior in natural systems, such as the soil organic carbon-partitioning coefficient, the bioconcentration factor, and the rate of biodegradation [24].

As the systems used to determine properties become more similar to environmental systems, they are more applicable for environmental applications, but a larger set of chemical parameters has to be considered.

The basic steps to estimate a property are shown in Table 6.

TABLE 6.

1. Draw structure of query compound
2. Identify similar compounds with known properties
3. Identify estimation techniques (manual or computerized)
4. Verify estimation techniques using similar compounds
5. Apply estimation techniques to query compound

Steps one and two involve drawing the structure of the query compound and identification of "similar" compounds with known properties. Step one is simple for compounds with a defined structure but represents a challenge for compounds, such as PCBs and nonylphenol ethoxylates, that occur as complex mixtures. The process of finding similar compounds with known property data often begins by removing functional groups and looking up the corresponding parent compound. The third step is to identify one or several estimation techniques that are appropriate for the compound class in question. The best approach is to consult a handbook (e.g., [24-26] and to retrieve data from published compilations [23]. An expedient way to retrieve published data is through the internet (e.g., www.ChemFinder.com). These data can then be used to familiarize oneself with the technique and to verify the appropriateness of the estimation method. At this point, the property of the query compound can estimated, and results can be evaluated in the context of data of similar compounds.

2.2. CLASSES OF PROPERTY ESTIMATION TECHNIQUES

The commonly used estimation techniques are highly empirical and rely on data sets developed for similar compounds. These estimation techniques can be classified as indicated in Table 7.

Table 7. Classes of Commonly Used Estimation Techniques
- Quantitative **property**-property relationships (QPPR)
- Quantitative **structure**-property relationship (QSPR)
- **Group** contribution models (GCM)
- **Similarity** based models

Mixed models

QPPRs are mathematical relationships that relate the query property to one or several other properties. QSPR models correlate molecular descriptors with property data. Molecular descriptors (also termed indicators) are functions that assign a numerical value to the structure of a chemical, based on structural information, physicochemical constants, or stereochemical factors. GCMs are based on the assumption that the contributions of individual parts (subgroups) to the overall property of a molecule are additive and independent of each other. The property of a molecule is given by the sum of the individual group contributions. Similarity based models are based on the observation that properties of similar compounds behave similarly in similar systems and have similar physicochemical values. Models one through four can be combined with elements of other models, leading to mixed models. Selected examples are discussed below.

Selected Examples

An example for a QPPR is the basic solvent regression equation (Colander equation), which correlates partitioning coefficients in different solvent systems. Equation 1 shows the Colander equation for 1-octanol-water (OW) and cyclohexane-water (CW) systems.

$$\log K_{ow} = a \log K_{cw} + b \qquad (1)$$

The coefficients a and b are known for some solvent systems [27]. In selecting a and b, the ability of the solute to donate or accept a proton needs to be considered [24]. For proton-donors, such as acids, phenols, alcohols, and nitriles, a and b are1.481 and 2.729, respectively.

For neutrals (aliphatic and aromatic hydrocarbon compounds) and proton acceptors such as amines, a and b are 0.941 and 0.690, respectively. As the complexity of the intra- and inter-molecular interactions increases, additional correction factors have to be considered. For ionized compounds, the partition coefficients are pH dependent (see, for example, [25]).

A straightforward example for a QSPR is derived from the observation that the aqueous solubility decreases linearly within several homologous series [28]. For n-alkanes, this relationship can be expressed

$$\log_{10}(C_W/\mu g{\cdot}mL^{-3}) = 4.416 - 0.569{\cdot}N_C \quad (2)$$

where N_C is a molecular descriptor that indicates the number of carbon atoms per molecule. Other examples for molecular descriptors are molecular mass, surface area, number of bonds, or indicators that are calculated based on theoretical considerations [25]. Although molecular descriptors are inherently precise, they only reflect a subset of a molecule's chemical properties, and complex molecular phenomena may not be captured. For instance, solubilities are often higher for odd values of N_C, causing the Equation 2 to zig zag upwards.

Many different GCMs have been developed. For illustration, we examine the approach developed by Hansch and Leo [27] for estimating K_{ow} values. In this approach, the log K_{ow} is calculated as the sum of fragments and the sum of structural factors (Equation 3).

$$\log K_{ow} = \text{sum of fragments (f)} + \text{sum of structural factors (F)} \quad (3)$$

f values have been calculated for over 100 atoms or atom groups and for 14 structural factors. For illustration, Cl-CH2-CH2-CH2-Cl is split into 3 carbon, 6 hydrogen, and 2 chlorine fragments. In addition, bond factor F_b is considered, which considers chain length effects.

Example: 1,3-dichloropropane

$3f_C$	= 3(0.20)		= 0.60
$+6f_H$	= 6 (0.23)		= 1.38
$+2f_{Cl-}$	= 2(0.06)		= 0.12
$+(4-1)F_b$		= 3 (-0.12)	= -.36
		log K_{OW}	= 1.74
		log K_{OW} observed.	= 2.00

The values for C, H, and Cl are 0.2, 0.23, and 0.06, and indicate the contribution of each atom to the log K_{ow} value. It is interesting to note that the contribution of hydrogen is largest although it is the smallest group whereas the contribution of chlorine is smallest although it is the largest group. The bond factor for 1,3-dichloroputane is 4 and considers two carbon-carbon bonds and two carbon-chlorine bonds. Its value is negative, which reflects the fact that flexible molecules can coil up and thus become more water soluble and less hydrophobic. The measured value is higher, however, suggesting that this effect may be overestimated. The method by Hansch and Leo is based on the contributions of small groups (atom and atom group), which adds flexibility. However, since the contributions of these relatively small atom groups are not independent from each other and depend strongly on the local structural environment, many different bond factors have to be taken into consideration. More recent methods for K_{ow} estimation such as the method by Broto et al. are based on larger fragments and consideration of their chemical environments [29]. Estimation for other properties have been reviewed by Reinhard and Drefahl [23].

Similarity based methods use one or several "similar" compounds as the starting point for the estimation of a query compound and evaluate the effect of minor structural modifications, for instance the replacement of one functional group with another, on the property. From the data set, a rule is developed that is employed to make estimate the property of the unknown. An example for the approach is the difference in the boiling point (ΔT_b) between n-alkanes and 2-oxaalkanes [25] (data from [30]). Balaban et al. authors evaluated boiling points at normal pressure for a set of 185 acyclic saturated compounds with one or two divalent oxygen or sulfur heteroatoms devoid of hydrogen bonding and correlated T_b with chemical structures using molecular descriptors. Analysis of these data reveals that replacing a methylene group with an oxa (-O-) group increases the boiling point by 0.57 °C. If the methylene group is replaced with an sulfa group (-S-), the increase is much more pronounced (48.7 °C), as expected based on the higher atomic weight of sulfur. Other examples illustrating the similarity approach as well as mixed approaches can be found elsewhere [25].

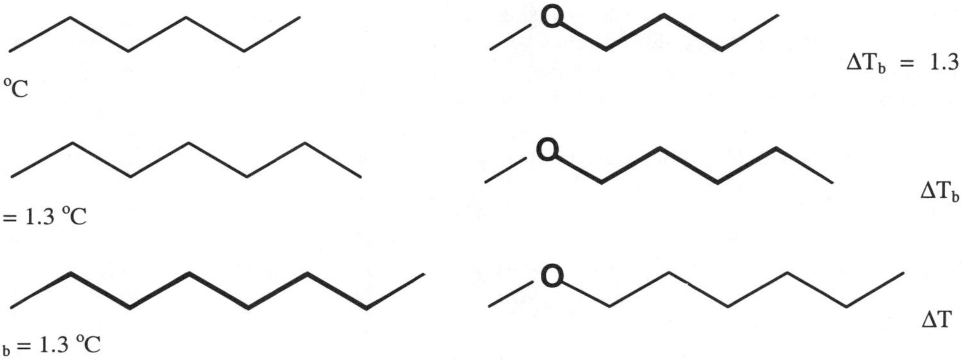

Figure 5. ΔT_b differences between n-alkanes and 2-oxaalkanes. For a large data set, the average ΔT_b is 0.57 °C [25].

Empirical relationships used to estimate properties are often based on statistical averages of relatively small data sets. Therefore, properties obtained by statistical correlation are always associated with an error. In some cases, the uncertainty associated with the estimation method is known (e.g., for the K_{ow} method by Broto et al. 1984) and can be considered for estimating the probable error of the estimated value. An alternate approach is to use different estimation techniques for the same property to obtain a range of values. The range obtained by these values may be used as an estimate for the error of the estimates. The fundamental problem is that it to difficult to predict changes in the chemical properties based on structural changes. Different approaches that lead to an estimate of the K_{ow} value are shown in Figure 6.

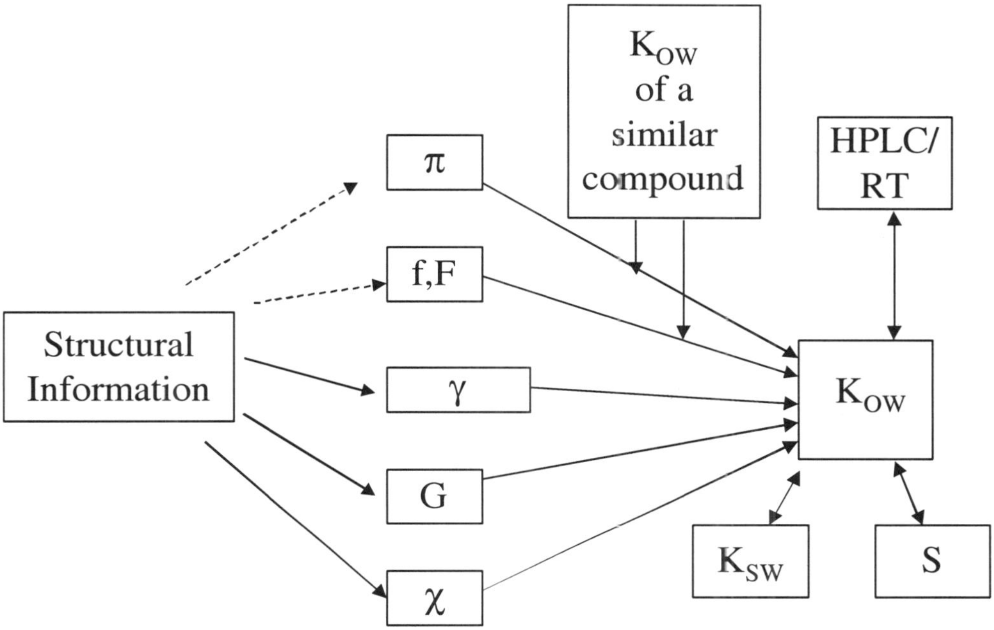

Figure 6. Different approaches to estimate K_{ow} (from Lyman et al [24]).

Starting from the compound structure, substituent constants (π) „ fragment (f) or structural factors can be used using the methods of Hansch and Leo [27], or related methods. Other approaches are based on the correlation between K_{ow} and the activity coefficient (γ), the free energy of solution, or the molecular connectivity index (χ), solvent water distribution coefficients other than 1-octanol-water, the solubility in water. or the retention time in hydrophobic chromatography columns.

Computer Assisted Property Estimation

Given the complexities of manual property estimation, the availability of computer assisted tools are highly desirable. The application of structure based estimation techniques requires that the structures are encoded in a computer readable form. One of the most frequently used encoding system for organic compounds is that by Weininger et al. [31-33]. The advantage of this system (termed the SMILES system for **S**implified **M**olecular **I**nput **L**ine **E**ntry **S**ystem) is that the rules are simple and intuitive. The most important rules are summarized in Table 8. More extensive tutorials and additional examples are given elsewhere ([25] or http://www.daylight.com/dayhtml/smiles/smiles-intro.html).

TABLE 8. Rules to Encode the Stucture of Organic Compound Using Smiles

H atoms are suppressed Methane CH_4 C	**Double bonds are not suppressed** Vinly chloride $ClHC{=}CH_2$ ClC=C
Single bonds are suppressed Methylamine CH_3NH_2 CN	**Triple bond not suppressed** Hydrogen cyanide HCN C#N
Ring closure at number following C Cyclohexane C_6H_{12} C1CCCCC1	**Aromatic bonds are suppressed** Benzene C_6H_6 c1ccccc1
2 fused rings closed at two different locations Naphthalene C_6 H_{10} c12ccccc1cccc2	**Aromatic heteroatoms lower case** Pyridine CN H5 c1cnccc1
Branching indicated by parentheses CH_3 t-Butylbenzene $C_6H_6CCH_3$ CC(c1ccccc1)(C)C CH_3	

Often there are several possible SMILES codes that can be generated from a single structure. Computer algorithms exist that convert different SMILES into a unique representation. A unique representation of a chemical structure in SMILES can be generated directly using computer programs. For instance, the "Toolkit for Estimating Physicochemical Properties of Organic Compounds" [23] converts chemical structures drawn in ChemDraw (CambridgeSoft, Cambridge, MA) into SMILES and allows the user to retrieve and estimate property data without prior knowledge of the SMILES code. In some cases, the user can choose among different estimation methods and compare the results obtained by different approaches. It is also possible to retrieve measured data and to compare measured with estimated values. Figure 7 shows the output of the Toolkit for estimating the K_{ow} of isobutylbenzene. The method of Broto et al [29] was selected. The output indicates the contributions of the different contributions and the total standard deviation, which is a measure for the uncertainty of the estimate.

List of Available Methods

n-Octanol/Water Partition Coefficient (GCM of Broto, Moreau and Vandyck

n-Octanol/Water Partition Coefficient (GCM of Viswanadhan, Ghose, Reva

Description

Estimation using Broto's fragmentation method

At.order	Code	At.contribution
1	C 551	0.095
2	C 55	0.311
3	C 55	0.311
4	C 55	0.311
5	C 55	0.311
6	C 55	0.311
7	C 111	0.029
8	C 11	0.456
9	C 1	0.631
10	C 1	0.631

log(p): 3.40
SD: 0.13
by Broto's method: Eur.J.Med.Chem.- Chim.Theor.,19,71(1984).

Figure 7. Output of the Toolkit for Estimating the K_{ow} of Isobutylbenzene [23].

In conclusion, methods are available for estimating the physicochemical properties of relatively simple compounds. Some of these methods have been implemented in computer programs such as the Toolkit. In general, the results of an estimation will be more reliable if experimental data from similar compounds are available.

3. Miscible Contaminant Transport And Availability

3.1. INTRODUCTION

A key factor in defining exposure, uptake and, ultimately, risk of contaminants to receptor organisms is the ability of the contaminant to partition to a mobile phase. The rate of migration of a contaminant from source to receptor is slowed by retardation associated with transient accumulation on the immobile soil or sediment phase. Uptake and risk at the point of contact is also dependent upon the quantity of contaminant present in the air or water phase that typically provides the pathway of exposure. Even contaminants that are ingested generally do not result in uptake or risk unless they can partition out of the solid carrier phase. Thus the partitioning of a contaminant between mobile and immobile phases, and particularly between the solid and water phases is a key indicator of the potential for exposure and risk. This section will quantify the influence of the partitioning process on transport and availability, specifically of organic contaminants.

3.2. REVERSIBLY SORBED CONTAMINANTS

For contaminants that exhibit reversible sorption and desorption, a single partition coefficient can be used to characterize the ratio of sediment concentration, W_s, to the adjacent porewater concentration, C_w, at equilibrium.

$$K_{sw} = \frac{W_s}{C_w} \tag{4}$$

For hydrophobic organic contaminants, the sorption is dominated by the organic carbon fraction of the soil or sediment which, to a first approximation, provides a uniform sorptive surface. Thus normalization of K_{sw} by the organic carbon fraction, f_{oc}, provides a soil or sediment independent partition coefficient that, in principle, is a function of only the hydrophobicity of the compound.

$$K_{oc} = \frac{K_{sw}}{f_{oc}} \tag{5}$$

It is this property that can be estimated by the techniques described above. This property relates directly to the mobility and availability of an organic contaminant subject to reversible sorption as can be seen by evaluation of the contaminant conservation equation.

$$\frac{\partial(\varepsilon C_w)}{\partial t} + \frac{\partial(\rho_b W_s)}{\partial t} = D\frac{\partial^2 C_w}{\partial z^2} - U\frac{\partial C_w}{\partial z} \tag{6}$$

This equation relates the accumulation per unit total volume in the porewater and solid phases (the first and second terms on the left hand side, respectively) to the migration processes of diffusion and advection that move the contaminant in the mobile phase (the first and second terms on the right hand side, respectively). Here D is an effective dispersion coefficient in the soil or sediment combining the effects of diffusion and dispersion and U is a superficial or Darcy velocity. For a reversibly sorbed contaminant in which local equilibrium defined by K_{sw} can always be assumed, this equation can be written

$$\frac{\partial(\varepsilon + \rho_b K_{sw})C_w}{\partial t} = D\frac{\partial^2 C_w}{\partial z^2} - U\frac{\partial C_w}{\partial z} \tag{7}$$

If the void fraction, bulk density and partition coefficient are independent of time, this can be written

$$\frac{\partial C_w}{\partial t} = \frac{D}{R_f}\frac{\partial^2 C_w}{\partial z^2} - \frac{U}{R_f}\frac{\partial C_w}{\partial z} \tag{8}$$

where the retardation factor, $R_f = \varepsilon + \rho_b K_{sw}$. Thus the effective dispersion or advection of a contaminant subject to reversible sorption onto an immobile solid phase is reduced by a factor proportional to the soil-water partition coefficient.

3.3. DESORPTION-RESISTANT CONTAMINANTS

The influence of sorption becomes more complicated when the local equilibrium cannot be assumed or when the contaminant does not sorb reversibly. Typically, some hysteresis in the sorption-desorption process is noted and desorption is reduced in rate or extent relative to sorption. The results of many laboratory and field observations indicate that a significant fraction of sediment-bound organic contaminants do not desorb linearly and reversibly as suggested by conventional partitioning models, are not biodegraded, and are difficult to remove by extraction with surfactants or cosolvents. For example, Pereira, et al. [34] found that the concentration of halogenated organic compounds in native water, suspended sediments, and biota was far below the values predicted with respect to concentrations in the contaminated bottom sediments collected from Bayou d'Inde, Louisiana. Similarly, McGroddy and Farrington [35] and Readman and Mantoura [36] observed a fraction of PAHs in river sediments not available for partitioning. Ghosh, et al. [37] has related highly sorbed concentrations and highly effective partition coefficients to the presence of soot in sediments. In addition, it is common for hydrophobic compounds to remain persistent in sediments even when conditions are appropriate for biodegradation to occur (Connaughton, et al. [38]; Hatzinger and Alexander [39]; Kan, et al. [40]). For most of these sediments, the contamination had ceased for periods of many years, yet the sediment-bound contaminants persisted over decades without significant concentration reduction or hydrocarbon fingerprint loss. Luthy et al. [41] has summarized evidence for sequestration in soils and sediments.

The sorption and desorption of organic chemicals to soils and sediments is a complex process, given the diversity, magnitude, and activity of chemical species, phases, and interfaces commonly present in contaminated subsurface environments. A research group at Rice University has made substantial progress in establishing the concept of irreversible adsorption or hysteresis as a fate mechanism in natural sediments (e.g., Kan, et al.[40]; Hunter, et al. [42]; Kan, et al. [43]; Kan, et al. [44]). It has been postulated that the observed phenomenon can be due to the occlusion of the chemicals from desorption by cooperative conformational changes of the organic phase during the adsorption process. The conformational rearrangement of the solid organic matter in the presence of adsorbed chemicals could cause the chemical environment of the adsorbate to be different and hence be the source of desorption resistance and hysteresis. At least one hypothesis has stated that the quality of organic matter determines the degree of hysteresis in sorption-desorption experiments (Huang and Weber [45]), suggesting that the size of the desorption resistant compartment can be related to organic matter age.

An alternative hypothesis is that an apparent hysteresis is only the result of slow desorption (Lick and Rapaka, [46] and Pignatello and Xing, [47]). Field sediments may have been contaminated for decades, allowing time for sorption into phases or portions of sediments that allow only equally slow desorption. Since the period of desorption can be very short compared to the period of sorption, the net effect is observation of an apparent hysteresis that does not reflect true equilibrium.

Regardless of the mechanism of desorption resistance and whether it is rate or extent (equilibrium) limited, recent work has shown that the porewater paradigm remains a good model for prediction of exposure and uptake in benthic organisms. That is, ultimately, the amount of, at least, organic contaminant that can accumulate in plants and animals in intimate contact with the sediment and porewater is controlled by partitioning into the porewater. Lu et al. [48] have shown that reductions in steady state biota-sediment accumulation factors in a freshwater oligochaete due to desorption resistance can be related to porewater concentration reductions. Gomez-Hermosillo et al. [49] has shown similar behavior in wetland plants.

Similarly, the effect of desorption-resistance on contaminant migration can be estimated by simply employing the retardation factor employing desorption partition coefficients. The limited desorption can reduce porewater concentrations producing a direct reduction in the potential for advective or diffusive transport of mass. A more accurate depiction would take into account the differing rates of adsorption and desorption and not simply assume local equilibrium but there exists little data to support this level of complexity.

This discussion has focused only on sorption as a mechanism for attenuation of the rate of migration of contaminants in soil and groundwater. The discussion has also focused only on the physicochemical processes of advection and diffusion of the miscible fraction of contaminants. A variety of other processes may influence contaminant migration, especially in the near surface environment where erosion of the solid phase may occur, where organisms may move soils or sediments (bioturbation), where microbial processes may be sufficiently fast to degrade contaminants or enhance transport through such processes as gas generation, and where organic material may maintain high colloidal carbon concentrations in the porewater. This may be especially important in contaminated sediments due to the variety of factors that influence contaminant fate and transport processes in sediments [50-52]

4. Availability Of Contaminants In Nonaqueous Phase Liquids: Dissolution Of Entrapped Nonaqueous Phase Liquids In Heterogeneous Aquifers

4.1. INTRODUCTION

The focus of this section is on problems dealing with potential contamination of groundwater by organic waste chemicals that are in the form of nonaqueous phase liquids or NAPLs. Many investigators (e.g. Kueper and Gerhard [53]; Illangasekare et al. [54,55]; Held and Illangasekare [56]; Illangasekare [57]) have demonstrated the complex flow and transport behavior of these chemicals in heterogeneous aquifers. Important to the site assessment and risk analysis involving NAPLs is the realization that the contamination will be distributed between free phase NAPL (pools, residual, and potentially in fractures); sorbed onto the matrix; and dissolved in aqueous phase. Unless completely removed, the mass present as free-phase and sorbed onto the media will provide ongoing sources of dissolved contamination in the aqueous phase---depending on the physical and chemical characteristics these sources can be expected to deliver mass to flowing groundwater for decades and perhaps centuries. In recognition of this risk, many remediation technologies have been developed for the removal of the separate phase from the entrapment zones. However, their effectiveness for complete removal has been limited by the heterogeneity of subsurface formations and the inability to locate all sources. Because of these limitations, it is certain that a fraction of the NAPL will remain undetected after source remediation, thus creating potential risk of contamination and subsequent long-term human exposure. The purpose of the fate-and-transport analysis is to estimate exposure point concentrations of the contaminant of concern.

The objective of this paper is to demonstrate the effects of aquifer heterogeneity on the entrapment of nonaqueous phase waste chemicals in the subsurface and their implications on remediation. The technology of surfactant enhanced dissolution is used to demonstrate that the uncertainty associate with the distribution affects the ability to predict the effectiveness of remediation.

4.2 ENTRAPMENT OF NAPLS IN HETEROGENEOUS FORMATIONS

It has been well recognized that the transport and transformation of contaminants in groundwater are influenced by the ubiquitous spatial variations of the properties of the aquifer material. These variations exist over a wide range of length scales, from the size of pores to regional scales. Many studies describe the basic process of how multiple fluids flow in homogenous porous media. However, all underground porous formations in their natural state will exhibit variability of properties in space. Natural heterogeneity resulting from the spatial variation of soil types and their properties affect the flow and entrapment behavior of NAPLs. The process of NAPL flow controls the ultimate entrapment distribution. The entrapped NAPLs may exist as discontinuous, stable pore-scale ganglia trapped in the porous media under capillary forces; also it may exist as an immobile continuous phase trapped by heterogeneities. We refer to this latter phenomenon as macroscale entrapment. NAPL immobilized by capillary forces is often referred to as a residual or irreducible saturation and has received the greatest attention in experimental mass transfer studies.

Two conceptual models, termed "by-passing" and "snap-off" have been proposed for NAPL entrapment in the microscopic scale. When the NAPL imbibes into water saturated pore space, the immiscible fluid phase becomes isolated in discrete volumes due to the tortuosity of flow paths and velocity differences in different size pores. These isolated fluid volumes are formed within the pores due to the bypassing of the aqueous phase. This mechanism of NAPL entrapment is referred to as "by-passing." Snap-off occurs when a volume of NAPL is squeezed through pores with small throat to body ratios (or aspect ratios), and inertial forces separate blobs from the main NAPL body. This mechanism is thought to give rise to small blobs occupying one or two pore bodies. Wilson et al. [58] demonstrated both of these microscale entrapment mechanisms using two-dimensional etched glass micromodels. Isolated single blobs occupying one or two pore bodies were seen to be formed due to snap-off, while by-passing gave rise to complex ganglia elongated in the direction of flow. In homogenous soils these blobs and ganglia produce NAPL entrapment at low saturations commonly referred to as *residual saturation*.

Hoag and Marely [59] have experimentally investigated microscale entrapment in sandy materials. Wilson et al. [58] showed that entrapped organic saturations are relatively insensitive to fluid properties but highly dependent on soil properties and heterogeneity. Powers [60] showed that well graded sands entrapped more NAPL and hold it in large ganglia, while uniform sands retained a smaller amount of NAPL and hold it in small blobs. These results are consistent with the hypothetical models of by-passing and snap-off. NAPL saturations due to microscale entrapment can reach as high as 25-30%.

Illangasekare et al. [54] have shown through laboratory spill experiments in large soils tanks that zones of residual saturations are not the only modes of NAPL entrapment in heterogeneous sand formations. The pore-scale processes that occur at the interfaces of different soils in heterogeneous formations result in NAPL entrapment in coarser pockets in saturated soils, and finer pockets in unsaturated soils (Illangasekare at al, [55]). The same interface effects also result in suspended NAPL pools at the transitions of soil formations. The NAPL saturations that result because of these transitions are much higher than the residuals that result from microscale entrapment. Because this type of entrapment occurs in the formation due to the macro-scale variability of soil properties, this mechanism of NAPL entrapment is referred to as macro-scale entrapment. In laboratory experiments it has been shown that the macro-scale saturations can reach as high as 75-80% in heterogeneous formations.

To demonstrate the combined effects of unstable fingering and capillary barrier effects of entrapment in heterogeneous formations, a series of experiments were performed by Campos [61]. These spill experiments were conducted in a 91-cm×61-cm test tank packed with five well-characterized sands to represent a spatially correlated random field (see Table 9 for sand properties). NAPL spills were conducted in five packing configurations to represent five realizations of the same random field. The stochastic parameters of the random field are: (1) Mean of the logarithm of hydraulic conductivity, $(\ln K)_{mean}$=3.46, (2) variance of logarithm of hydraulic conductivity, $\sigma^2_{\ln K}$ =1.10, (3) vertical correlation length, λ_v=5.08, and (4) horizontal correlation length, λ_h=2.54. In all the parameters the hydraulic conductivity is measured in m d^{-1}.

Figures 8 and 9 show the packing configuration based on one of the realizations and the corresponding distribution of the NAPL after a spill. The final spatial distribution of NAPL saturation was measured using a dual-gamma attenuation system (Illangasekare et al [62]). This test shows how the NAPL has moved through the heterogeneous field through fingering and got entrapped within the coarser formations. The NAPL has pooled at the coarse-fine sand interfaces due to the capillary barrier effects. Similar pooling phenomena is expected at sand-clay interfaces due to the high entry pressures and low permeability of the clay.

TABLE 9. Properties of sand used in NAPL spill experiments

Sand Type	#16	#30	#50	#70	#110
Saturated hydraulic conductivity, K_{sat} (m d^{-1})	694.1	171.4	35.1	21.0	4.56
Mean grain size, d_{50} (mm)	1.08	0.5	0.32	0.2	0.12
Brooks-Corey, pore size index, λ	3.5	2.1	2.0	1.9	1.8
Immobile saturation, S_r	0.07	0.26	0.29	0.3	0.26

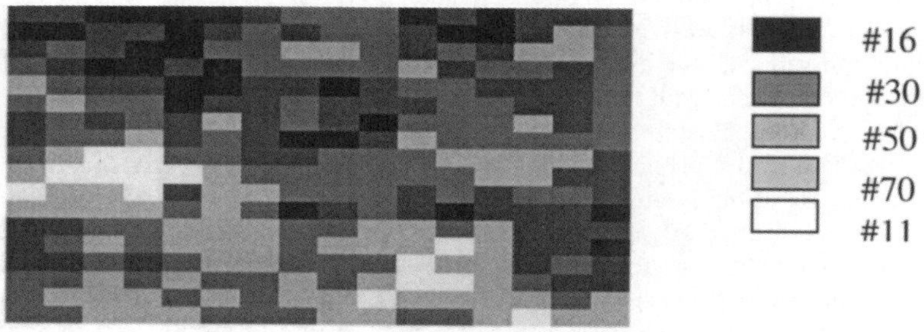

Figure 8. Packing configuration of a randomly heterogeneous test aquifer.

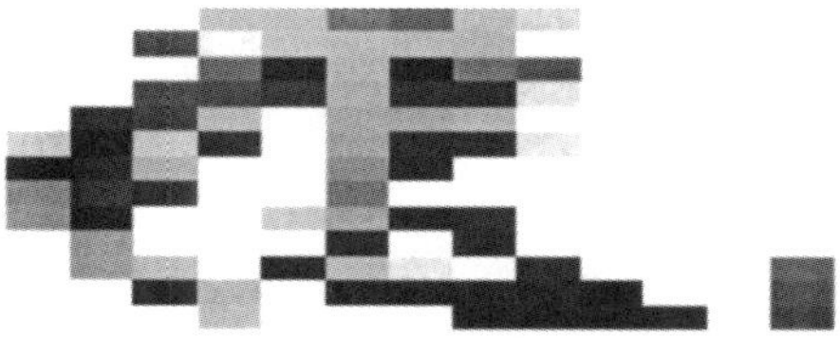

Figure 9. Distribution of NAPL saturation in a randomly heterogeneous test aquifer.

4.3. DISSOLUTION MODELING

Dissolution of the entrapped NAPL to the flowing water is a result of the mass transfer that occurs across the NAPL-water interfaces. The rate of mass transfer depends on the maximum solubility and the groundwater flow velocity. Also, when the NAPL sources are treated, the rate of mass transfer is expected to change with the physical, chemical and biological transformation that occur as a result of treatment. In this section, the basic process of mass transfer under natural and surfactant enhanced dissolution is discussed. A numerical model that uses mass transfer relationships that are up-scaled to two-dimensional flow fields is also discussed.

4.3.1. *Mass Transfer*

Mass transfer phenomena have been the focus of considerable attention in the literature, but there are still many unresolved issues. While dimensional analysis has shown the mass transfer rate to be a function of several non-dimensional groups (Bird et al. [63]), most laboratory work to date has only investigated a few of these. A majority of the laboratory studies have used a first-order rate model to simulate mass transfer in homogeneous porous media with length scales on the order of centimeters (e.g., Miller et al. [64]; Powers et al. [60]; Powers et al. [65]; Imhoff et al. [66]; Imhoff and Miller [67]). Results from these studies support the establishment of near equilibrium conditions over relatively short length and time scales. However, field-scale observations do not typically show groundwater plumes to be near equilibrium conditions.

Miller et al. [64] proposed that mass transfer occur based on a linear driving force between the aqueous and organic phases described by:

$$J = \hat{k}\left(C_s - C\right) \tag{9}$$

where J [$ML^{-3}T^{-1}$] is the solute mass flux from the immiscible liquid phase to an aqueous phase, $\hat{k}$ [T^{-1}] is a lumped mass transfer coefficient, C_s [ML^{-3}] is the aqueous solubility of the organic phase, and C [ML^{-3}] is the solute concentration in the bulk aqueous phase. Methods to determine values of $\hat{k}$ are usually based on correlations containing a modified Sherwood (Sh) number as shown in equation (10).

$$\hat{k} = \frac{D_m}{d_{50}^2}(\mathbf{Sh}) \tag{10}$$

where D_m [L^2T^{-1}] is the molecular diffusion coefficient for the soluble constituent, and d_{50} [L] is the mean particle diameter. Several researchers have proposed dimensionless correlations of interphase mass transfer (Sh) to be a function of various dimensionless groups such as Reynolds (Re), Schmidt (Sc), and Peclet (Pe) numbers etc. (Miller et al. [64]; Powers et al. [60]; Powers et al. [65]; Imhoff et al. [66]). These empirical correlations are referred to as Gilland-Sherwood models. Although these models developed from data generated in 1-d columns allows for the estimation of mass transfer coefficients to predict NAPL dissolution, they may not provide accurate mass transfer predictions in larger field scales where the groundwater flow is three-dimensional. Saba [68], Saba and Illangasekare [69] and Saba et al. [70] developed up-scalable mass transfer coefficients for both natural and surfactant-enhanced dissolution. A series of experiments were conducted in which externally prepared NAPL (para-xylene) samples with known saturations were placed in a test tank. Several NAPL-source samples with different lengths were used to investigate the scale dependency on mass transfer rate. Aqueous phase para-xylene concentrations were measured downstream of the NAPL source and breakthrough curves were analyzed to estimate the rate of mass depletion. Following a tracer test to determine media dispersivity, dissolution behavior under different groundwater velocities was investigated for three entrapped source lengths. Through inverse modeling the transient mass transfer rate coefficients were determined based on the transient values of Reynolds number (Re), Schmidt number (Sc), and volumetric NAPL content during the simulation of the dissolution process. The resulting equation was of the form;

$$\mathbf{Sh}_{\mathbf{natl}} = \alpha(\mathrm{Re})^{\beta}(\mathbf{Sc})^{\gamma}\left(\frac{\theta_n d_{50}}{\tau L^*}\right)^{\eta} \tag{11}$$

The parameter τ is the tortuousity factor of the porous medium, and L^* [L] is the dissolution length. L^* accounts for the path length of the contaminated numerical soil block in the flow direction. This characteristic length parameter allows for the correction of the scale-dependent mass transfer coefficient. The Schmidt (Sc) and Reynolds (Re) numbers are the properties of flowing fluids where $\mathbf{Sc} = \nu D_m^{-1}$ and $\mathrm{Re} = d_{50}U''\nu^{-1}$. U'' [LT^{-1}] and ν [L^2T^{-1}] are the average linear pore velocity and the kinematic viscosity of the flowing fluid, respectively. The constants $\alpha, \beta, \gamma, \eta,$ and ξ are empirical fitting parameters that are determined from a multiple, non-linear regression analysis of the experimental data. Table 10 shows the empirical fitting parameters used for the numerical simulations; these values were determined from the 2-d tank experiments reported in Saba [68].

Using similar laboratory and modeling procedures that were used in the development of correlations for natural dissolution, a relationship was developed by Saba et al. [70] for surfactant enhanced dissolution. This relationship is given as:

$$\mathrm{Sh}_{\mathrm{surf}} = \alpha(\mathrm{Re})^{\beta}(\mathrm{Sc})^{\gamma}\left(\frac{\theta_n}{1-\theta_n}\right)^{\eta}\left(\frac{d_{50}}{\tau L^*}\right)^{\xi} \tag{12}$$

The parameters used for the simulations were determined from the tank experiments reported in Saba [68]. As the rate limited mass transfer depends on the NAPL during mass removal, a dual-gamma system was used in the experiments to monitor the changes in NAPL saturation in the source zone during surfactant enhanced dissolution The parameters of the model in equation (12) are given in Table 10. In contrast to previous mass transfer models, this proposed relation was determined using two-dimensional experimental data and has the ability to be up-scaled to different numerical block sizes in simulation models.

TABLE 10. Values of empirical fitting parameters used in natural and surfactant-enhanced dissolution models

Parameter	Natural	Surfactant
α	11.34	0.4727
β	0.2767	0.2793
γ	0.3333	0.5
η	1.037	1.642
ξ	-	0.1457

4.3.2 *Numerical Model*

A simplified model that has the capability of upscaling the mass transfer processes in heterogeneous systems was developed in this study to demonstrate the effect of heterogeneity on surfactant delivery efficiency. More complete analysis could be conducted using complex models such as UTCHEM by Delshad et al. [71].

The mixing of the dissolved NAPL in the source zone and the subsequent downstream migration of the dissolved NAPL solute are controlled by the local groundwater flow field within the entrapment zone, regional groundwater flow field, advection, dispersion and reactive processes. In this study, a three-dimensional finite difference groundwater flow model, MODFLOW (McDonald and Harbaugh [72]; Harbaugh et al. [73]) was used to model the local and regional groundwater flow fields. The solute transport is simulated using MT3D (Zheng [74]; Zheng and Wang [75]).

The dissolution modeling package developed by Okeson [76] and Saba [68] was modified and coupled to MT3D to account for the dissolution of NAPL under natural and surfactant-enhanced conditions. This modified package allows MT3D to solve for NAPL dissolution based on equations (1) and (2) and the phenomenological models given in equations (11) and (12). Given an initial NAPL saturation in each grid block, MODFLOW solves the groundwater flow equation based on modified hydraulic conductivity resulting from NAPL entrapment. The partially water saturated or modified hydraulic conductivity (K [LT^{-1}]) can be expressed in terms of the relative permeability (k_{rw}), and saturated hydraulic conductivity (K_{sat} [LT^{-1}]) as;

$$K = k_{rw} K_{sat} \tag{13}$$

Wyllie [77] proposed the relationship between the relative permeability of the wetting phase and NAPL saturation using equation (14) as:

$$k_{rw} = \left(\frac{1 - S_n - S_r}{1 - S_r} \right)^3 \tag{14}$$

where S_n is the NAPL saturation, and S_r is the residual saturation of the aqueous phase. Wyllie's equation was chosen in the model because it was found to best describe the relative permeability measurement data for the test sands used in our past laboratory investigations (Saba [68]; Saba and Illangasekare [69]).

The value of hydraulic conductivity is time-dependent since NAPL content decreases during surfactant-enhanced dissolution. Consequently, during the simulation, the flow field within the NAPL entrapment zone will change accordingly with the updated hydraulic conductivity.

Injection of water or surfactant solution is simulated using the well package of MODFLOW. The delivery of surfactant to the NAPL entrapment zone is simulated by solving the advection-dispersion equation using MT3D for surfactant concentrations. For simplicity, the aqueous phase and the micro-emulsion phase were combined, treating them as a single aqueous phase. In addition, the viscosity and density of the surfactant solution were assumed to be the same as for water. At the end of each surfactant transport simulation step, finite difference blocks having surfactant concentration greater than the critical micelle concentration (CMC) are flagged and identified as zones of enhanced dissolution. Within these flagged computational blocks, the mass transfer coefficients are calculated using equation (12). In the un-flagged blocks where the surfactant concentrations have not reached the CMC, the mass transfer is modeled using equation (11) that estimates the mass transfer coefficient for natural dissolution.

When the modified MT3D solves for the mass transfer of NAPL to aqueous phase, it keeps track of the rate of dissolution of the NAPL into the flowing aqueous phase and hence the rate of depletion of the NAPL mass. Temporal variation of NAPL saturation, effective porosity for the aqueous phase, and relative permeability are updated at the end of each simulation time step. With an updated hydraulic conductivity field, MODFLOW simulates the new groundwater flow field within the NAPL source zone.

In order to take the surfactant-NAPL interaction into consideration, the surfactant concentrations within the blocks where the surfactant is used to enhance dissolution were updated at the end of each time step for the subsequent surfactant transport calculation. Previous research (Ewing [78]) found the solubility of para-xylene (the NAPL that is used in this study) is proportional to the amount of surfactant (Tween-80) in groundwater. Based on laboratory batch studies Ewing [78] proposed the following relationship:

$$\frac{C}{C_s} = 2637.3 C_{\text{surf}} + 1 \tag{15}$$

where C is the para-xylene concentration, C_s is the solubility of para-xylene in pure water, and C_{surf} is the surfactant concentration in g cm^{-3}. The surfactant concentrations were updated using equation (15).

4.4 SOURCE ZONE REMEDIATION

Because of the enormous concerns for long-term groundwater contamination from DNAPL sources and the long-term costs associated with pump-and-treat schemes, interest has increased in the application of aggressive free-phase removal technologies. These include, air-sparging, surfactant or co-solvent flushing, chemical oxidation and biostabilization/ bioisolation. These technologies cannot completely remove DNAPLs, especially in complex, heterogeneous aquifers (USEPA [79,80], Grubb and Sitar [81]). Several factors contribute to incomplete DNAPL removal including insurmountable viscous resistance of entrapped DNAPL, incomplete contact of reactive agents (e.g. surfactant, microorganisms, chemical oxidants) with DNAPL at all locations, complexity of DNAPL morphology (pools or residual fingers and blobs), and changes in DNAPL morphology (e.g. remobilization) and chemical composition during application of remediation technologies. The technology of surfactant enhanced dissolution is used to demonstrate that the uncertainty associate with the distribution affects the ability to predict the effectiveness of remediation.

Surfactant enhanced source removal technologies are based on the ability of surfactant molecules to create low water/oil interfacial tension to reduce capillary forces and thus mobilize entrapped NAPLs. The mobilized NAPLs are surrounded by surfactant molecules forming stable micelles when surfactant concentration is above the critical micelle concentration (CMC). As a result, the apparent aqueous solubility of a NAPL can significantly increase in the presence of surfactants allowing a more rapid cleanup than is typically observed in conventional pump-and-treat techniques. SEAR processes have been tested in a small scale (usually 1-d column experiments), and have been proven to be successful in cleaning up contaminated soils (Pennell et al., [82]; Pennell et al.[83]). However, the effectiveness of SEAR in field applications is usually hindered by the inability of the surfactant to access all the entrapped free-phase NAPL in the source zone.

4.4.1 *Simulation Problem*

The example problem that is selected for a model demonstration to show the uncertainty of mass removal through source zone treatment is based on proposed laboratory simulations that will be conducted in an intermediate-scale test tank. The test tank is 9.75 m (32 ft.) in length and 1.22 m (4 ft) in height (Figure 10). The tank is constructed using 2.44 m (8-ft) long prefabricated plexiglas wall units. Two constant-head end tanks are used to create a sloping water table along the length of the tank. Three monitoring wells are placed downstream of the NAPL entrapment zones to act as the points of compliance. In the planned experiments, a NAPL entrapment zone will be created by spilling a test NAPL into a heterogeneous formation produced by using five well-characterized test sands (Table 9).

The NAPL source distributions used here are based on experimental NAPL spills in past intermediate-scale tank experiments (Campos [61]). These NAPL source distributions could also be obtained from screening models (e.g., Willson et al. [84]), invasion percolation models (e.g., Miller et al. [85]) or multiphase, multicomponent models (e.g., Delshad et al. [71]).

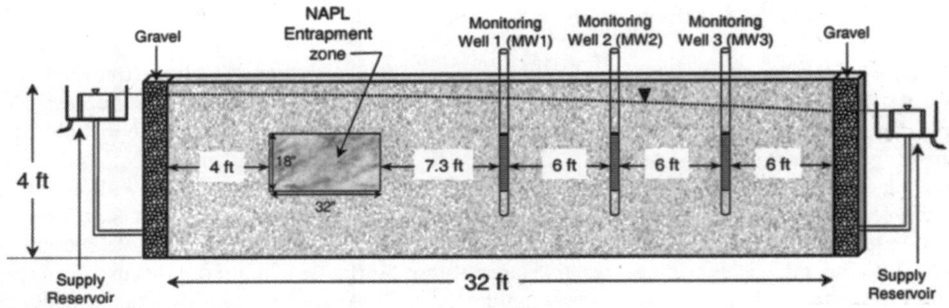

Figure 10. Schematic of the intermediate-scale dissolution experiment

To simulate the solute breakthrough concentrations at the downstream monitoring wells, it is necessary to provide the parameters for the flow, solute transport and dissolution models. Table 11 summarizes the model parameters for natural and surfactant enhanced dissolution simulations. The transport parameters are based on laboratory measured values obtained for the test sands in previous intermediate-scale experiments.

TABLE 11. Model parameters used in natural and enhanced dissolution

Parameter	Units	Natural dissolution	Enhanced dissolution
Cell size along raw (Δx)	ins	2	2
Cell size along column (Δy)	ins	2	2
Layer thickness (Δz)	ins	1	1
Longitudinal dispersivity (α_L)	cm	0.1	0.1
Transverse dispersivity(α_T)	cm	0.01	0.01
Average groundwater velocity	m/d	2.53	2.53
NAPL partition coefficient (K_p)	$cm^3\ g^{-1}$	6410	60
NAPL density (ρ_{nw})	$g^{-1}\ cm^3$	0.8611	0.8611
Solution viscosity	cP	1.000	1.177

4.2 SIMULATION RESULTS

Figure 11 shows the simulated velocity distribution within the entrapments zone. The velocity vector field depends on the heterogeneity of the system and the saturation of the entrapped NAPL. During enhanced dissolution, the velocity vector field changes due to the reduction in NAPL saturation.

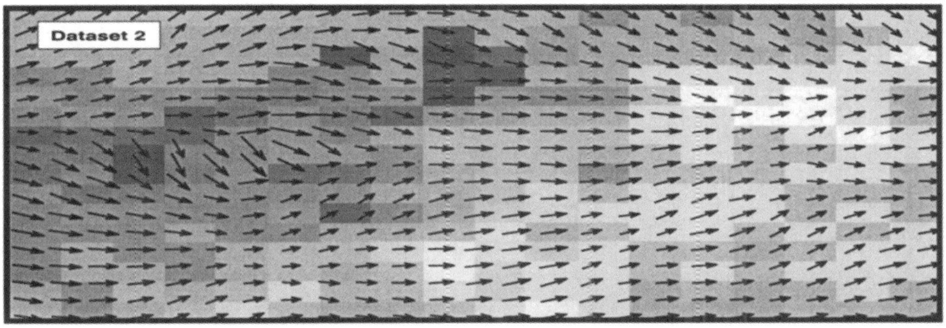

Figure 11. Vector field of velocity during NAPL dissolution

Figure 12 shows the simulated downstream concentration breakthrough curves at the three monitoring wells for one of the data sets (observed NAPL distribution 26 days after spill in realization #2.) These results show that, under natural dissolution, the downstream concentration reaches a steady-state value of 10.7 mg l^{-1}. The steady-state values obtained from the other data sets ranged between 9.2 mg l^{-1} and 17.3 mg l^{-1}. However, these steady-state concentrations are still much lower than the aqueous (equilibrium) solubility of para-xylene (156 mg l^{-1}) indicating rate-limited mass transfer.

Figure 13 shows surfactant-enhanced dissolution breakthrough curves at monitoring well #2 (MW2) assuming both complete and incomplete delivery of surfactant for the same data set (data set #2). In the case of complete delivery, the observed para-xylene concentration remains initially high but dramatically decreases as NAPL mass is depleted. Even though the breakthrough curve is smooth, it shows different zones of transitions reflecting the model's ability to simulate dissolution in zones of varying entrapment saturations. That is, surfactant will first remediate zones where NAPL saturations are low (high relative permeability, k_{rw}) and then deplete mass in high saturation zones. In this case, dissolved NAPL concentration eventually reaches zero, indicating that the source zone is fully remediated.

In practical situations, it is not possible for surfactant to reach all of the entrapment zones, and some locations of NAPL entrapment may not undergo remediation within a reasonable length of time. This is shown by the incomplete surfactant delivery cases. As surfactant delivery is controlled by the heterogeneity of the entrapment zone, the breakthrough curve is not smooth (see Fig. 13). This is due to the fact that the surfactant concentration will not reach CMC in all of the computational blocks containing free-phase NAPLs and that the flow field is continuously changing with the NAPL depletion.

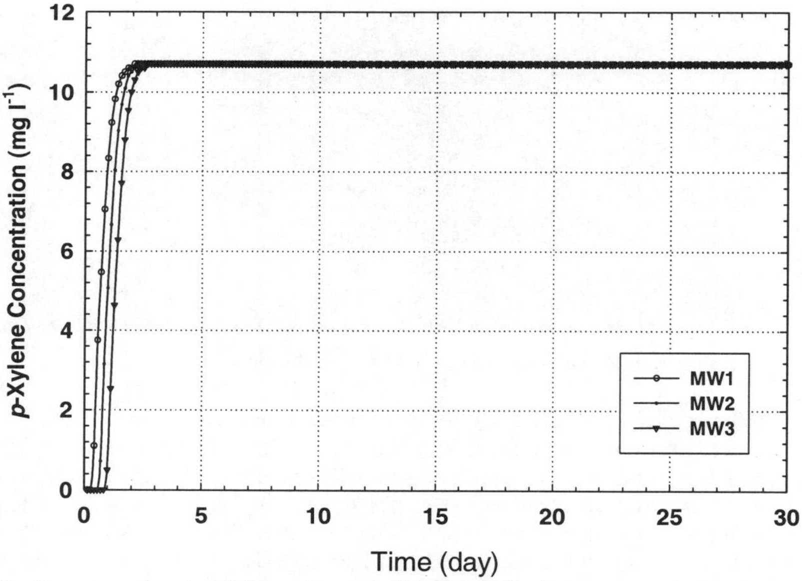

Figure 12. Concentration breakthrough at monitoring wells due to natural dissolution (Data Set #2).

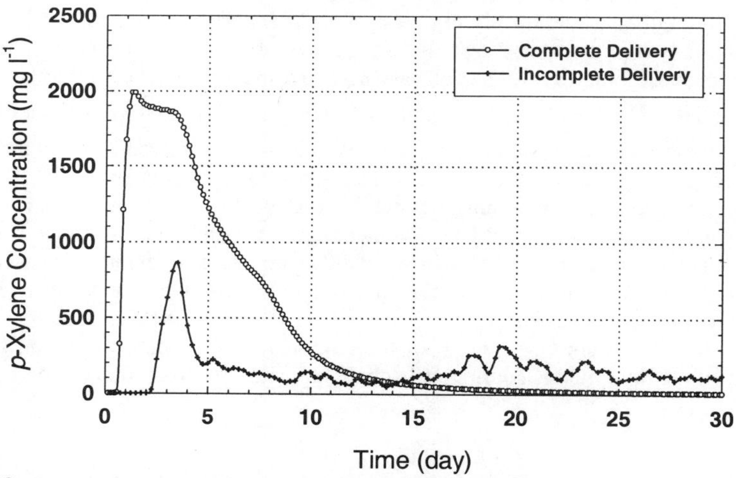

Figure 13. Concentration breakthrough at monitoring well #2 due to enhanced dissolution: complete vs. incomplete delivery (Data Set #2).

The NAPL mass depletion in the source zone as a function of time for the ten simulations are plotted in Figure 14 for complete and incomplete surfactant delivery, respectively. Mass depletion during surfactant-enhanced dissolution for the ten cases show a range of variability due to the uncertainty of NAPL distribution and soil heterogeneity. For complete surfactant delivery, although the variability of mass depletion is evident, the total mass of NAPL will eventually decrease to zero for all cases. In reality, however, soil heterogeneity as well as high NAPL saturation can create flow bypassing of the injected surfactant solution. As a result, the cleanup takes place slowly in the NAPL entrapment zones where bypassing occurs. This is shown in Fig. 14 for the incomplete delivery cases. In some extreme cases, the surfactant will not access the entrapment zones and the NAPL will never be remediated (e.g. data set #3 in Fig. 14 shows mass of NAPL in the source zone remains constant after ~13 days of surfactant injection). A much greater mass depletion variability is observed for incomplete surfactant delivery compared to the case where complete delivery is assumed.

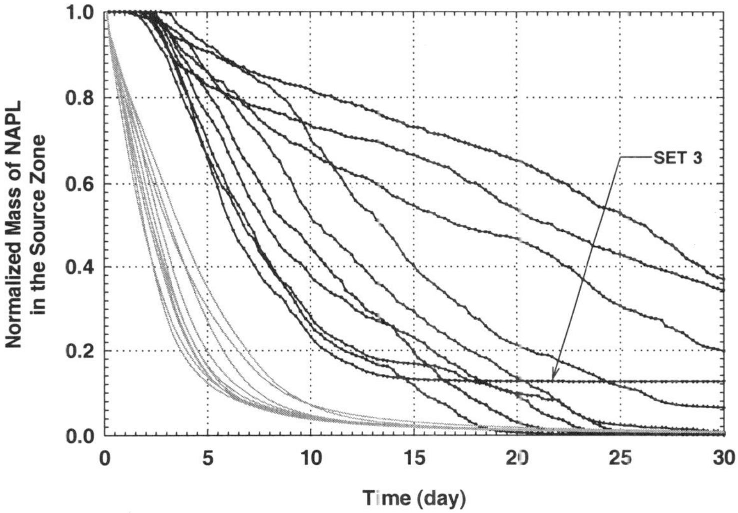

Figure 14. Normalized mass depletion as a function of time for surfactant-enhanced dissolution (incomplete delivery). The gray lines represent complete delivery cases.

4.5. CONCLUSIONS

This study illustrates the effect of heterogeneity on the effectiveness of surfactant delivery to the source zone of NAPL undergoing remediation. A comparison of complete and incomplete surfactant delivery as controlled by heterogeneity and relative permeability effects shows a much greater variability in the mass depletion when incomplete surfactant delivery is accomplished. The primary hypothesis that the heterogeneity controls the delivery of the surfactant was tested. The results establish that when the heterogeneity cannot be fully characterized, there exists significant uncertainty regarding the achievement of cleanup goals and the reduction of risk.

Even though a simplified model was used in the analysis, the model captures the basic processes adequately to arrive at these conclusions. An experimental study in an intermediate-scale test tank is planned to generate an accurate data set to validate the model in the laboratory before applying it to field situations. These results will be used to conduct similar analysis to determine remediation end points for risk analysis.

5. Acknowledgements

The support of the US EPA Hazardous Substance Research Center, South and Southwest, Western Region and Great Plains and Rocky Mountains is acknowledged as is the support of the Mechanical and Environmental Science Division of Army Research Office, the Department of Defense DURIP funds and National Science Foundation (MRI). The contributions of S. Saenton, T. Saba and R. Compas, Dr. K. Soga at Cambridge University and Dr. Clint Wilson at Louisiana State University are gratefully acknowledged.

6. References

1. Fairbrother A., Glazebrook P.W., Van Straalen, Tarazona J.V. (1999) Test methods for hazard determination of metals ans sparingly soluble metal compounds in soils: summary of a SETAC Peelston Workshop. SETAC, ed A. Fairbrother. pp1-24.
2. Janssen CR, De Schamphelaere K, Heijerick D, Muyssen B, Lock K, Bossuyt B, Vangheluwe M, Van Sprang P (2000) Uncertainties in the environmental risk assessment of metals *Human Ecolog. Risk Assessment.* **6,** 1003-1018.
3. Hill SJ Speciation of trace metals in the environment (1997) *Chem. Soc. Reviews* **26,** 291-298.
4. Murphy AP, Coudert M, Barker J. (2000) Plants as biomarkers for monitoring heavy metal contaminants on landfill sites using sequential extraction and inductively coupled plasma atomic emission spectrophotometry (ICP-AES). *J. Env. Monitoring* **2,** 621-627.
5. Oste LA, Dolfing J, Ma WC, Lexmond TM (2001) Cadmium uptake by earthworms as related to the availability in the soil and the intestine *Environ. Toxicol. Chem.* **20,** 1785-1791.
6. Peijnenburg WJGM, Baerselman R, de Groot AC, Jager T, Posthuma L, Van Veen RPM (1999) Relating environmental availability to bioavailability: Soil-type-dependent metal accumulation in the oligochaete *Eisenia andrei. Ecotox. Environ. Safety* **44,** 294-310.
7. Ellickson KM, Meeker RJ, Gallo MA, Buckley BT, Lioy PJ (2001) Oral bioavailability of lead and arsenic from a NIST standard reference soil material. *Arch. Environ. Cont. Toxicol.* **40,** 128-135.
8. Lin SH, Lai SL, Leu HG (2001) An improved method for determination of heavy metal bioavailability in contaminated soil. *Environ. Techn.* **22,** 731-739.
9. Lock K, Janssen CR (2001) Test designs to assess the influence of soil characteristics on the toxicity of copper and lead to the oligochaete *Enchytraeus albidus. Ecotoxicol.* **10,** 137-144.
10. Davison W, Zhang H (1994).In-Situ speciation measurements of trace components in natural-water using thin-film gels. *Nature* **367,** 546-548
11. Zhang H, Zhao FJ, Sun B, Davison W, McGrath SP (2001) A new method to measure effective soil solution concentration predicts copper availability to plants. *Env. Sc. Technol.* **35,** 2602-2607.
12. Lebourg A, Sterckeman T, Ciesielski H, Proix N. (1996) Suitability of chemical extraction to assess risks of toxicity induced by soil trace metal bioavailability. *Agronomie* **16,** 201-215.
13. Collard JM, Corbisier P, Diels L, Dong Q, Jeanthon C, Mergeay M Taghavi, S, van der Lelie D, Wilmotte A, Wuertz S. (1994) Plasmids for heavy metal resistance in *Alcaligenes eutrophus* CH34: mechanisms and *applications FEMS Microbiol. Rev.***14,** 405-414.

14. Corbisier P. Mergeay M., Diels L. (1998).Fused genes, vectors containing them, process for preparing them and their use for determining the presence of metals or of xenobiotic compounds *US patent* No. 5 786 162.
15. Corbisier P. van der Lelie D., Borremans B., Provoost A., de Lorenzo V., Brown N., Lloyd J., Hobman J., Csöregi E., Johansson G, Mattiasson B (1999) Whole cell- and protein-based biosensors for the detection of bioavailability heavy metals in environmental samples. *Anal. Chim. Acta* **387,** 235-244.
16. Corbisier P., E. Thiry, A. Masolijn and L. Diels (1994) in Campbell, LJ Kricka and P.E Stanley (eds.) Bioluminescence and Chemiluminescence: Fundamentals and Applied Aspects, New York, Brisbane, Toronto, Singapore pp 151-155.
17. Tibazarwa C., Corbisier P., Mench M., Bossus A., Solda P., Mergeay M., Wyns L., van der Lelie D. (2001) A microbial biosensor to predict bioavailable nickel in soil and its transfer to plants. *Environmetal Pollution*. **113,** 19-26.
18. van der Lelie D. and P. Corbisier (2000) Use of bacterial tests (the VITOTOX® genotoxicity test and the BIOMET heavy metal test) to analyse chemicals and environmental samples *In* Microbiotests for routine toxicity testing and biomonitoring, pp. 197-207.
19. Corbisier P., Thiry E. and Diels L. (1996) Bacterial Biosensors for toxicity assessment of solid wastes *Env. Toxicol. Water Qual.* **11**, 171-177.
20. Mergeay M., Nies D., Schelegel HG, Gerits J., Charles P. and Van Gysegem F.(1985) *Alcaligenes eutrophus* CH34 is a facultative chemolithotrop with plasmid-bound resistance to heavy metals. *J. Bact* **162,** 328-334.
21. Kaiser K.L.E. and Devillers J. (1994*) Ecotoxicity of chemicals to photobacterium phosphoreum*, Gordon and Breach Science Publishers, Switzerland.
22. Vangronsveld J., Spelemans N., Clijsters H., van der Lelie D., Noë L., Mergeay M., Corbisier P., Bierkens J., Adriaensens E., Carleer R. and Van Poucke L. (2000) Physico-chemical and biological evaluation of the efficacy of in situ metal inactivation in contaminated soils In: *11th Annual International Conference on Heavy Metals in the Environment* (J. Nriagu, Editor), Contribution #1043. University of Michigan, School of Public Health, Ann Arbor, MI (CD-ROM).
23. Reinhard, M. and A. Drefahl. (1999) *Toolkit for Estimating Physicochemical Properties of Organic Compounds.*
24. Lyman, W.J., W.F. Reehl and D.H. Rosenblatt. (1982) *Handbook of Chemical Property Estimation Methods. Environmental Behavior of Organic Compounds.*
25. Reinhard, M. and A. Drefahl. 1999. *Estimating Physicochemical Properties of Organic Compounds*, First Edition. New York.
26. Boethling, R.S. and D. MacKay. (2000) *Handbook of Property Estimation for Chemicals: Environmental Health Sciences*, CRC Press, Boca Raton, FL.
27. Hansch, C. and A. Leo. (1979) Substituent Constants for Correlation *Analysis in Chemistry and Biology*, John Wiley & Sons, New York.
28. Coates, M., D.W. Connell and D.M. Barron. (1985) Aqueous Solubility and Octan-1-ol to Water Partition Coefficients of Aliphatic Hydrocarbons. *Environ. Sci. Technol. 19*:628-632.
29. Broto, P., G. Moreau and C. Vandycke. (1984) Molecular structure: perception, autocorrelation descriptor and sar studies. System of atomic contributions for the calculation of the n-octanol/water partition coefficients. *Eur. J. Med. Chem. - Chim. Ther. 19*:71-78.
30. Balaban, A.T., L.B. Kier and L.H. Hall. (1992) Correlations between Chemical Structure and Normal Boiling Points of Acyclic Ethers, Peroxides, Acetals, and Their Sulfur Analogues. J. *Chem. Inf. Comput. Sci. 32*:237-244.

31. Weininger, D., A. Weininger and J.L. Weininger. (1989) SMILES. 2. Algorithm for Generation of Unique SMILES Notation. *J. Chem. Inf. Comput. Sci. 29*:97-101.
32. Weininger, D. (1988) SMILES, a Chemical Language and Information System. 1. Introduction to Methodology and Encoding Rules. *J. Chem. Inf. Comput. Sci. 28*:31-36.
33. Weininger, D. (1990) SMILES. 3. Depiction of Graphical Depiction of Chemical Structures. *J. Chem. Inf. Comput. Sci. 30*:237-243.
34. Pereira, W.E.; C.E. Rostad; C.T. Chiou; T.I. Brinton; L.B. Barbar II. (1988) "Contamination of Estuarine Water, Biota, and Sediment by Halogenated Organic Compounds: A Field Study." Environ. Sci. Technol., 22, pp. 772-778.
35. McGroddy, S.E. and J.W. Farrington. (1995) "Sediment Porewater Partitioning of Polycyclic Aromatic Hydrocarbons in Three Cores from Boston Harbor Massachusetts." Environ. Sci. Technol., 29, pp. 1542-1550.
36. Readman, J.W. and R.F.C. Mantoura. (1987) "A Record of Polycyclic Aromatic Hydrocarbon (PAH) Pollution Obtained from Accreting Sediments of the Tamar Estuary U.K.: Evidence for Nonequilibrium Behaviour of PAH." Total Environ., 66, pp. 73-94.
37. Ghosh, U.; R.G. Luthy; S.J. Clemett; R.N. Zare. (2000) "Microscale Location, Characterization, and Association of Polycyclic Aromatic Hydrocarbons on Harbor Sediment Particles." Environ. Sci. Tech., 34, pp. 1729.
38. Connaughton, D.F; J.R. Stedlinger; L.W. Lion; and M.L. Shuler. (1993) "Description of Time-Varying Desorption Kinetics: Release of Naphthalene from Contaminated Soils." Environ. Sci. Technol., 30, pp. 1145-1151
39. Hatzinger, P.B. and M. Alexander. (1996) "Biodegradation of Organic Compounds Sequestered in Organic Solids or in Nanopores within Silica Particles." Environ. Tox. and Chem. 15, 2215-2221.
40. Kan, A.T.; G. Fu; and M.B. Tomson. (1994) Environ. Sci. Technol., 28(5): 859-867
41. Luthy, R.G.; G.R. Aiken; M.L. Brusseau; S.D. Cunningham; P.M. Gschwend; J.J. Pignatello; M. Reinhard; S.J. Traina; W.J. Weber; J.C. Westall. (1997) "Sequestration of Hydrophobic Organic Compounds by Geosorbents." Environ. Sci. Technol., 31, pp. 3341-3347.
42. Hunter, M.A.; A.T. Kan; M.B. Tomson. (1996) "Development of a Surrogate Sediment to Study the Mechanisms Responsible for Adsorption/Desorption Hysteresis." Environ. Sci. Technol. 30. pp. 2278-2285.
43. Kan, A.T.; G. Fu; M.A. Hunter; W. Chen; C.H. Ward; M.B. Tomson. (1998) "Irreversible Sorption of Neutral Hydrocarbons to Sediments: Experimental Observations and Model Predictions." Environ. Sci. Technol., 32, pp. 892.
44. Kan, A.T.; G. Fu; M.A. Hunter; M.B. Tomson. (1997) "Irreversible Adsorption of Naphthalene and Tetrachlorobiphenyl to Lula and Surrogate Sediments." Environ. Sci. Technol., 31, pp. 2176-2185.
45. Huang, W. and W.J. Weber. (1997) "A Distributed Reactivity Model for Sorption by Soils and Sediments 10. Relationships between Desorption, Hysteresis, the Chemical Characteristics of Organic Domains." Env. Sci. Technol., 31, pp. 2562.
46. Lick, W. and V. Rapaka. (1996) "A Quantitative Analysis of the Dynamics of the Sorption of Hydrophobic Organic Chemicals to Suspended Sediments." Environmental Toxicology and Chemistry, 15(7), pp. 1038-1048.
47. Pignatello, J.J. and B. Xing. (1996) "Mechanisms of Slow Sorption of Organic Chemicals to Natural Particles." Environ. Sci. Technol., 30, pp. 1-11.
48. Lu, X., D. Reible, J. Fleeger, Y. Chai, (2002) "Bioavailability of Desorption-resistant Phenanthrene to the Oligochaete, Ilyodrilus templetoni," Environmental Toxicology and Chemistry, in press.

49. Gomez-Hermosillo, H., J. Parude, D. Reible. (2002) "Wetland Plant Uptake Of Desorption-Resistant Organic Compounds From Sediments," submitted to Environ. Sci. Tech.
50. Reible, D.D.; K.T. Valsaraj; and L.J. Thibodeaux. (1991) "Chemodynamic Models for Transport of Contaminants from Sediment Beds." In Handbook of Environmental Chemistry, O. Hutzinger (ed.), Springer-Verlag, Heidelberg, p. 187-228.
51. Danny D. Reible; Richard H. Jensen; Samuel J. Bentley;Mimi B. Dannel; Joseph V. DePinto; James A. Dyer; Kevin J. Farley; Marcelo H. Garcia; David Glaser; John M. Hamrick; Wilbert J. Lick; Robert A Pastorok; Richard F. Schwer; C. Kirk Ziegler (2002) "The Role of Modeling in Managing Contaminated Sediments" in Environmental Modeling and Management: Theory, Practice, and Future Directions,p. 63-110.
52. Reible, D.D. (2002) "Contaminant Processes in Sediments," in Sedimentation Engineering- ASCE Manual Volume 54, Ed. Marcel Garcia (2nd Edition in preparation)
53. Kueper, B.H. and Gerhard, J.I. (1995) Variability of point source infiltration rates for two-phase flow in heterogeneous porous media. Wat. Resour. Res.. 31, 2971-2980..
54. Illangasekere, T.H., Armbruster, E.J., Yates, D.N. (1995a) Non-aqueous Phase Fluids in Heterogeneous Aquifers: An Experimental Study, J. Envir. Engin., 121, 571-579.
55. Illangasekare, T.H., Ramsey, J.L., Jensen, K.H., and Butts, M. (1995b) Experimental Study of Movement and Distribution of Dense Organic Contaminants in Heterogeneous Aquifers, J. Contam. Hydrol., 20, 1-25.
56. Held, R.J. and Illangasekare, T.H., (1995) Fingering of Dense Non-aqueous Phase Liquids in Porous Media, 1. Experimental Investigation, Wat Resour. Res., 31, 1213-1222.
57. Illangasekare, T.H., (1998) Flow and entrapment of nonaqueous phase liquids in heterogeneous soil formations, In Physical Nonequilibrium in Soil, (ed. H.M. Selim and L. Ma) (Ann Arbor Press) 417-423.
58. Wilson, J. L., S. H. Conrad, W. R. Mason, W. Peplinski, and E. Hagan. (1990) Laboratory Investigation of Residual Liquid Organics. EPA/600/6-90/004.
59. Hoag, G.E. and M. Marley. (1986) Gasoline Residual Saturation in Unsaturated Uniform Aquifer materials. Journal of Environmental Engineering. v. 112, p. 586-604.
60. Powers, S.E., Abriola, L.M., and Weber, Jr., W.J. (1992) An experimental investigation of nonaqueous phase liquid dissolution in saturated subsurface systems: Steady state mass transfer rates. Water Resour. Res., 28(10), 2691-2705.
61. Compos, R. (1998) Hydraulic Conductivity distribution in a DNAPL entrapped zone in a spatially correlated random field, M.S. Thesis, University of Colorado, Boulder.
62. Illangasekare, T.H., Znidarcic, D., Walser, G. and Weaver, J. (1994) An experimental evaluation of two sharp front models for vadose zone non-aqueous phase liquid transport, Report for United States Environmental Protection Agency, Office of Research and Development, EPA/600/R-94/197
63. Bird, R.B., Stewart, W.E., Lightfoot E.N. (1960) Transport Phenomena, John Wiley, New York, 780 pages.
64. Miller, C.T., Poirier-McNeill, M.M. and Mayer, A.S. (1990) Dissolution of trapped nonaqueous phase liquids: mass transfer characteristics, Wat. Resour. Res., 26, 2783-2796.
65. Powers, S.E., Abriola, L.M., and Weber, Jr., W.J. (1994) An experimental investigation of nonaqueous phase liquid dissolution in saturate subsurface systems: Transient mass transfer rate. Water Resour. Res., 30(2), 321-332.
66. Imhoff, P.T., Jaffe, P.R., and Pinder, G.F. (1994) An experimental study of complete dissolution of a nonaqueous phase liquid in saturated porous media. Water Resour. Res., 30(2), 307-320.

67. Imhoff, P.T., and Miller, C.T. (1996) Dissolution fingering during the solubilization of nonaqueous phase liquids in saturated porous media: 2. Experimental observations. Water Resour. Res., 32(7), 1929-1942.
68. Saba, T.A. (1999) Upscaling of mass transfer from entrapped NAPLs under natural and enhanced conditions. Ph.D. Dissertation, University of Colorado.
69. Saba, T. and Illangasekare, T.H. (2000) Effect of groundwater flow dimensionality on mass transfer from entrapped nonaqueous phase liquid contaminants. Wat. Resour. Res., 36(4), 971-979.
70. Saba, T., Illangasekare, T.H., and Ewing, J. (2000) Surfactant enhanced dissolution of entrapped NAPLs in multi-dimensional flow fields, Submitted for publication in J. of Contaminant Hydrology, in revision.
71. Delshad, M., Pope, G.A., and Sepehrnoori, K. (1996) A compositional simulator for modeling surfactant enhanced aquifer remediation: 1. Formulation, J. Contam. Hydrol. 23(4), 303-327.
72. McDonald, M.G. and Harbaugh, A.W., (1988) A modular three-dimensional finite difference groundwater flow model, Modeling Techniques, Techniques Water Resour. Invest. U.S.Geol. Surv., 6.
73. Harbaugh, A.W., Banta, E.R., Hill, M.C., and McDonald, M.G. (2000). MODFLOW-2000, The U.S. Geological Survey Modular Ground-Water Model - User guide to modularization concepts and the ground-water flow process. Open file report 00-92. U.S. Geological Survey.
74. Zheng, C. (1990) MT3D: A modular three-dimensional transport model for simulation of advection, dispersion, and chemical reactions of contaminants in groundwater systems, Rep. 74280, U.S. EPA Robert S. Kerr Environ. Res. Lab., Ada, OK.
75. Zheng, C., and Wang, P.P. (1999) MT3DMS: A modular three-dimensional multispecies transport model for simulation of advection, dispersion, and chemical reactions of contaminants in groundwater systems; Documentation and user's guide. Contract report SERDP-99-1. U.S. Army Engineer Research and Development Center
76. Okeson, S.J. (1995) A study of dissolution of multicomponent nonaqueous phase liquids in a two dimensional flow field and model development. M.S. Thesis, University of Colorado.
77. Wyllie, M.R.J. (1962) Relative Permeability in Petroleum Production Handbook. Reservoir Engineering, vol. II. McGraw-Hill, New York.
78. Ewing, J.E. (1996) Effects of dimensionality and heterogeneous on surfactant-enhanced solubilization of nonaqueous phase liquids in porous media. M.S. Thesis, University of Colorado.
79. USEPA. (1989) Evaluation of ground-water extraction remedies. Vol. 1-2 of EPA/540/2-89/054b. U.S. Environmental Protection Agency Offce of Emergency Response, Washington, D.C.
80. USEPA (1992) Dense nonaqueous phase liquids – a workshop summary, United States Environmental Protection Agency, Office of Research and Development, Dallas, TX Workshop, April 16-18, 1991. EPA/600/R-92/030
81. Grubb, D.C. and Sitar, N. (1994) Evaluation of technologies for in situ cleanup of DNAPL contaminated sites, Report for United States Environmental Protection Agency, Office of Research and Development, EPA/600/R-94/12
82. Pennell, K.D., Abriola, L.M., and Weber, Jr., W.J. (1993) Surfactant enhanced solubilization of residual dodecane in soil columns: 1. Experimental investigation. Environ. Sci. Technol., 27(12), 2332-2340.

83. Pennell, K.D., Abriola, L.M., and Pope, G.A. (1994) Surfactant enhanced remediation of soil columns contaminated by residual tetrachloroethylene. J. Contam. Hydrol., 16(1), 35-53.
84. Willson, C.S., Weaver, J.W., Illangasekare, T.H., and Charbeneau, R.J. (1996) Framework for a Screening Model for DNAPL Contamination of Porous Media, Non-Aqueous Phase Liquids (NAPLs) in the Subsurface Environment: Assessment and Remediation, edited by L.N. Reddi, 407-418.
85. Miller, C.T., Willson, C.S., Gleyzer, S.N., Farthing, M.W., McBride, J.F., and Imhoff, P.T. (1998) NAPL-aqueous phase mass transfer in heterogeneous porous media. Groundwater Quality: Remediation and Protection, Proceedings of the GQ '98 Conference, Tubingen, Germany, September 1998, IAHS Pbl., no 250, 141-148.

BIODEGRADATION AND BIOREMEDIATION

Principles and Applications

P. ADRIAENS[1], P.J.J. ALVAREZ[2], L. BASTIAANS[3], L. DIELS[3], D. MAJOR[4], Z. FILIP[5], D. SPRINGAEL[3]
[1]*Environmental and Water Resources Engineering, 1351 Beal Ave., The University of Michigan, Ann Arbor, Michigan 48109-2125, USA*
[2]*Civil and Environmental Engineering, The University of Iowa, 5317 Seaman's Center, Iowa City, Iowa 52242-1527, USA*
[3]*Flemish Technological Institute, Boeretang 200, B-2400 Mol, Belgium*
[4]*GeoSyntec Consultants, 160 Research Lane, #206, Guelph, Ontario N1G5B2*
[5]*Federal Environment Agency, Langen Building, Paul Ehrlich Strasse 29, 63225 Langen, Germany.*

1. Abstract

During the last decade, advances in our understanding of biodegradation principles, and lessons learned from bioremediation applications, have increased our knowledge base to help transition laboratory-based processes to the field. Within the group of organic anthropogenic compounds, polycyclic aromatic hydrocarbons (PAH), chlorinated volatile organic compounds (VOC), and the alkylbenzene (benzene, toluene, ethyl benzene, and the xylenes, BTEX) hydrocarbons are among the most frequently encountered pollutants in soils, groundwater aquifers and sediments. This chapter aims to describe the challenges posed by, and methodologies used during, the transfer of biodegradation principles to field application by addressing the following specific objectives: (i) principles of technology transfer and application to a chlorinated solvent-contaminated site, (ii) principles and applications of PAH, VOC, and BTEX bioremediation, and (iii) principles and applications of soil health assessment. Whereas the fundamental processes of biodegradation have been well established, harnessing these processes for in situ application is fraught with substantial uncertainty. These uncertainties are mainly due to site specificities associated with expression of the desired biodegradation activity, the difficulty with establishing causal relationships between biological activity and geochemical (organic and inorganic) changes, and the lack of data for a convincing cost-benefit analysis associated with bioremediation applications (i.e. does the investment of time and money substantially increase the overall health of the impacted environmental compartment?). In addition to the requirement for substantial technical expertise, these uncertainties are major impediments for site managers, problem owners, and technology adopters and implementers to make effective use of bioremediation technologies. The results indicate the need of more research on the use of *in situ* biomonitoring techniques at various temporal and spatial scales of environmental impact, and in optimization strategies for technology implementation within a quantitative uncertainty-based decision framework.

D. Reible and K. Demnerova (eds.), Innovative Approaches to the On-Site Assessment and Remediation of Contaminated Sites, 67–113.

2. Technology Transfer for Environmental Remediation Systems

2.1. INTRODUCTION

For clarification, the use of the terms biodegradation and bioremediation throughout this manuscript should be defined. According to figure 1 below, **biodegradation** is the culmination of the dependent research variables that determine microbial activity in a given system bounded by organic and inorganic geochemical characteristics (concentrations, fluxes, physical form and molecular structure) on the one hand, and the microbial ecology and its associated intrinsic activity on the other. The optimization of these boundary conditions for a given (class of) organic compounds – typically reported in the literature – then results in effective biodegradation process control. The capitalization on these processes during *in situ* or on site application of electron donor amendments, redox manipulation, and microbial inoculation or stimulation strategies with the aim to mitigate contaminant fluxes or concentrations is **bioremediation**. The transition between both levels of optimization is neither trivial, nor logical, and requires a substantial investment of time and funding to address issues of scaling, which can only be accomplished via successful integration of a wide range of expertise in a team-effort.

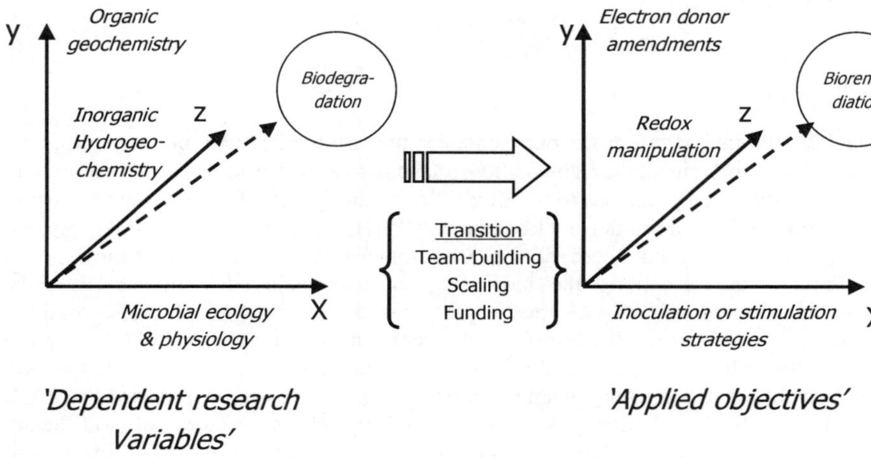

Figure 1. Multi-dimensional space for biodegradation and bioremediation optimization.

A conceptual illustration of the transition phase between process demonstration in the laboratory, and its implementation in the field is shown in figure 2. The timeline assumes that the fundamental processes (e.g. halorespiration of VOC, degradation of BTEX) are understood, and starts with laboratory feasibility studies ('microcosms') applying the process of interest to the specific medium /soil which needs bioremediation. The pilot study focuses on a contained location in the field, and under field conditions to be evaluated for full-scale application. The cost escalation is in part due to the number of investigators involved, the initial capital investment, and disposal costs, but tapers off if the equipment acquired for the pilot study can be easily scaled for full-scale application. In that case, operations and maintenance comprise the majority of the investment. It should be noted that the costs for pre- and post-operational environmental health assessment are not included, as success of a bioremediation application is usually measured in terms of flux or concentration reduction.

Note also that the transition from a closed and defined laboratory system to a relatively open field system increases the uncertainty associated with decisions and success, mainly due to the decrease in reliability on the achievable causal relationships.

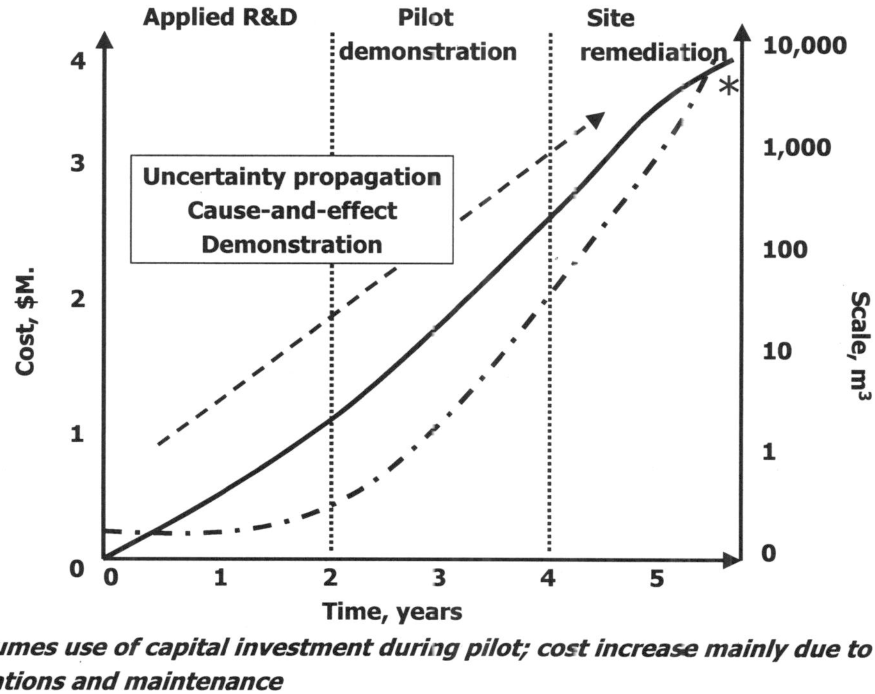

Figure 2. Multi-dimensional parameter propagation during technology transfer transition from biodegradation to bioremediation (cost: solid line; scale: dotted • —).

2.2. A CASE IN POINT: SOURCE AND PLUME CONTROL STRATEGIES

The water quality of the Great Lakes watershed is impacted by nutrient, organic and microbial contaminant loadings resulting from coastal urban and industrial development. Owing to their ubiquitous domestic and industrial usage, chlorinated solvents are among the dominant and most frequently observed organic contaminants discharged. Typically, coastal sites contaminated with chlorinated solvents exhibit two distinct phases: a non-dissolved phase (the source), and a dissolved phase (the plume) that emanates from the source and migrates in the direction of surface waters such as Great Lakes.

Under a phased contract (Figure 3) with the Michigan Department of Environmental Quality (M-DEQ) and with The University of Michigan as the lead institution in a 7-institute collaboration, two innovative source and plume control technologies were evaluated and implemented to prevent or at least minimize discharge of chlorinated solvents to Lake Huron (in Oscoda, MI). The estimated discharge prior to the use of control technology is on the order of 30 kg/year at this site alone. The technologies are:

Surfactant-enhanced aquifer remediation (SEAR): This technology capitalizes on the properties of surfactants to either solubilize or mobilize the non-dissolved source material, which is trapped in the soil matrix. Through targeted injection of the surfactant, the surfactant-solvent mixture is recuperated after 'sweeping' the contaminated area.

Dechlorinating biocurtains: This technology capitalizes on a biodegradation process whereby anaerobic microorganisms utilize the dissolved chlorinated compounds in the plume, resulting in their conversion to harmless by-products. Injection of nutrients and microorganisms, followed by circulation of the groundwater creates a 'reactive biocurtain' in the subsurface through which the plume must pass.

Research Phases and Deliverables 1997-2002

Phases	Basic & Applied Research 1997-99	Technology Demonstration (field research-pilot) 2000-01	Technology Integration 2001-02	Site-wide Remediation (full scale) 2002-
Deliverables	Site biogeochemistry (Halo and SEAR) Pure and mixed culture isolations (Halo) Microbial characterization (rates, mass balances, efficiency) (Halo) Surfactant treatability studies (SEAR) Modeling (SEAR)	Test plot operational organic geochemistry (Halo and SEAR) Design and implementation of pilot-scale systems and inoculum scale-up (Halo) Proof-of-concept -Source control (SEAR) -Plume control (Halo) plume-specific designs for full-scale 50 % (Halo)-70 (SEAR) %	Site-wide integration of source reduction and plume control Delineation of contamination 90% design	Construction and implementation of source reduction and plume control

Figure 3. The GLMAC-HSRC-MDEQ technology transfer initiative for innovative source and plume control technologies

The approach includes a basic and applied research phase, a field demonstration, a technology integration phase, ultimately resulting in full-scale remediation. The first phase, using laboratory microcosms and column or sand box studies, is designed to screen a wide range of operational parameters relevant to the process on material collected from the field. The screening approach not only results in a laboratory-based optimization protocol, but more importantly includes sufficient variables such that the proposed technology can be transferred to sites other than the site under consideration. The second phase represents the proof-of-concept stage with sufficient confidence in the protocol to allow siting procedures, equipment design, and economic analysis. The third phase will integrate both source and plume control technologies on the same site (the second phase is demonstrated on separate sites to eliminate operational interferences) and scale the application to the entire site, which will then be remediated in the fourth phase.

2.2.1. *Site Description*

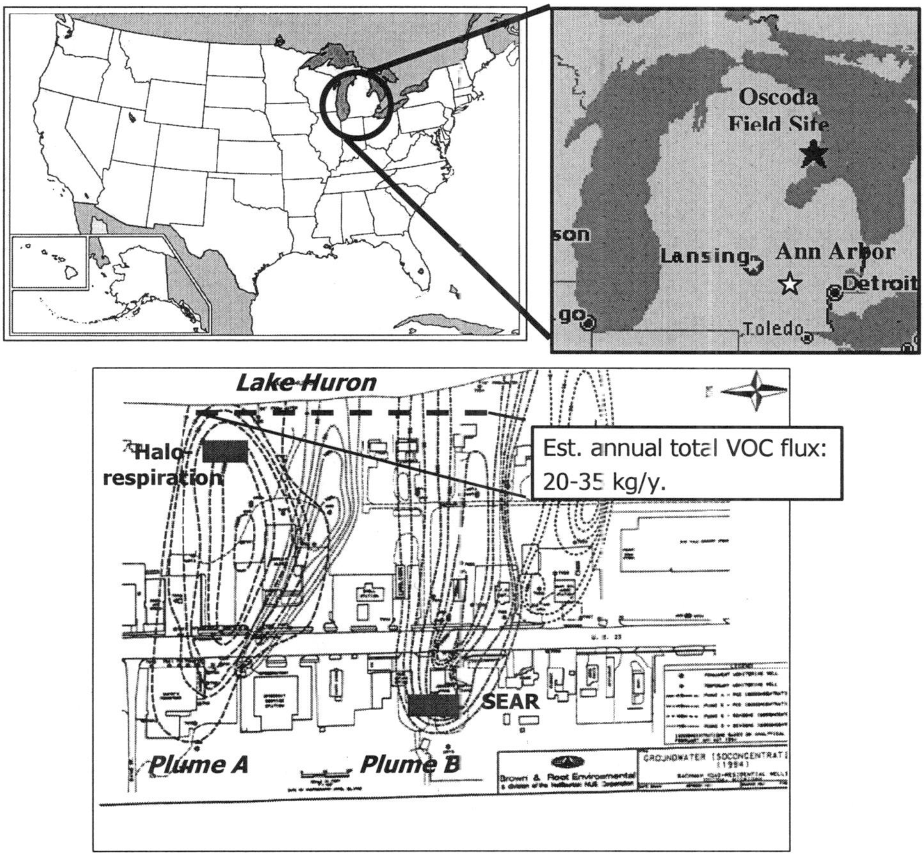

Figure 4: Geographical location of the Bachman Road Residential Wells site in Oscoda, Michigan. Both demonstration sites are bounded by Lake Huron to the East.

2.2.2. *SEAR Technology Application*

Simulation results and site constraints dictated that the final pilot-scale test design (Figure 5) consist of a single extraction well (19.7 l min^{-1}), a row of three water injection wells (W1-W3: 3.8 l min^{-1} each) to establish a flow field through the source zone, and a gallery of three surfactant injection wells (S1-S3: 1.9 l min^{-1} each) positioned between the water supply and extraction wells. The selected pumping schedule involved start-up of the extraction well, followed shortly by all injection wells. Injection of a 6% surfactant solution over the entire aquifer thickness was initiated 3 weeks later for 5 days. Operation of well S1 was then discontinued and injection of surfactant continued for an additional 5 days in wells S2 and S3, screened over the top and bottom 1.2 m of saturated depth.

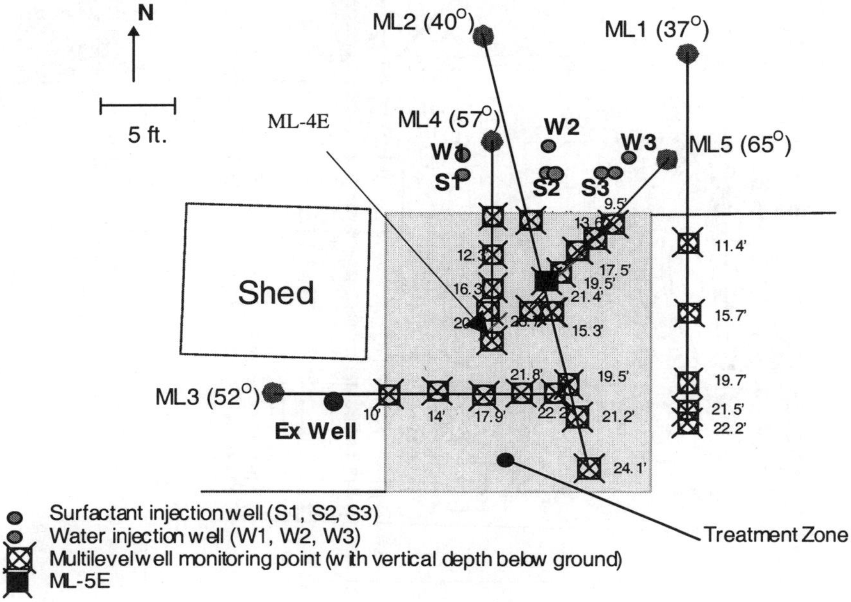

Figure 5: Bachman Road SEAR pilot-scale test site. SEAR treatment zone is shaded.

Simulations predicted that this targeted injection scheme would efficiently deliver surfactant to suspected highly contaminated regions and reduce overall surfactant cost. Water injection and extraction were then maintained for an additional month to ensure surfactant and plume capture. Simulations using this final design predicted a sweep of the entire treatment zone and over 95% recovery of injected surfactant.

Field data Figure 6 presents PCE and Tween 80 concentrations measured at the extraction well. As expected, the concentration of PCE rises with the arrival of Tween 80. A second increase of PCE concentration due to the targeted injection scheme is also evident. PCE concentrations at the extraction well prior to water injection varied from 15.36 to 19.48 ppm. The PCE concentration tailing at the end of the test may indicate that soluble PCE originating outside the treatment zone was arriving at the extraction well. This would be possible because the extraction well zone of influence is significantly larger than the targeted treatment zone. Another possible explanation is that solubilized PCE was still present in the treatment zone at the end of the test.

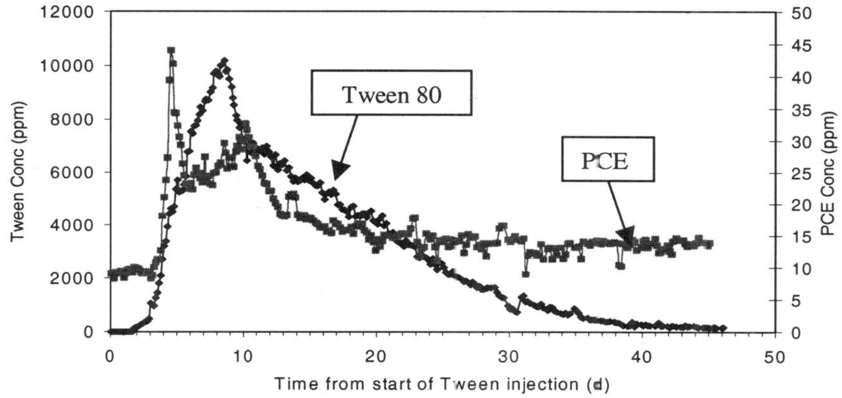

Figure 6. Measured PCE and Tween 80 concentrations at extraction well

Overall, approximately 95% of the injected surfactant was recovered, and 19 L of PCE, illustrating the effectiveness of the Tween 80 flush to enhance the solubility of PCE.

2.2.3. Halorespiration Barrier Technology Application

The test and control plots were spaced normal to groundwater flow (Fig. 7). Groundwater flow is toward Lake Huron, located 80 meters east of the plots.

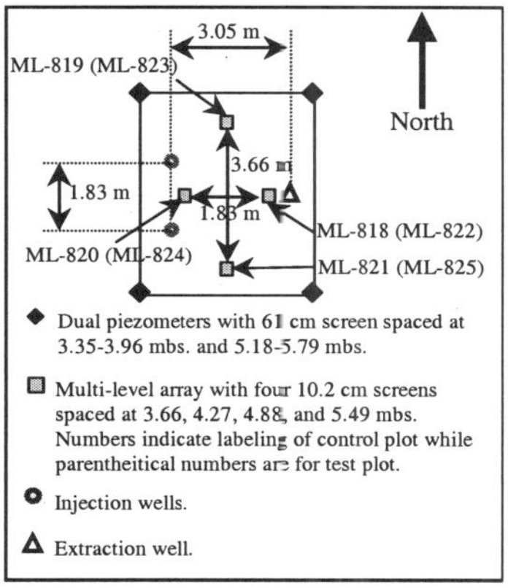

Figure 7. Well and sampling points layout each plot consists of dual piezometers with 61cm screens at each corner. Screened depths are at 3.4 – 4.0 meters and 5.2 – 5.8 meters below surface (mbs) for the shallow and deep piezometers, respectively. The plot size is 5.5 meters north to south (normal to groundwater flow) and 4.6 meters east to west. Two 10cm diameter injection wells, screened at 3.4 – 5.8mbs, are located on the west side of the plots and a single extraction well (screened at 4.6 – 5.8mbs) located toward the eastern border of the plot. The injection wells are screened over most of the saturated thickness of the aquifer and the extraction well only in the deeper portion.

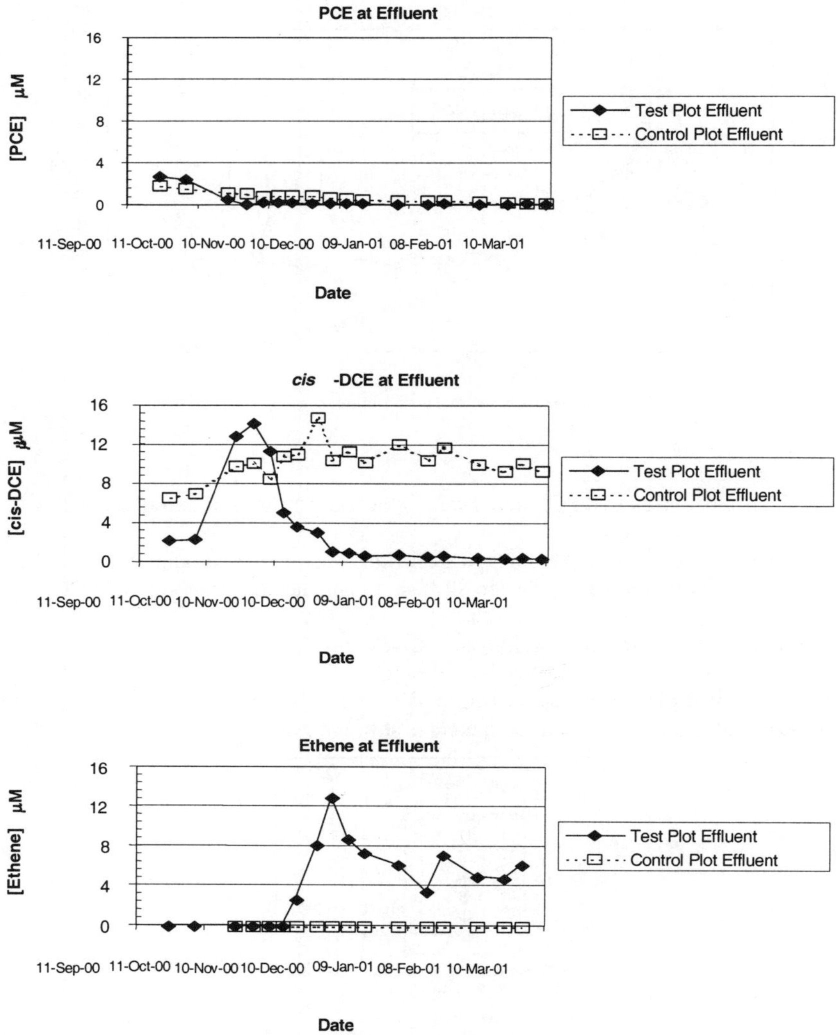

Figure 8. Aqueous concentrations of PCE, cis-DCE and Ethene with time in the test and control plot effluents. TCE and VC data were also sampled but were minor in concentration relative to values shown.

This design resulted from initial characterization of the aquifer microbial cultures indicating that the cis-DCE to ethene degraders were predominantly in the deep portion of the aquifer. It was necessary for them to be present in sufficient numbers in both the shallow and deep portions for this study to effectively dechlorinate PCE to ethene in the shallow zone as well as the deep zone. Finally, four separate multi-level arrays are positioned as shown with 10cm screens at 3.7, 4.3, 4.9, and 5.5mbs to provide discrete monitoring points with depth over the center of each plot.

Evidence of contaminant transformation following the start of the test compared with that of the control plot is shown in Fig. 8 for water sampled from the extraction wells. These data show a sharp increase in cis-DCE concentration relative to the control plot between the start of injection of lactate and nutrients and inoculation of halorespiring consortium in the test plot. This is potentially a result of biostimulation of native microorganisms in the test plot due to addition of carbon source and nutrients. Following inoculation with halorespiring microorganisms, there is essentially complete dechlorination in the test plot effluent to ethene in less than 50 days. Over the same time period, there was an increase of aqueous cis-DCE concentration in the control plot.

2.3. CONCLUSIONS

Due to the characteristics of increased uncertainty and cost associated with scaling and transferring technologies to the field, a clear demonstration of cause-and-effect have to accompany pilot demonstrations. This added requirement relies heavily on mathematical modeling to evaluate alternative scenarios as to how field observations can be explained, as well as on exhaustive analysis of chemical (e.g. contaminants, surfactants) and molecular (e.g. gene probes, total community DNA) parameters. Hence, the SEAR application for source control carefully compared computer simulations for PCE and Tween 80 recovery with observations in both the extraction and multilevel monitoring wells. Thus, it is known how much of the PCE was recovered, and how this was correlated to the surfactant sweep. The main unknown in this case is the field assessment of how much contamination was originally there; this quantification is order of magnitude accuracy. The effectiveness of the halorespiration biocurtain was evaluated based on the analysis of the entire suite of transformation products from PCE (i.e. TCE, DCE, VC, and ethene), on the qualitative relative occurrence of two indigenous species Desulfuromonas and Dehalococcoides which catalyze PCE→ cDCE and cDCE→ ethene dechlorination, respectively, as well as based on the total DNA analysis of aquifer microbial communities. The latter technique, transverse restriction fragment length polymorphism (tRFLP), allows for the qualitative discrimination between 'native' and 'inoculated' populations. It was clearly demonstrated (data not shown, [1,2]) that the occurrence of inoculated bacteria was correlated to the observed dechlorination activity.

3. Principles and Applications of VOC Bioremediation

Chlorinated solvents are some of the most commonly encountered chemicals in groundwater, and it is a generally held view that there is little that can be done to clean up sites contaminated with these compounds. As a result, containment and long-term risk management is usually selected as the only reasonable approach. However, there is growing evidence that biodegradation, either naturally-occurring (intrinsic) or enhanced may change this view.

3.1. INTRODUCTION TO THE DNAPL PARADIGM

Widespread production and use of chlorinated ethenes, ethanes and methanes started during World War II and increased rapidly in the following decades, finding applications in a variety of chemical, electrical, manufacturing, cleaning and transport industries. These chemicals are referred to as dense, non-aqueous-phase liquids (DNAPLs) and the physical and chemical properties that makes them good solvents also complicates their distribution, ensures their long-term persistence in the subsurface, and limits the effectiveness of many remedial technologies to cleanup a site.

When a sufficient amount of a DNAPL is released, it can migrate through the water table, leaving behind residual phases in the vadose and saturated zones, and pool on low hydraulic conductivity (K) soils. Small differences in K can have a dramatic effect on the distribution of a DNAPL, vastly increasing it distribution and even distributing it hydraulically upgradient from its point of release. How the DNPAL is released will also affect the depth and lateral distribution, with slow chronic releases at the same location penetrating deeper into the subsurface, and rapid, catastrophic releases tending to spread out further from the point of release prior to migrating downward. and its distribution will be governed. Figure 1 demonstrates the complexity of DNAPL release, with the presence of gaseous, dissolved, residual, and free-phases.

Once a DNAPL has entered into the subsurface, the ability of any technology will be affected by the dominance of one or more of the following factors:

- fracture flow and dual porosity (key in fractured bedrock systems);
- matrix diffusion and matrix counter-diffusion; and
- distribution and kinetics of dissolution of non-aqueous phases

Each of the above processes limits the feasibility of any remedial strategy and remedial technology.

3.1.1. Fracture Flow and Dual Porosity.

Fractured bedrock typically has both a fracture and a bulk porosity, with groundwater movement normally being much faster and dominated by advective processes in the fractures (i.e., fractured or secondary porosity), and much slower and dominated by diffusion in the unfractured rock matrix (bulk porosity or primary porosity). The degree that each contributes to the bulk groundwater flow is dependent on the rock type. For example, fracture flow will dominate groundwater flow in igneous and metamorphic rock, whereas matrix porosity contributes more to groundwater flow in sedimentary rock. Fractures may also be discontinuous and form "dead-ends" where groundwater flow is essentially zero. The combination of fracture and matrix flow is often considered to be to be "equivalent to porous media" such as sand at some scale, and this assumption is used to estimate the average groundwater flow for design purposes. However, often a small percentage of fractures are responsible contributing the most to the bulk groundwater flow, and groundwater will be travelling at a much higher velocity (10 to 1000 times faster than the "average" groundwater velocity) within these fractures. It is through these fractures that both dissolved and free-phase chemicals will also preferentially migrate.

3.1.2. Matrix and counter diffusion

Both the DNAPL and free-phase chemicals will flow through paths of least resistance (i.e., fractures, high K layers) and avoid lower permeable zones. However, both phases can enter into dead-end fracture spaces and into unfractured rock and low K layers, dependent on the concentration gradient. Most remedial technologies will treat chemicals in the prominent fractured or permeable zones first, and thus reverse the concentration gradient resulting in the counter-diffusion of chemicals from the dead-end fractures and rock matrix into the prominent fracture zones. This process often results in a "rebound" of concentrations after treatment and the flux from the rock matrix into the fractures, or from low K to high K layers, and thus repeated or continuous application of the technology may be required to address rebound effects.

3.1.3. Dissolution Rate of Non-Aqueous Phase Chemicals

Most organic contaminants have a low solubility and thus the dissolution rate of the residual or trapped DNAPL is the limiting step of any technology that relies on removing or directly treating the residual DNAPL phases for site clean-up. The dissolution rate is affected by the surface area to volume ratio of the non-aqueous phase and the groundwater velocity nearby. Because there is usually poor mixing of groundwater near the residual phase of a chemical, dissolution becomes controlled by diffusion instead of advection. Under diffusive conditions, the chemical reaches its solubility limit near the water/free-phase interface. This creates a small concentration gradient between the source materials and the bulk groundwater and thus the remaining source dissolves slowly. Thus, a significant amount of the residual non-aqueous phase must be removed before there is a permanent decrease in the total mass flux of the chemical. Technologies that attempt to treat the source must promote advection rather than diffusion, increase the chemical's solubility, or increase the concentration gradient between the source and bulk groundwater.

These issues exhibit implications for the application of remedial technologies:

complex distribution: the distribution of residual sources is complex and likely impossible to predict with sufficient confidence to prevent them from acting as long term sources. Because there is a limited ability to absolutely define the location of residual sources, a larger volume of the potential source area will require treatment to provide confidence that the source areas are encompassed by the treatment;

inaccessible mass: a significant mass of chemicals may be present in the rock matrix, "dead-end fractures", and low K layers, that are relatively inaccessible to extraction/removal technologies;

significant removal of mass to impact flux from the source: this is required to permanently reduce the mass flux from remaining sources (and hence shrink the dissolved phase plume);

matrix counter diffusion: repeated application or longer-term operation of a treatment or technology may be required to address "rebound" due to matrix counter diffusion;

source dissolution rate: slow dissolution increases the time until the source is removed;

fracture flow velocities and preferential pathways: chemicals may be transported at much higher velocities in primary fractures than would be estimated from the equivalent porous media assumption, and thus their residency time in treatment zones may be too short for effective treatment, and reactants added to treat source materials will flow along similar fractures and not come into contact with chemicals in "dead-end" fractures or that have diffused in to the rock;

3.2. SOLUTIONS

There are various methods being demonstrated to remove or treat DNAPL sources, including oxidation (e.g., permanganate), various in situ heating (steam stripping, 6-phase radio-frequency heating, and surfactant or alcohol flooding. However, the technical issues identified above all impact the successful remediation of DNAPL sources. One very promising approach is the application of bioremediation. Previously it was believed that microorganisms could not degrade chlorinated solvents. However, a significant body of laboratory and field research over the past 10 years has shown that microorganisms that naturally exist in groundwater environments possess the capability to biodegrade chlorinated solvents under a variety of reduction-oxidation (redox) conditions for food and energy. In fact there are several microorganisms that only use chlorinated ethenes as their electron acceptor and hydrogen as their electron donor (i.e., they breath on chlorinated solvents)! It is now known that microorganisms can and can metabolize high dissolved concentrations of chlorinated solvents. The application of bioremediation to treat chlorinated solvents and their sources has many advantages (Table 1) .

TABLE 1. Technical issues and the advantage and limitation of applying bioremediation.

Technical Issue	Advantage/Limitation Bioremediation
Complex distribution	Relatively easy and inexpensive to enlarge the zone of biological activity to overcome the uncertainty of the source zone dimensions or heterogeneities through increasing the nutrient feed concentration and/or the number of nutrient injection points
Inaccessible mass	Active degradation in the source zones will increase the mass flux by enhancing the diffusion rate of chemicals from low K layers, dead-end fractures, and rock matrix, and result in faster cleanup times of sources than can be achieved through simple dissolution and diffusion;
Matrix counter diffusion	See above, and will degrade those chemicals as they diffuse into the primary fractures or from low to high K layers
Significant removal of mass to impact flux from the source	Same constraint as other technologies, and will require longer-term operation until permanent reduction in mass flux. However, because of enhanced dissolution due improved concentration gradients, more mass will be removed faster and thus cleanup should be quicker
Source dissolution rate	Bioremediation near the source/water interface will increase the flux and thus enhance the dissolution rate and thus removal of the source term. Fast degradation rates could provide "biological containment"
Fracture flow velocities and preferential pathways	Microorganisms will migrate toward areas with higher concentrations of chemicals. However, fracture orientation and preferential pathways will affect effective mixing of dissolved nutrients, and may sweep nutrients away too fast. Slow release nutrients may be considered as they could persist longer in fracture zones because they are not as water-soluble.

The three metabolic routes, through which chlorinated solvents can be biodegraded and that can occur naturally or be can be induced are:

Direct aerobic: the chlorinated solvent serves as a carbon and energy source(the electron donor), while respiring on oxygen (the electron acceptor);

Direct anaerobic: which is the same as direct aerobic, but the electron acceptor is another inorganic species that will be reduced such as iron, manganese, nitrate, sulfate , and carbon dioxide; and

Co-metabolic: a "fortuitous" process, where the microorganism gains no energy or carbon. This occurs when the microorganism is metabolizing a primary food source for food and energy (the cometabolite) that induces enzymes that react with the chlorinated solvent transforming it into another compound (which may undergo other reactions, or be metabolized by another microorganism).

There has been much interest in the assessment and application of intrinsic biodegradation, the main component of monitored natural attenuation, as a remedial alternative. Generally, the intrinsic biodegradation of these compounds is assessed by correlating the of distribution of parent chlorinated solvents, to their degradation daughter products, and the changes in geochemistry that are associated with known microbial activities and the biodegradation of these compounds. However, these natural degradation processes are not ubiquitous, and not always able to meet remedial goals. Fortunately these processes can be enhanced through the addition of the appropriate nutrients and, in some cases, the requisite microorganisms.

Let's use one of the most common class of chlorinated solvents, the chlorinated ethenes, to illustrate how these chemicals biodegrade, and how this information is applied to assess if chlorinated solvents are undergoing intrinsic biodegradation or to how to promote biodegradation. Chemicals in this class include tetrachloroethene (PCE), and trichloroethene (TCE), 1,2-dichloroethene (1,2-DCE) and vinyl chloride (VC) can all be biodegraded to non-chlorinated, environmentally-acceptable end products such as ethene, carbon dioxide, water and chloride. In anaerobic-reducing environments, the known biodegradation mechanisms for chlorinated ethenes are reductive dechlorination to ethene and anaerobic oxidation to CO_2 (for 1,2-DCE and VC). In aerobic-oxidizing environments, the known degradation mechanisms for chlorinated ethenes are oxidation to CO_2 (for VC and 1,2-DCE) and cometabolism to CO_2 (for TCE, 1,2-DCE and VC). At many sites, combinations of these anaerobic and aerobic biodegradation processes operate in sequence to destroy chlorinated ethenes. In natural sequential anaerobic-aerobic bioremediation of chlorinated ethenes in groundwater and wetland systems the highly-chlorinated parent VOCs released to the subsurface are first transformed under anaerobic-reducing conditions to less-chlorinated degradation intermediates. The degradation intermediates are then biodegraded to environmentally-acceptable end products (CO_2, chloride and water) under aerobic-oxidizing conditions. This pattern of chlorinated ethene degradation has been observed at multiple sites and suggest that this type of sequential degradation activity is robust and widespread in subsurface environments.

The following sections provide background information on each of the known biodegradation processes: reductive dechlorination, anaerobic oxidation, aerobic oxidation and cometabolism. The sections identify the conditions under which the processes are known to proceed, the common degradation products of the reactions, and the data that can be measured during field investigations to determine if the biodegradation reactions are occurring and to assess whether site conditions are appropriate for the reactions to occur or be promoted.

3.2.1. Reductive Dechlorination

In anaerobic environments, reductive dechlorination is the most common biodegradation mechanism for chlorinated ethenes. Reductive dechlorination involves the sequential replacement of chlorine atoms on the alkene molecule by hydrogen atoms. In the case of chlorinated ethenes, PCE is sequentially dechlorinated via TCE, 1,2-DCE (preferentially the cis-isomer of 1,2-DCE) and VC to ethene, as shown in Figure 2. For this type of reaction to be thermodynamically favorable, the redox potential of the groundwater must be very low (i.e., negative oxidation-reduction potential or Eh), thereby excluding the presence of dissolved oxygen.

The complete dechlorination of chlorinated ethenes requires the concerted efforts of a number of different groups of anaerobic bacteria including halo-respirers (the specific anaerobic microorganisms that mediate the dechlorination reactions) and supporting bacteria such as methanogens (methane-producers), acetogens (acetate-producers) and sulfate-reducers. To date, a number of distinct types of halo-respiring bacteria have been identified as being capable of dechlorinating chlorinated ethenes, including *Dehalospirillium multivorans*, *Dehalobacter restrictus*, and *Dehalococcoides ethenogenes*. Of these, *Dehalococcooides ethogenes* is able to cause complete dechlorination to ethene. These halo-respiring microorganisms, or other microorganisms possessing similar metabolic capabilities, are thought to be relatively widespread in subsurface environments, although they may not be present at every site.

To mediate dechlorination reactions, halo-respiring microorganisms require an ample supply of electron donors. Hydrogen is the ultimate electron donor used by these bacteria, and can be derived from the metabolism of a wide variety of organic compounds including alcohols (e.g., methanol and ethanol), volatile fatty acids (VFAs; acetate, propionate), sugars (e.g., glucose), edible oils, complex polymers, natural organic matter, and petroleum hydrocarbons. Hydrogen is also required to sustain the activity of other microorganisms (e.g., methanogens, acetogens) that supply the halo-respirers with essential growth nutrients (halo-respirers are typically present in mixed anaerobic consortia containing methanogens and acetogens).

3.2.2. Anaerobic Oxidation

The lesser chlorinated ethenes, VC and 1,2-DCE, can be anaerobically-oxidized to CO_2 and chloride under iron- and manganese-reducing conditions, respectively. In these anaerobic reactions, the chlorinated ethene serves as the electron donor (rather than the electron acceptor) and is oxidized to CO_2 (Figure 2); iron or manganese serves as the electron acceptor. For this type of reaction to occur, redox conditions must be anaerobic and reducing. No evidence exists at this time for the direct anaerobic oxidation of either TCE or PCE.

3.2.3. Aerobic Oxidation

VC can be aerobically oxidized to CO_2 and chloride, where VC serves as the electron donor and is oxidized to CO_2 (Figure 2); oxygen serves as the electron acceptor. This reaction occurs can control the migration of VC at contaminated sites. There is limited evidence that 1,2-DCE may be similarly oxidized to CO_2; however the extent of this reaction in subsurface environments is unknown. Aerobic oxidation reactions require the input of oxygen, and therefore 1,2-DCE and VC oxidation occurs under aerobic-oxidizing conditions. No evidence exists at this time for the direct aerobic oxidation of either TCE or PCE.

3.2.4. Cometabolism

TCE, 1,2-DCE and VC can be fortuitously oxidized to CO_2 (and chloride and water) by non-specific microbial oxygenase enzymes produced by a variety of aerobic microorganisms. These enzymes are produced to metabolize specific growth substrates (e.g., methane, propane, toluene, phenol, ammonia and ethene) termed cometabolites that serve as the primary food source for these microorganisms. However, TCE, 1,2-DCE and VC can fortuitously react with these enzymes, producing unstable epoxide intermediates that spontaneously decompose to CO_2, chloride and water. The microorganisms that produce the enzymes and mediate the cometabolic biodegradation reactions do not derive benefit from degradation of the chlorinated ethenes. Specific oxygenase enzymes that have been documented to promote cometabolism of TCE, 1,2-DCE and VC include methane monooxygenase, ammonia monooxygenase, toluene monooxygenase and dioxygenase, phenol dioxygenase, propane monooxygenase, and several alkene monooxygenases. The fully-chlorinated molecular structure of PCE prevents its cometabolic biodegradation (i.e., PCE does not react with the enzymes).

The rate at which a chlorinated ethene can be cometabolized in an aerobic environment varies depending on: the availability of the primary substrate and oxygen; the ratio between the primary substrate, oxygen and the chlorinated ethenes; and the metabolic capabilities of microbial populations that are present. In subsurface environments, the active zone where cometabolic biodegradation reactions occur can be relatively small because cometabolic reactions are often very rapid. Aerobic cometabolic reactions tend to occur in small zones at the boundaries of plumes where redox conditions transition from anaerobic to aerobic. The cometabolites (e.g., methane, ammonia, ethene) are often produced through metabolic activities within the anaerobic zone. The cometabolites are then consumed at the boundary where redox conditions transition to aerobic. At some sites, cometabolite is present in the groundwater as a result of co-release (e.g., for toluene) or production in an anaerobic source area (e.g., for methane or ethene). At these sites, intrinsic cometabolic biodegradation of chlorinated ethenes has occurred naturally where conditions are appropriate (i.e., correct ratio of cometabolite, oxygen and VOC).

3.3. APPLICATION OF BIOREMEDIATION AND FUTURE DIRECTIONS

Our understanding of how chlorinated solvents biodegrade has lead to the application of Monitored Natural Attenuation and accelerated bioremediation. MNA incorporates intrinsic biodegradation processes, with other naturally occurring physical and chemical processes such as dilution, dispersion, and stabilization, to meet remedial goals. Key to evaluating the applicability of MNA is correlating the distribution of parent and breakdown daughter products with changes in geochemistry known to occur when biodegradation is occurring. If these processes are able to mitigate risks, and/or cleanup a site within a reasonable time frame, than monitoring the site over time to verify performance can be an acceptable remedy. However, natural attenuation processes may not be sufficient, and the degradation processes need to be enhanced. This may involve the addition of electron donors, acceptors or co-metabolites (i.e., nutrients), and even the requisite microorganisms.

For example, researchers at Stanford University have conducted a field trial of toluene addition to enhance the cometabolic degradation of TCE in California; lactate addition to stimulate reductive dechlorination was demonstrated by Dupont. However, these studies required that the requisite microorganisms be present at the site. Using TCE as an example, it is clear that this is not always the case.

Cultures have been isolated that can be used to stimulate complete dechlorination of TCE to ethene at sites that are lacking the requisite dechlorinating microorganisms. Through continued research, several stable, natural microbial consortia have been isolated that are capable of mediating complete dechlorination of TCE to ethene. General Electric Corporate Research & Development have isolated one such culture, referred to as the Pinellas culture, from a Department of Energy facility in Pinellas, Florida. This culture was recently added to groundwater (a process termed bioaugmentation) at Dover Air Force Base in Delaware as part of a field demonstration conducted by the Remediation Technologies Development Forum (RTDF). At this test site, TCE dechlorination had stalled at cis-1,2-DCE despite continued electron donor addition. Shortly following bioaugmentation, complete dechlorination of cis-1,2-DCE via VC to ethene was observed in the demonstration area.

GeoSyntec, working with U of T, has isolated a similar stable, natural microbial consortia (referred to as KB-1) capable of stimulating rapid dechlorination of TCE to ethene at sites where this activity is otherwise deficient. Bioaugmentation of microcosms in which TCE dechlorination had stalled at cis-1,2-DCE despite continued electron donor addition with KB-1 immediately stimulated dechlorination of cis-1,2-DCE via VC to ethene; cis-1,2-DCE was completely transformed to ethene in a matter of days. Development of these and similar microbial cultures now provides the ability to accelerate bioremediation of TCE and related chlorinated solvents at sites where complete dechlorination reactions do not otherwise occur. Bioaugmentation studies with KB-1 have been completed at Kelly Air Force Base in Texas which showed that half-lives on the order of hours (not days) was achieved in the field. Furthermore, the study used molecular probes (based on the 16S rRNA gene) to detect the presence of *D. ethenogens*. These probes showed that this microorganism was absent from the site. However, the strain of *D. ethenogens* that is associated with the KB-1 culture was shown to spread from the point of its introduction. At the end of the test, *D. ethenogens* was only detected within the test area.

However, it is generally believed that high concentrations (100 ppm to aqueous solubility limits) of chlorinated solvents were toxic to the dechlorinating microorganisms. However, exciting laboratory research and field demonstrations have shown that dechlorinating microorganisms do function at high concentrations of chlorinated ethenes. Complete dechlorination of TCE to ethene has been demonstrated to be occurring naturally at a field site in Auburn NY when TCE was at 80% of its solubility limit. An anaerobic bacterial culture enriched from natural sources by researchers at Cornell University has been shown to completely dehalogenate PCE to ethene at concentrations of 55 mg/L, which is 36% of water saturation for PCE. KB-1has been shown to dechlorinate TCE at 100 ppm.. These concentration levels are higher than typically observed in the field, even when large quantities of DNAPL are present. These results suggest that a variety of dehalogenating microorganisms are tolerant of high levels of chlorinated aliphatics and will be active in proximity to DNAPL sources. Further work is focussing on how microbial activity, either occurring naturally, or through bioaugmentation, in the source zones will enhance the dissolution rates by orders of magnitude and significantly shorten the clean-up time. New nutrient amendments (e.g., slow release compounds) and strategies are being investigated to further reduce the cost of cleanup.

3.4. CONCLUSIONS

The complex behavior of DNAPLs has reduced most remedial efforts of contaminated sites to containment and long-term management strategies. New approaches aimed at attacking the root problems of DNAPL persistence are being developed, and bioremediation is one technology that has many potential advantages. Providing adequate microbial sensing and control technologies can be economically and reliably scaled to the field, bioremediation will rapidly become a preferred technology to treat chlorinated solvent-impacted environments.

4. Principles and Applications of PAH Bioremediation

4.1. INTRODUCTION

Polycyclic aromatic hydrocarbons (PAHs) are hydrophobic pollutants that can persist in the environment for many years due to their low water solubility and their strong tendency to sorb to soil particles as well as their low degradability. Due to its abundance and toxicity 16 PAHs have been classified as priority pollutants by the U.S. Environmental Protection Agency [3].

Bioremediation of polluted soil mostly depends on indigenous micro-organisms. The principle of engineered bioremediation is to stimulate those micro-organisms, which are able to degrade the xenobiotic substrates. In response to the introduction of electron acceptors (like oxygen) or nutrients to soil, specific degrader organisms will multiply. However, despite of a potentially metabolicaly active biomass, *in-situ* bioremediation of soils polluted with hydrophobic organic pollutants such as PAHs frequently results in slow pollutant degradation rates, high residual concentrations and, as a consequence, in limited clean-up efficiencies [4]. The unequal spatial distribution of micro-organisms and pollutants in combination with physicaly retarded substrate diffusion are nowadays generally accepted as key limiting factors for efficient biodegradation of hydrophobic contaminants in soil.

Extensive mechanical mixing of the soil, the addition of surfactants, and electrokinetic stimulation are considered as bioavailability-enhancing engineering solutions. Although many bacteria have been found to degrade and to live on PAHs, still very little is known about the strategies of bacteria themselves to improve their access to hydrophobic pollutants in soil. Indeed, the fact that despite of the generally poor bioavailability of PAHs some bacteria obviously subsist on these compounds, has not been looked at in detail and might be of major interest for bioremediation of PAH-polluted matrices. Bacterial features may enhance the transfer of poorly bioavailable substrates (i) by reducing the mean distances between pollutants and bacteria, e.g. by adhesion to sorbents, (ii) by causing active transfer of pollutants into the aqueous phase, (iii) by uncoupling the development and maintenance of active biomass from pollutant bioavailability, and (iv) by using uptake systems with high specific affinity [5,6].

4.2. RESULTS AND DISCUSSION

4.2.1. Biodegradation and isolation of strains

PAHs are found in fosil fuels and formed during the pyrolysis of organic material (at 2000°C non-substituted; between 400 and 800°C non- and alkylsubstituted; between 50 and 150°C alkylsubstituted PAHs are formed). They are hydrophobic, recalcitrant, have a low solubililty and are in some cases genotoxic. An overview of the important characteristics is presented in table 2.

PAHs are biodegradable under aerobic conditions and the biodegradability decreases with increasing ring number. The first ring is degraded to a salicylate form (upper pathway) and than further degraded in the lower pathway to water and carbon dioxide. The biodegradation follows a defined pathway. In a first step the ring is prepared by the introduction of hydroxyl groups on two adjacent (ortho) or opposing (meta) carbons by a dioxygenase or monooxygenase respectively followed by a dehydrogenase. Afterwards the ring fission proceeds via an ortho or intradiol cleavage or a meta or extradiol cleavage in case of methylated PAHs. Afterwards the ring fission products are further degraded.

TABLE 2. Characteristics of the 16 PAHs of EPA.

PAH	Molecular weight	Water solubility (ppb)	Carcinogenicity
Naphtalene	128	31700	NR
Acenaphalene	152	-	NR
Acenaphtene	154	3930	NR
Fluorene	166	1980	-
Fenantrene	178	1290	-
Anthracene	178	73	-
Fluoranthene	202	260	-
Pyrene	202	135	-
Benz(a)anthracene	228	14	+
Chrysene	228	2	±
Benzo(a)pyrene	252	4	++++
Benzo(b)fluoranthene	252	-	++
Benzo(k)fluoranthene	252	-	+
Benzo(ghi)perylene	276	0.26	±
Indeno(123cd)pyrene	276	-	++
Dibenz(ah)anthracene	278	0.5	+++

NR : not reported ; - : not carcinogenic ; ± : carcinogenicity unsure ;
+ : low carcinogenicity ; ++ : mean carcinogenicity ; +++ : strong carcinogenicity ;
++++ : very strong carcinogenicity.

Lower molecular weight PAHs such as naphtalene and phenanthrene [7], acenaphtene and acenaphtylene [8] and fluorene [9] are relatively easy to degrade and a large number of strains able to metabolize or cometabolize these compounds has been described. Anthracene, although identical to phenanthrene in number of aromatic rings, seems much more difficult to degrade which is probably due to its lower solubility in water [10]. Until 1990, there have been no reports of axenic microbial cultures utilizing PAHs containing four or more fused rings as sole source of carbon and energy. Since then, a number of pure cultures able to (co)metabolize fluoranthene [11] and pyrene [12] have been reported. Literature data describing microbial growth on chrysene and benz(a)anthracene are rather scarce [13], whereas no micro-organisms capable of utilizing five-ring PAHs, such as benzo(a)pyrene, as sole carbon and energy source are known.

Anaerobic PAH biodegradation is only reported for low molecular weight PAHs. Fungal biodegradation is possible via extracellular H_2O_2-ligninases leading to PAH-quinones and finally to ring fission. Also fungal cytochrome P450 monooxygenases can form arene oxides which are further non-enzymatically rearranged into phenols or by hydrolases transformed into trans-dihydrodiols. Also combined processes between fungal attack and futher bacterial degaradation are possible detoxification routes. Several unspecific peroxidase type exoenzymes can oxidise PAH. These oxidated PAH can form under certain conditions polymeric substnaces of PAH or can be bound via an oxidative coupling into the humus structure. Both polymerization reactions lead to a decrease in availability and toxicity of the PAHs and can also be used as a remediation process.

4.2.2. Bioavailability

Some studies suggest that certain bacteria are adapted to the utilization of poorly bioavailable compounds. These bacteria may either be very oligotrophic or they may apply strategies to improve the bioavailability of their potential substrates. Such bacteria may enhance the substrate transfer by shortening their distance to a substrate source, for instance by attaching to it, or by developing high-affinity uptake systems that efficiently reduce the aqueous phase substrate concentration. In order to isolate such bacteria a habitat directed isolation protocol was designed.

One of these methods was based on a solid/liquid biphasic system [6]. In this case highly hydrophobic teflon membranes were covered with a solid phase of PAHs. This teflon coated piece was introduced in a soil slurry system containing a mixture of historically contaminted soil samples. After three to four weeks months the teflon membrane was recoverd from the slurry system, rinsed and placed on top of an agar plate with or without supplying extra PAH crystals on top of the membranes. Some bacteria started growing on the membrane and from these several PAH biodegrading bacteria could be isolated. It turned out that all belonged to the *Mycobacterium* genus. Isolates from classical liquid enrichment cultures did selected for *Sphingomonas spp.* (e.g. *Sphingomonas LB126* could biodegrade fluorene) and many gram-negative bacteria. The isolated *Mycobacterium* showed to have very specific membrane characteristics which allowed them to stick to the surface of PAH crystals. One strain, *Mycobacterium sp.* LB501T displayed an unusual combination of high adhesion efficiency and an extremely high negative charge. This anthracene degrading strain may represent a new bacterial species as suggested by 16SrRNA gene sequence analysis. *Mycobacterium* LB307T was isolated on phenanthrene and *Mycobacterium* LB208 was isolated on pyrene. The new isolation method is appropriate to select for adherent PAH-degrading bacteria, which might be useful to biodegrade sorbed PAHs in soils and sludge. This method favors to isolate PAH-degrading bacteria that are strongly attached to the soil particles and grow very slowly or not in suspension.

4.2.3. Biodegradation at lab scale

In some first tests the biodegradation of single PAH compounds was tested in batch experiments by incubating soil, spiked with one or more PAH compounds, with cultures of the isolated specific PAH-degrading bacteria. In that way the ability of the strains to biodegrade its specific PAH and others as well as the combination (inhibition of stimulation) of several PAHs was evaluated. Later bigger lab experiments were done in a dry slurry reactor. In this reactor (cylinder with rotating plows on a central axis) of 50 kg soil more real conditions of landfarming could be simulated. Figure 9 presents the results of the treatment of a soil spiked with fluorene before and after the addition of *Mycobacterium* LB208. The figure shows a full fluorene biodegradation. Only when the inoculum of LB208 was high enough (more than 10^5 cfu/ml) the biodegradation of the PAH could be started. It was also observed that once the fluorene was degraded the concentration of fluorene degrading bacteria decreased.

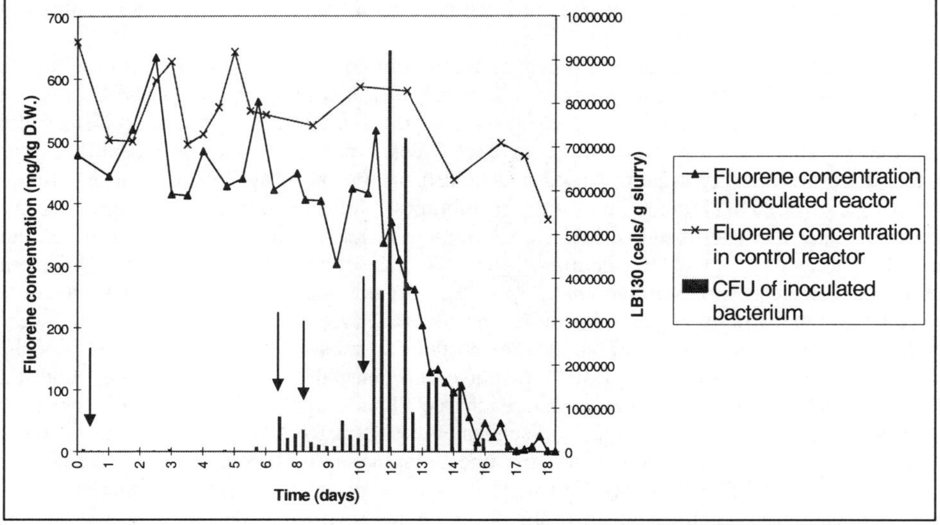

Figure 9. Fluorene biodegradation in a dry slurry reactor

4.2.4. Biodegradation at pilot scale

At a manufactured gas plant in the city of Lier in Belgium biotreatment in biopiles was demonstrated. Three biopiles were set up in the following way: 1A: unamended soil; 1B: soil amended with nutrients; 2A: soil with nutrients and a liquid inoculum; 2B: soil with nutrients and an inoculum grown on soil; 3: soil amended with nutrients and inoculated after a pretreatment with Fentons reagens. Each biopile contained 20 ton of soil, adjusted to 60% WRC. The inoculum consisted only of PAH-degrading *Sphingomonas sp.* (3×10^6 cfu/g soil). The PAH-degrading *Mycobacterium sp.* could not be added due to contamination problems during the 'industrial' cultivation. Even after 17 months in the inoculated soils the concentration of phenanthrene degraders remained higher compared (1×10^3 cfu/g soil) to the control soils (table 3). In contrast to the slurry experiments, chemical analyses on the soil samples indicated that none of the treatments had a significant effect on the concentration of the PAHs. Therefore it was decided to restart the biopile demonstration in the field on a smaller scale, using aerated big bags, containing 1.5 tons of soil per bag.

During the new field (big bag) experiment, special attention was given to the monitoring of the PAH-degrading bacterial inoculum, which consisted of two *Sphingomonas* species (LB126 and LH128) and three different *Mycobacterium* species (LB307T, LB501T and KV27). These strains have been isolated on their capacity to use several PAHs as sole source of carbon and energy.

TABLE 3. Bacterial counts in the biopile experiment.

Time	Media	Biopile 1A	Biopile 1B	Biopile 2A	Biopile 2B	*Biopile 3*
14 days	Rich medium	$6.7\ 10^5$	$1.6\ 10^6$	$8.7\ 10^5$	$1.2\ 10^7$	$3.2\ 10^7$
	MM gl + phenanthrene	$< 10^3$	$< 10^3$	$1.0\ 10^3$	$1.0\ 10^5$	$1.0\ 10^4$
5 months	Rich medium	$1.6\ 10^6$	$4.2\ 10^5$	$1.8\ 10^5$	$1.4\ 10^5$	$3.6\ 10^5$
	MMgl + phenanthrene	$1.3\ 10^3$	$4\ 10^2$	$1\ 10^2$	$< 1\ 10^2$	$2.3\ 10^3$
10 months	Rich medium	$1.5\ 10^5$	$4.4\ 10^5$	$2.7\ 10^5$	$2.0\ 10^5$	$4.7\ 10^5$
	MM gl + phenanthrene	$1.4\ 10^3$	$1.0\ 10^3$	$1.0\ 10^2$	$1.5\ 10^3$	$2.9\ 10^3$
17 months	Rich medium	$4.5\ 10^5$	$5.0\ 10^5$	$7.3\ 10^5$	$3.2\ 10^5$	$3.8\ 10^5$
	MM gl + phenanthrene	$<1.0\ 10^2$	$<1.0\ 10^2$	$1.6\ 10^3$	$2.2\ 10^3$	$2.7\ 10^3$

Some first three conditions (see figure 10) were initiated with bag 1 as control. Bags 2 and 3 were inoculated with a mixture of the mentioned *Sphingomonas* and *Mycobacterium* strains at $2x10^6$ cfu/g soil. Bag 3 was pretreated with Fentons reagens. Bags 4 and 5 were inoculated 3 months later with a commercial *Pleurotus ostreatus* mushroom substrate added at 10%. To bag 5 also a bacterial inoculum was added at a concentration of $6x10^5$ cfu/g soil.

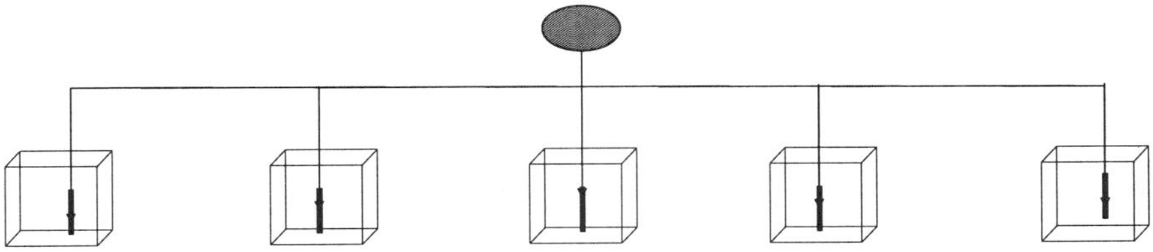

Figure 10. Configuration of a big bag experiment

The viable bacterial counts and PAH-(co)metabolizing strains was determined by microbiological platings. In the first three bags (table 4a), total viable counts were rather constant during the monitoring period, i.e. 10^5 to 10^6 cfu/g soil, and were not influenced by the inoculations. In bag 5 and 6 (table 4b), higher total viable counts were obtained, i.e. 10^7 cfu/g soil, probably due to the inoculation of fungi which were grown on straw.

TABLE 4a. Bacterial counts in the big bag (1-3) experiment

Time	Sampling	Rich medium	Phenan-threne	Pyrene	Fluorene	Fluoran-thene	Anthracen e
Before	Untreated	1.6 10^5	2.0 10^2	1.0 10^3	1 10^2	1 10^2	< 1 10^2
Inoc.	Fentons	9.5 10^5	7.6 10^3	< 1 10^2	< 1 10^2	< 1 10^2	< 1 10^2
1 day	Bag 1-1	4.6 10^5	< 1 10^2	3.0 10^2	< 1 10^2	< 1 10^2	< 1 10^2
	Bag 2-1	5.8 10^5	2.3 10^4	8.0 10^4	8.0 10^4	4.6 10^4	4.0 10^4
	Bag 3-1	7.2 10^6	2.3 10^5	1.4 10^5	2.0 10^4	1.5 10^5	4.5 10^3
3 days	Bag 1-2	1.6 10^5	1.7 10^3	4.0 10^4	< 1 10^2	1.8 10^3	< 1 10^2
	Bag 2-2	3.2 10^5	1.7 10^5	2.4 10^5	1.2 10^5	1.8 10^6	2.0 10^6
	Bag 3-2	2.8 10^6	6.2 10^5	8.0 10^5	8.0 10^4	6.3 10^5	1.7 10^5
3 wk	Bag 1-3	1.4 10^4	4.0 10^2	< 1 10^2	< 1 10^2	< 1 10^2	< 1 10^2
	Bag 2-3	3.1 10^5	3.7 10^5	3.9 10^5	2.5 10^5	4.3 10^5	6.2 10^5
	Bag 3-3	1.7 10^6	1.9 10^5	3.4 10^5	1.5 10^5	8.3 10^5	4.0 10^5
5 wk	Bag 1-4	1.6 10^5	< 1 10^2	< 1 10^2	< 1 10^2	< 1 10^2	< 1 10^2
	Bag 2-4	3.5 10^5	2.6 10^5	1.7 10^5	1.8 10^5	3.4 10^5	1.3 10^5
	Bag 3-4	8.0 10^5	5.1 10^5	4.2 10^5	8.0 10^4	3.6 10^5	2.7 10^5
16 wk	Bag 1-6	2.1 10^5	6.0 10^3	4.0 10^2	< 1.0 10^{2*}	< 1.0 10^{2*}	< 1.0 10^2
	Bag 2-6	3.3 10^5	3.5 10^4	> 1.3 10^{4*}	<1.0 10^{2*}	>4.0 10^{3*}	2.4 10^4
	Bag 3-6	1.2 10^6	7.0 10^4	8.0 10^4	6.0 10^{2*}	7.8 10^4	2.1 10^4
25 wk	Bag 1-7	6.1 10^5	2.0 10^3	<1.0 10^2	7.2 10^3	<1.0 10^2	1.4 10^4
	Bag 2-7	3.0 10^5	6.3 10^3	3.0 10^4	2.2 10^4	2.0 10^4	1.7 10^4
	Bag 3-7	2.5 10^6	1.0 10^5	4.0 10^4	1.1 10^5	1.4 10^5	5.0 10^4
30 wk.	Bag 1-8	5.3 10^5	1.0 10^2	2.8 10^4	2.1 10^4	5.7 10^4	3.3 10^4
	Bag 2-8	2.8 10^5	7.9 10^4	2.4 10^4	2.6 10^4	1.1 10^4	1.5 10^4
	Bag 3-8	1.5 10^6	2.1 10^4	2.2 10^4	5.3 10^4	2.1 10^4	1.5 10^4
38 wk.	Bag 1-9	1.3 10^5	8.0 10^2	3.0 10^2	<1.0 10^2	<1.0 10^2	1.0 10^2
	Bag 2-9	3.7 10^5	1.0 10^2	1.0 10^2	<1.0 10^2	6.0 10^3	1.0 10^3
	Bag 3-9	5.9 10^6	4.0 10^3	2.0 10^3	<1.0 10^2	3.0 10^3	2.0 10^3
46 wk	Bag 1-11	1.3 10^6	4.5 10^3	2.0 10^3	1.1 10^4	3.6 10^3	1.0 10^3
	Bag 2-11	2.4 10^5	1.0 10^3	3.6 10^3	3.4 10^3	2.6 10^3	4.910^3
	Bag 3-11	1.4 10^6	2.0 10^4	3.9 10^3	8.3 10^3	2.2 10^3	4.810^3

TABLE 4b. Bacterial counts in the big bag (4-5) experiment

Time	Sampling	MED	PHE	PYR	FLUE	FLUA	Anthracene
	Before	9.4 10^4	3.0 10^3	4.0 10^3	4.6 10^3	1.[illegible] 10^3	3.7 10^3
1 day	Bag 4-1	1.4 10^7	2.0 10^4	5.0 10^3	<1.0 10^2	<1.0 10^2	1.2 10^4
	Bag 5-1	4.2 10^7	2.0 10^5	1.2 10^5	<1.0 10^2	3.4 10^4	1.3 10^5
12 days	Bag 4-2	3.8 10^7	< 1.0 10^2	< 1.0 10^2	< 1.0 10^2	< 1.0 10^2	< 1.0 10^2
	Bag 5-2	3.6 10^7	< 1.0 10^2	< 1.0 10^2	< 1.0 10^2	< 1.0 10^2	< 1.0 10^2
8 weeks	Bag 4-3	8.7 10^7	4.0 10^3	3.0 10^4	3.6 10^3	2 10^2	4.5 10^3
	Bag 5-3	4.3 10^7	1.1 10^5	1.2 10^5	< 1.0 10^2	< 1.0 10^2	1.1 10^5
12 weeks	Bag 4-4	9.7 10^6	< 1.0 10^2	1.0 10^2	< 1.0 10^2	< 1.0 10^2	< 1.0 10^2
	Bag 5-4	2.0 10^7	< 1.0 10^2	< 1.0 10^2	< 1.0 10^2	< 1.0 10^2	< 1.0 10^2
16 weeks	Bag 4-5	1.9 10^6	2.0 10^2	< 1.0 10^2	< 1.0 10^2	4.0 10^3	1.0 10^2
	Bag 5-5	2.5 10^6	3.6 10^4	7.3 10^4	1.0 10^3	6.7 10^3	1.0 10^2

Viable phenanthrene, pyrene, fluorene, fluoranthene and anthracene (co)metabolizing bacteria were quantified based on clearing zones which appeared on minimal mineral media supplied with glucose and sprayed with the appropriate PAH-compound. Up to 25 weeks, the number of PAH-degrading bacteria in the inoculated bags 3 and 4 stayed significantly higher than in the uninoculated bags 1 and 2. Afterwards, the difference was less pronounced and even absent. Immediately after inoculation, a significant higher amount of PAH-degrading bacteria is found in bag 5 in comparison with bag 4, which is only inoculated with fungi (table 3). Afterwards, the values fluctuate somehow. Fluorene degrading bacteria were in most cases below the detection limit. A relatively high percentage of the clearing zone forming colonies were yellow, which indicates that *Sphingomonas* and *Mycobacterium* may indeed be present.

A number of the colonies was purified and REP-PCR was used to verify if it concerns the inoculated PAH-degrading strains. PCR-detection using en *Mycobacterium* genus specific primer set revealed that *Mycobacterium* species were present in all the soil samples. DGGE-separation of the PCR-product further allowed to distinguish the inoculum *Mycobacterium* strains from the native *Mycobacterium* sp. LB307T and KV27 could not be distinguished, as their 16 SrRNA gene sequence is identical. *Mycobacterium sp.* LB501T could be detected directly using a *Mycobacterium* cluster specific PCR primer set. LB501T was clearly detected in samples up to 3 weeks.

Although there were strong indications that the inoculated PAH-degrading strains survived relatively well in the MGP-soil, chemical analyses revealed that in none of the tested conditions significant PAH-removal was realized.

4.3. CONCLUSIONS

Many hydrophobic organic compounds are priority soil contaminants because of their toxicity and their tendency to persist in soils and to escape biological degradation even in engineered bioremediation. Availability for biodegradation can be defined as the extent of mass transfer to microbial cells relative to the intrinsic catabolic potential of these organisms. However many soil bacteria have been found to degrade and to live on such hydrophobic organic compounds as PAHs despite their generall low bioavailability. Some specific genera could be isolated.

Interestingly, a relation can be observed between the complexity and recalcitrance of PAH-compounds and the number of *Sphingomonas* and Gram positive bacterial isolates which have been described to biodegrade these PAHs. Most bacteria selected on napthalene, a 2-ring PAH, belong to the fluorescent *Pseudomonads* [14]. Many phenenathrene and anthracene utilizing isolates are *Pseudomonas sp.*, *Sphingomonas* sp. or Gram positives of the *Nocardia-Rhodococcus-Mycobacterium* group [16]. Above 3 aromatic rings, mainly Gram positive PAH-degrading bacteria have been reported, but also some *Sphingomonas sp*. The genera *Sphingomonas* and especially *Mycobacterium* therefore seem to be specialized in degrading such less bioavailable compounds. They both have a particular outer cell wall layer, i.e. glycosphingolipids for *Sphingomonas* [16] and glycolipids such as mycolic acids for *Mycobacteria* [17], wich may be important for the interaction/uptake with/of hydrophobic compounds.

Based on microbiological platings as well as on PCR-detection, there are strong indications that at least some of the inoculated stains survived relatively well (up to 25 weeks) in the MGP soil during the aerated bigbag experiment. PCR-based detections using specific primer sets were found useful for the monitoring of well-defined inocula. Although the inoculum seems to survive in the soil for at least some weeks, chemical analysis indicated that none of the treatments led to significant PAH-removal. This can be contributed to the limited bioavailability of the PAHs in the weathered tar drops and soil. The added strains are also specialized in degrading lower PAHs (up to 3 and 4 rings), while higher PAHs may only be cometabolized very strongly.

5. Principles and Applications of BTEX Bioremediation

5.1. INTRODUCTION

Monoaromatic hydrocarbons such as benzene, toluene, ethylbenzene, and the three isomers of xylene (BTEX) are ground water pollutants commonly associated with petroleum product releases (Figure 11). All six BTEX compounds are depressants to the central nervous system, and chronic benzene exposure can cause leukemia. Thus, contamination of potential drinking water sources by BTEX represents a serious threat to public health (Table 5). To put the magnitude of this problem in perspective, 412,392 confirmed releases have been reported in the US from leaking underground tanks alone.

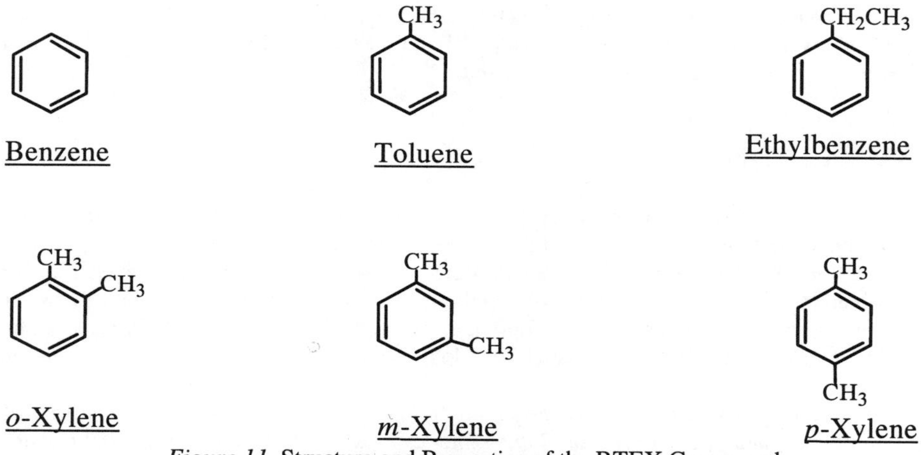

Figure 11. Structure and Properties of the BTEX Compounds

TABLE 5. Physical-Chemical Characteristics of BTEX compounds

MCL = Maximum Contaminant Level allowed in drinking water by U.S. regulations; v.p.=vapor pressure; H=Henry's constant.

Compound	MW (g/mole)	v.p. (mm Hg)	H (atm.l/mol)	ρ (g/mL)	log K_{ow}	Solubility (mg/L)	MCL* (μg/L)
Benzene	78	95.2	5.43	0.8737	2.13	1,791	5
Ethyl-benzene	106	4.5	8.44	0.8670	3.15	161	700
Toluene	92	28.4	5.94	0.8660	2.73	535	1,000
o-Xylene	106	6.6	5.10	0.8802	3.12	175	10,000
m-Xylene	106	8.3	7.68	0.8684	3.20	146	10,000
p-Xylene	106	3.2	7.68	0.8611	3.15	156	10,000

Traditional processes for removing BTEX from aquifers involve pumping the contaminated groundwater for above ground treatment with activated carbon or air strippers. BTEX compounds, however, are moderately hydrophobic and sorb to the aquifer material. This makes them difficult to withdraw by pumping, and serves as a slow-release mechanism for sustained groundwater contamination. Pump-and-treat technologies alone can result in prohibitively long time for the removal of the residual contamination, and are often economically unfeasible. In addition, treatment with activated carbon or air strippers merely transfer the contaminants from one phase to another, rather than transforming it into less harmful compounds. In situ bioremediation, which involves the use of indigenous microorganisms to degrade the target compounds within the aquifer, is receiving increasing attention due to its potential cost-effectiveness. Other advantages of bioremediation include minimum land disturbance, reduce volume of generation waste, reduced potential of human exposure to contaminated media, reduced potential of cross-media contamination, it does not dewater the aquifer due to excessive pumping, and it is environmentally sound because it ultimately transforms the target BTEX into harmless products such as CO_2 and H_2O. Nevertheless, the successful exploitation of natural degradation processes is site-specific and requires fulfillment of the following conditions.

5.2. REQUIREMENTS FOR BTEX DEGRADATION

5.2.1. Presence of microorganisms with potential to biodegrade the target compounds.
Hydrocarbons have a natural pyrolytic origin and have been in contact with microorganisms throughout evolutionary periods of time. Thus, it is not surprising that many bacteria have acquired the ability to utilize hydrocarbons as food. The ability of microorganisms to degrade BTX has been known since 1908, when Stormer isolated the bacterium *Bacillus hexabovorum* by virtue of its ability to grow on toluene and xylene aerobically. In an early review, over 100 microbial species from 30 genera were identified that could degrade hydrocarbons. The existence of BTX degraders is a widely accepted fact. Furthermore, they are widely distributed. The ubiquity of soil bacteria capable of degrading BTX was first demonstrated in 1928 when 146 out of 245 uncontaminated soil samples contained bacteria capable of metabolizing hydrocarbons.

5.2.2. Accessibility of target pollutants to the microorganisms.
A common limitation of natural degradative process is the lack of adequate contact between pollutants and microorganisms. The target pollutants must be accessible in various aspects, including physicochemical bioavailability (e.g,. desorption from aquifer solids, dissolution from pure phase, or diffusion through nano-pores to reach the cell surface), structurally (e.g., bonds requiring cleavage must be exposed and not be sterically blocked by large atoms such as chlorine), and biochemically (e.g., the target pollutant must be able to pass through the cellular membrane – and the uptake may be regulated). In regards to BTX, these requirements are generally met.

5.2.3. Induction of appropriate degradative enzymes.
This process involves activation of specific regions of the bacterial genome. When some target substrates are present, they initiate a cascade of biochemical reactions that result in the transcription of genes that code for the synthesis of the necessary degradative enzymes. In regards to BTX degradation, many enzymes require induction, and the inducer (e.g., toluene) must be present at a higher concentration than the minimum threshold for induction. In general, this threshold is very low and enzyme induction is rarely a limiting factor in BTX bioremediation. Furthermore, BTX contamination is often discovered several years after the fact, and significant microbial acclimation and enzyme induction can occur during this time. Nevertheless, the presence of easily degradable substrates could exert diauxic effects (i.e., preferential substrate utilization) which would hinder the induction of BTX degrading enzymes. This may be an important factor in gasoline amended with ethanol, a common oxygenate that is added to ensure cleaner combustion. In this case, a lag period may be observed during which ethanol is degraded before any significant BTX degradation occurs.

5.2.4. Availability of electron acceptors.
Hydrocarbons are in a reduced state, and their oxidation is thermodynamically very feasible. Microorganisms mediate their oxidation using electron acceptors during natural respiratory processes. The following preferential utilization has been observed, and reflects a decreasing oxidation potential of the potential electron acceptor:

oxygen > nitrate > ferric iron > sulfate > carbon dioxide.

In general, the kinetics of hydrocarbon oxidation is faster for electron acceptors with higher oxidation potential.

Biodegradation using molecular oxygen. Aerobic BTX degradation is usually fastest. Using toluene as an example, half lives associated with its aerobic degradation typically vary between 1 and 20 days. Yet, it is not uncommon that half lives fall out of this range, depending on the active microbial concentration and mass transfer limitations. Aerobic toluene mineralization follows the following stoichiometry:

$$C_7H_8 + 9\ O_2 \rightarrow 7\ CO_2 + 4\ H_2O \qquad (1)$$

Therefore, the complete mineralization of toluene requires about 3.1 mg/L for each mg/L of toluene (or the other BTX compounds). The extent of aerobic BTX degradation is controlled by the amount of contamination released, the rate of oxygen transfer into the subsurface, the background oxygen concentration of the groundwater (usually 6 to 12 mg/L), and the occurrence of alternate substrates. The presence of ethanol in gasoline represents a significant additional oxygen demand by the soluble components, and is likely to decrease the extent of aerobic BTX degradation in aquifers.

Biodegradation using nitrate. Once oxygen is depleted, some facultative denitrifiers can replace oxygen with nitrate as the terminal electron acceptor during BTX degradation. Half lives for toluene degradation under denitrifying conditions are a little longer, typically ranging from 20 to 50 days, but degradation rates could be as fast as under aerobic conditions if the denitrifier concentration is sufficiently high. The mineralization of toluene coupled to the complete nitrate reduction to nitrogen gas is represented by the following reaction:

$$C_7H_8 + 7.2\ NO_3^- + 7.2\ H^+ \rightarrow 7\ CO_2 + 7.6\ H_2O + 3.6\ N_2 \qquad (2)$$

Numerous laboratory and field studies have shown that toluene, *m*-, *p*-, and *o*-xylene, ethylbenzene, and naphthalene can be degraded under strictly anaerobic denitrifying conditions. However, benzene, which is the most toxic of the BTX, is often recalcitrant under denitrifying conditions.

Biodegradation using ferric iron. When oxygen and nitrate have been used up, some indigenous microorganisms can use ferric iron [Fe(III)] as the electron acceptor during BTX degradation. Large amounts of ferric iron are present in mineral forms in most aquifers, which constitute a large potential electron acceptor pool for hydrocarbon oxidation. For example, the mineralization of toluene coupled to the reduction of ferric hydroxide can be represented as follows:

$$C_7H_8 + 36\ Fe(OH)_3 + 72\ H^+ \rightarrow 7\ CO_2 + 36\ Fe^{+2} + 94\ H_2O \qquad (3)$$

Although the mechanisms of hydrocarbon degradation under iron-reducing conditions are not fully understood, field evidence suggests that this is an important mechanism in the subsurface biodegradation of dissolved BTX.

Biodegradation using sulfate. Laboratory and field studies have shown that BTX can be degraded under sulfate reducing conditions. Anaerobic toluene mineralization under sulfidogenic conditions is described by the following stoichiometric equation:

$$C_7H_8 + 4.5\ SO_4^{-2} + 3\ H_2O \rightarrow 7\ HCO_3^- + 2.25\ H_2S + 2.25\ HS^- + 0.25\ H^+ \qquad (4)$$

This process is relatively slow, and its extent and significance in aquifers has not yet been quantified. When it occurs, it is usually near the center of the plume, farthest from the surrounding uncontaminated (aerobic) groundwater.

Biodegradation using carbon dioxide. Laboratory studies have also shown that BTX can be degraded under methanogenic conditions. The reaction for toluene can be represented as:

$$C_7H_8 + 5\ H_2O \rightarrow 2.5\ CO_2 + 4.5\ CH_4 \qquad (5)$$

Similar to BTX degradation under sulfate reducing conditions, this process is relatively slow and its significance as an attenuation mechanism in aquifers has not been proven. This may be due, in part, to the high sensitivity of methanogens and sulfate reducers to a wide variety of environmental conditions, including pH, temperature, and toxicants (including oxygen in this case). An imbalance in any of these factors could easily inhibit these anaerobic BTX degradation processes.

5.2.5. Availability of inorganic nutrients.

Microorganisms need macronutrients to synthesize cellular components, such as nitrogen for aminoacids and enzymes, phosphorus for ATP and DNA, sulfur for some coenzymes, calcium for stabilizing the cell wall, and magnesium for stabilizing ribosomes. A C:N:P ratio of 30:5:1 is generally sufficient to ensure unrestricted growth in aquifers. Microbes also need micronutrients to perform certain metabolic functions. For example, trace metals such as Fe, Ni, Co, Mo, and Zn are needed for some enzymatic activities. In general, aquifer minerals contain sufficient nutrients to support microbial activity. Nevertheless, geochemical analyses and laboratory biodegradation assays should be performed to verify that the presence of inorganic nutrients is sufficient for the success of intrinsic bioremediation.

5.2.6. Adequate pH.

Enzymes are polymers of amino acids, and their activity requires the proper degree of aminoacid protonation. This is controlled by the pH. Optimum groundwater pH is usually near the neutral value of 7.0, but most aquifer microorganisms can perform well between pH values of 5 and 9. Groundwater is typically well buffered within this range, so that the microbial physiological requirement for adequate pH is generally met in aquifers. Nevertheless, aquifers contaminated by municipal landfill leachates may contain elevated concentrations of organic acids and pH's as low as 3.0. In these cases, pH may represent a significant environmental problem to the indigenous bacteria.

5.2.7. Adequate temperature.

Temperature is one of the most important environmental factors influencing the activity and survival of microorganisms. Low temperatures reduce the fluidity and permeability of the cellular membrane, which hinders nutrient (and contaminant) uptake. Higher temperatures are associated with higher enzymatic activity and faster biodegradation rates, up to an optimum value which is species specific. In this range, BTX degradation rates can triple as a result of a temperature increase of 10°C. If the temperature rises much beyond the optimum value, proteins, enzymes, and nucleic acids become denatured and inactive. The temperature of the upper 10 m of the subsurface may vary seasonally; however, that between 10 and 100 m approximates the mean annual air temperature of a particular region.

5.2.8. Absence of toxic substances.

Some contaminants can be present in aquifers at sufficiently high concentrations that inhibit microbial activity. For example, it is not uncommon for aquifer microorganisms to encounter potentially toxic heavy metals such as Pb, Hg, Cd, and Cr. While heavy metals are required in trace quantities for nutritional purposes, they can be bactericidal if present in soluble form at concentrations greater than about 1 mg/L. Hence, geochemical analyses and laboratory degradation assays with representative aquifer samples should be performed to verify the absence of inhibitory substances.

The primary mechanisms that limit BTX migration are biodegradation and, to a lesser extent, volatilization. Abiotic reactions such as hydrolysis are not significant because BTX are stable under environmental conditions. In intrinsic BTX bioremediation, the rate-limiting attenuation mechanism is frequently the influx of oxygen, which in turn limits aerobic BTX degradation kinetics. Nevertheless, the presence of a free phase is a critical factor influencing whether a plume will expand or recess. For example, a fluctuating water table can continue to flush BTX into the plume, which would hinder its recession. Consequently, the removal of the free-phase and the sorbed hydrocarbons from the source area source is an important prerequisite for the successful implementation of intrinsic bioremediation.

5.3. ENGINEERED BTEX BIOREMEDIATION SYSTEMS

The ubiquity of microorganisms capable of degrading BTEX has been established. Thus, BTEX bioremediation often involves stimulating microbial activities by supplying nutrients and electron acceptors (e.g., O_2 and sometimes NO_3^-), with success often limited by the ability to distribute the stimulating materials throughout the contaminated zone. Bioremediation works best for homogeneous aquifers of high permeability ($>10^{-5}$ m/s) after the free product has been removed. The most common engineered bioremediation technologies are summarized below:

5.3.1. Bioventing

This approach is used to stimulate aerobic degradation processes above the water table by the action of vacuum pumps that pull air through the unsaturated zone. Bioventing is often used with infiltration galleries that deliver (with nutrients) to prevent desiccation in the unsaturated zone.

5.3.2. Water Circulation Systems

This method is based on extracting contaminated groundwater for above-ground treatment and reinjecting it into the ground with stimulatory amendments (e.g., H_2O_2 as oxygen source, and nutrients). Clogging near injection well screens and infiltration galleries can occur due to bacterial growth and mineral precipitation. In general, pulsing nutrients results in less clogging than continuous delivery. Occasional pulsing of Cl_2 or H_2O_2 to control biofouling can also prevent clogging.

5.3.3. Air Sparging

This engineered system involves injection of compressed air into the contaminated subsurface to deliver oxygen and strip the BTEX into a vapor-capture system. Air sparging can be a relatively effective and inexpensive BTEX bioremediation approach, but it is not effective when low-permeability soil traps or diverts the airflow.

5.3.4. Biobarriers

This term refers to biologically active zones that are placed in the path of narrow BTEX plumes, often incorporating air spargers (i.e., air-curtains) or oxygen-releasing compounds to enhance oxidative biodegradation processes. Hydraulic or physical controls on groundwater movement may be required to ensure that BTEX pass through the barrier.

5.4. PASSIVE BTEX BIOREMEDIATION SYSTEMS

In many cases, natural conditions at contaminated sites meet all the essential environmental factors so that bioremediation can occur without human intervention. This process is called *natural attenuation*, and differs from no-action alternatives in that it requires thorough documentation and monitoring of the role of microorganisms in eliminating the target contaminants.

Natural attenuation can be defined as the combination of natural biological, chemical and physical processes that act without human intervention to reduce the mass, toxicity, mobility, volume, or concentration of the contaminants (e.g., intrinsic bioremediation, dispersion, dilution, sorption, and volatilization).

Natural attenuation is typically used in conjunction with active remediation measures (e.g., source control) or as a follow-up to such measures, and is most appropriate for sites experiencing low exceedences of cleanup levels. The success of natural attenuation as a remedial option depends on the following: 1) adequate site characterization; 2) a long-term monitoring plan consistent with the level of knowledge regarding subsurface conditions at the site; 3) evaluation and (if needed) control of the contaminant source; and 4) a reasonable time frame to achieve the remedial objectives.

5.5. EFFECT OF GASOLINE OXYGENATES

Understanding the factors that affect the fate and transport of BTEX compounds in aquifers is of paramount importance for risk assessment and corrective action purposes. While significant advances have been made towards understanding the genetic and biochemical bases of BTEX biodegradation, little attention has been given to how differences in gasoline formulation affect such natural attenuation processes. In this regard, there is a recent initiative being considered to phase out methyl *tert*-butyl ether (MtBE) as a gasoline oxygenate, due to its recalcitrance, ability to rapidly impact drinking water sources, and low taste and odor thresholds (<5-40 ppb). The most likely candidate to replace MtBE (which accounts for 80% of current oxygenate use) is ethanol (currently accounting for 15% of oxygenate use). Ethanol is also a renewable resource that can serve as a substitute fuel for imported oil. Therefore, an increase in the use of EtOH as a gasoline additive seems imminent, and a better understanding of its effects on BTEX migration and natural attenuation is warranted.

In theory, both MTBE and ethanol could enhance BTEX mobility by exerting a cosolvent effect that favors BTEX partitioning into the aqueous phase and decreases sorption-related retardation. Such effects, however, are concentration-dependent and are unlikely to be observed when oxygenate concentrations are below 10,000 mg/L. Since this concentration is unlikely to be exceeded in gasohol-contaminated sites, adding ethanol to gasoline (e.g., at 10% v/v) should not enhance BTEX dissolution from the fuel phase or decrease BTEX retardation factors. However, column breakthrough experiments showed significantly decreased retardation and enhanced BTEX mobility when ethanol was added at 50%, suggesting that neat ethanol spills in a bulk terminal can exacerbate groundwater pollution by pre-existing petroleum product releases.

Experiments were also conducted with large aquifer columns (120-cm long). Benzene and toluene (BT) were completely degraded within the first 3 cm of the inlet in columns fed BT alone or fed BT plus MtBE (Fig. 12A). MtBE did not affect the degradation of EtOH either. On the other hand, BT degradation was adversely affected by the degradation of EtOH, which was preferentially utilized within 3 cm of the column inlet (Fig. 12B). The degradation of EtOH used up the available oxygen and exacerbated the consumption of anaerobic electron acceptors such as nitrate and sulfate (data not shown). The high electron-acceptor demand exerted by EtOH created strongly reducing conditions (-350 mV near the inlet), which are not conducive to rapid BTEX degradation (Fig. 13).

Because of the high oxygen demand exerted by gasohol spills, EtOH is likely to be degraded predominantly under anaerobic conditions. None of the products of anaerobic EtOH degradation is toxic, although some metabolites (e.g., butyrate) could have adverse aesthetic impacts. In addition, acetate and other volatile fatty acids can decrease the pH if they accumulate at high concentrations. This is illustrated in Fig. 14A, which shows acetate accumulation (1.1 g/L) only in the columns fed BT plus EtOH. Acetate accumulation caused a decrease in pH from 7.5 to as low as 6.2 (Figure 4B), even though this was a well-buffered system (1 g/L as $CaCO_3$). Poorly buffered systems could experience a greater decrease in pH that could inhibit microbial activity and the further degradation of EtOH and other compounds. Another potential concern is the accumulation of ethanol-derived methane, which could represent an explosion hazard. However, the highest methane concentration measured in the column fed BT plus EtOH was only 1.8 mg/L (Figure 14A), which is considerably below the solubility of methane (~21 mg/L at 25°C). Therefore, this experiment suggests that gasohol spills should not pose a significant explosion risk, notwithstanding the possibility that some site-specific conditions could favor more extensive methanogenesis.

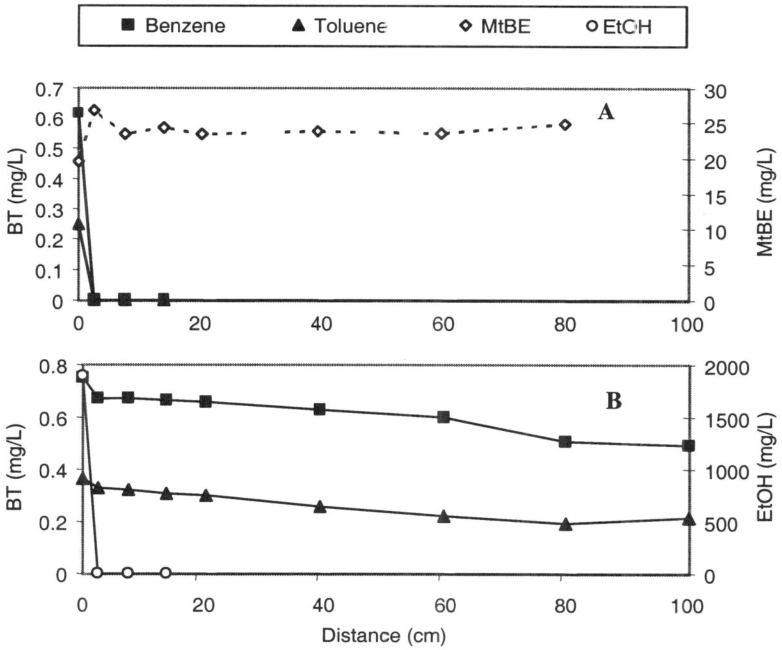

Figure 12. Benzene and toluene concentration profiles in columns fed MtBE (Panel A) or EtOH (Panel B) after 110 days of operation.

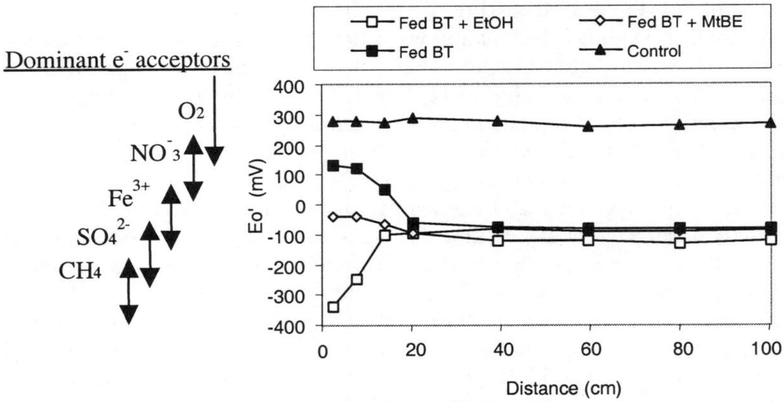

Figure 13. ORP profiles after 110 days of operation.

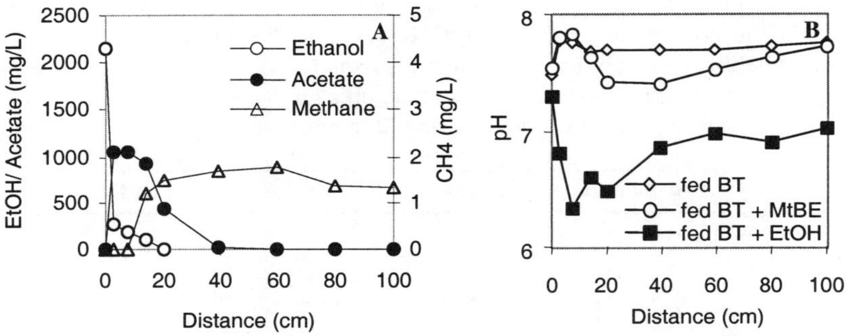

Figure 14. Acetate and methane concentration along the column fed BT plus EtOH (Panel A), and pH profiles (Panel B) after 226 days of operation.

Microcosms were used to determine how EtOH and MtBE affect BTEX degradation rates under different electron-acceptor conditions commonly found in contaminated aquifers experiencing natural attenuation. Response variability was also investigated by preparing microcosms with aquifer material from different sites (Table 6). BTEX and EtOH were typically degraded earlier in microcosms with previously contaminated aquifer material, although previous exposure did not always result in high degradation activity. MtBE was not degraded within 120 days under any conditions, and did not affect BTEX or EtOH degradation patterns. EtOH was typically degraded before BTEX compounds, but had a variable effect on toluene degradation as a function of electron-accepting conditions and bacterial source. In some cases, EtOH retarded toluene degradation (e.g., under methanogenic conditions for Site 1 or sulfate-reducing conditions for Site 3; Table 6), but it occasionally enhanced toluene degradation (e.g., under denitrifying conditions for Sites 3 and 4; Table 6). Enhancement of toluene degradation by EtOH may be attributable to a low initial concentration of BTEX degraders and their incidental growth during EtOH degradation.

The preferential degradation of ethanol by aquifer microorganisms, the accompanying depletion of oxygen and other electron acceptors, and the observed decrease in pH as a result of acetate accumulation in anaerobic zones suggest that ethanol could hinder natural attenuation of BTEX plumes. This is particularly important for the fate of benzene, which is the most toxic and the most recalcitrant of the BTEX under anaerobic conditions. Thus, ethanol could increase the distance that benzene migrates before attenuating processes stabilizes the plume. Nevertheless, the impacts associated with the use of ethanol seem to be less significant and more manageable than those associated with the use of MTBE.

TABLE 6. Time (days) required for 50% toluene degradation in microcosms with material from different sites under different electron acceptor conditions.

		Site*			
		1	**2**	**3**	**4**
	BTEX alone	1	8	11	4
Aerobic	With EtOH	1	6	14	>13
	With MtBE	1	NA	NA	7
	BTEX alone	4	32	>48	51
Denitrifiying	With EtOH	4	30	14	31
	With MtBE	4	NA	NA	36
	BTEX alone	5	29	10	NA
Iron-reducing	With EtOH	18	>70	17	NA
	With MtBE	5	NA	NA	NA
	BTEX alone	11	>70	24	30
Sulfate-reducing	With EtOH	7	>70	33	11
	With MtBE	11	NA	NA	8
	BTEX alone	5	57	>54	38
Methanogenic	With EtOH	34	35	>56	26
	With MtBE	6	NA	NA	16

***Note:** Site 1 = Travis Air Force Base, BTEX and MtBE exposure history.
Site 2 = Tracy (CA), no previous BTEX exposure.
Site 3 = Northwest Terminal site, EtOH (average 160 mg/L) and BTEX exposure history.
Site 4 = Sacramento (CA), BTEX and MtBE exposure history.
NA = Data not available (i.e., not tested).

6. Ecological, Legal and Methodical Approaches to Biological Indication of Soil Quality

Soil quality represents an integral value of the compositional structures and functions of terrestrial soils in relation to their different uses and to long-term environmental conditions on site. Among the indigenous soil structural components, edaphon, and especially microorganisms, play a key role in different ecologically important functions of soil. This applies to maintaining of matter and energy transfer in terrestrial ecosystems, including the cycling of plant nutrients. Under stress conditions caused by chemical pollutants or other adverse anthropogenic effects, soil microorganisms and their biochemical activities may undergo severe alterations. In order to avoid negative consequences of anthropogenic disturbances to soil functions, legal regulations should be adopted internationally. In Germany, a Federal Soil Protection Act, and a Federal Soil Protection Ordinance have been put in force in 1999 for this reason. The aim of these legal instruments is to protect or to restore different soil functions on a sustainable basis. In order to evaluate soil quality for this purpose, not only abiotic parameters but also microbiological and biochemical soil characteristics, should be used. In our respective approach more than twenty different parameters were tested in an joint effort with scientists from the Czech Republic, Hungary, Russia and Slovakia. A general evaluation of results obtained from repeated sampling and analysing of 49 differently anthropogenically affected soils, it has been concluded, that total microbial biomass, dehydrogenase activity, N_2-fixing bacteria, nitrification activity, soil respiration (CO_2 release), and also a microbial humification of plant organic material could be preferentialy used as indicators of soil quality. The setting up of appropriate limits for different soils, however, remains open until yet.

6.3. INTRODUCTION

Human beings and other living creatures represent an important part of the existing natural heterogeneity and contribute to the dynamics of environmental processes. Simultaneously, all living organisms modify directly or indirectly their environment and this is true also for humans. Human alteration of Earth's ecosystem based on agriculture, industry, recreation and commercial activities became a substantial factor for land transformation with consequences for a global environmental change [18]. That change may include alteration of soil quality with severe consequences for soil fertility, human and animal health, and different environmental issues.

In our understanding, soil quality is an integral value of the compositional structures and functions of terrestrial soils in relation to their different uses and to long-term environmental conditions on site. It is hardly possible to assess and evaluate soil quality in a simple way. From the agricultural point of view, fertility of soils served as an indicator of soil quality for decades, and its development was mainly positive. For example, Matson et al. [19] reported an average increase in yield of wheat by 100 %, and of corn up to 500 % from 1940 until 1990 in Colorado, USA. However, possible hazards to soil and its functions other than fertility, which could be caused, e.g., by an extensive use of agro-chemicals have been not co-evaluated. Similarly, the density of industrial emissions and other adverse human activities might not be reflected if using soil fertility as a parameter of soil quality.

Anthropogenically caused risks to soil environmental functions has been recognized to exist many years ago [20, 21], but the believe in natural self-purification capacity of soils persisted for many years [22]. The first moves towards the development of soil protection in Europe were made in Germany and the Netherlands, and later in the European Community [23-25]. Since then, the necessity to evaluate physical, chemical and biological soil characteristics as a prerequisite of an effective assessment of soil quality has been stressed repeatedly, and different physical and chemical parameters have been already adopted internationally [26,27]. In difference to that, biological parameters remained a matter of discussion, and focussed again on consequences for soil fertility, water and air quality and on sanitary aspects [28]. In a general way, some attempts to indicate the usefullness of biological methods in evaluation of soil health were made [29-32], but an ecologically based conceptual approach has not been fully developed yet. Instead, numerous individual approaches were presented (see e.g. [33]) which, however, have only a little chance to contribute to the soil protection practice and policy. The attempt of this paper is (i) to address some of the particular concerns of soil protection in Germany; (ii) to underline the necessity of a process-linked biological approach in the assessing of soil quality; (iii) to indicate the usefulness of some ecologically based parameters of soil quality.

6.4. SOME GLOBAL ENVIRONMENTAL EFFECTS OF SOIL ORGANISMS

On a global scale, soils contain by far the highest numbers and the greatest diversity of organisms. In a trivial comparison, there are at least twice as much or even more microorganisms in a teaspoon of soil than the total population of humans on the Earth. For s the main groups of soil organisms the following numbers of species were estimated by Pankhurst [34]: Bacteria 30,000; Fungi 1 500,000; Algae 60,000; Protozoa 10,000; Nematodes 500,000; Earthworms 3,000. According to restriction analysis of DNA extracted from soil samples, between 6,000 and 18,000 microbial species might exist in a 1 g soil [35]. All these organisms, i.e., edaphon, do not only inhabit soils but they also actively contribute to different transformations of their habitats. Those organisms affect soil environments more quickly than abiotic processes can do. For comparison, the time scales for biological processes involved in the development of terrestrial ecosystems take only 1-100 years, while those for physical processes may take up to 10,000 years [36].

Approximate numbers of soil organisms in arable soils and their main functions are schematically shown in Fig. 15. Bacteria (including actinomycetes) and fungi undoubtedly represent the most active groups of soil organisms. They are involved in both degradative and synthesis biogeochemical processes in soil. The decomposer community including organisms capable of degrading lignin, cellulose, pectin, chitin, protein and other organic compounds contribute to the integrity of the ecosystem by converting different remnants of plant, animal and microbial tissues. In this way, soil microorganisms maintain the predominant part of the matter and energy transfer in terrestrial ecosystems.

Since the life on Earth is based upon carbon, CO_2 is the main final product of microbial mineralization of organic compounds. Soils represent a major source in liberating CO_2 to the atmosphere. Under steady state conditions in an ecosystem, the CO_2 emission should roughly equal the respective demand of photosynthetically active plants. However, long-term measurements combined with an interpolation of data available from the past indicate that due to different activities, humanity added CO_2 to the atmosphere resulting in a concentration increase of about 30 % relative to the pre-industrial era. That increase has been assumed as driven primarily by fossil fuel combustion [18], but according to Bertram [37] non less than two-thirds of the increased atmospheric CO_2 may come from an increase in soil respiration. Usually, about 74 % of soil CO_2 is directly related to microbial activity and 26 % to a root respiration, which again strongly depends on the activity of rhizospheric microorganisms inhabiting plant roots [38]. Conclusively, the mineralization activity of soil microorganisms expressed in CO_2 release, should be considered as an important parameter of soil quality from the ecological point of view.

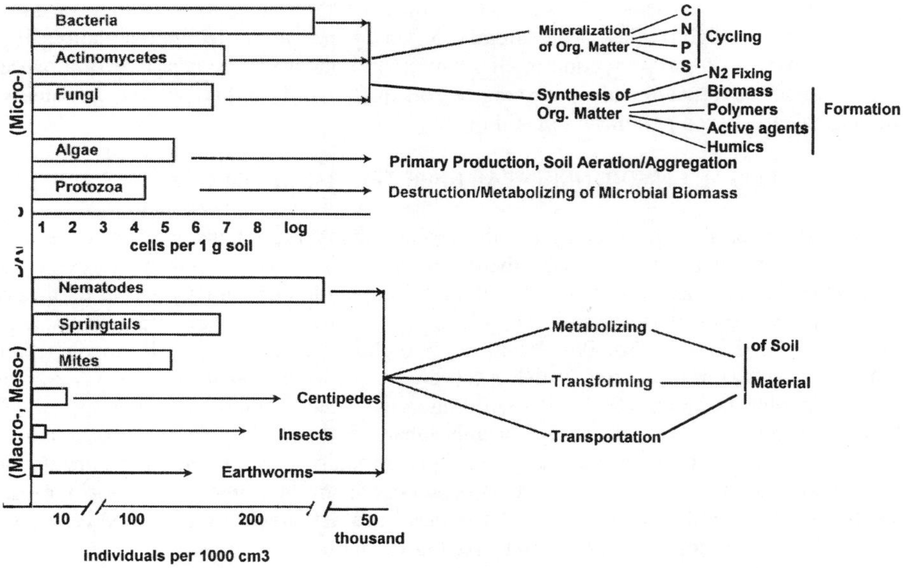

Figure 15. Numbers, Functions and Interactions of Soil Organisms

It has been reported that up to 50 % of plant residues in soil could be digested by invertebrates [39]. However, those soil animals are mainly capable only of an uncomplete utilizing of some plant polysaccharides; thus, their contribution to the mineralization of organic matter in soil could be rather neglected [40].

From the energetic point of view, mineralization of organic matter represents a catabolic process releasing energy for anabolic (synthesizing) activities of soil organisms. One can assume that anabolic activities of soil microorganisms may affect the global cycling of carbon. Norby [41] calculated, that in fact only about 45 % of the globally increased CO_2 amounts remains in the atmosphere while the main part is apparently missing. That missing carbon is proposed to be assigned to soil environments. Deliberated CO_2 could be transferred below ground either into a labile, short-lived pool such as microbial biomass, or into a refractory soil organic matter such as humic substances. For this reason, the assessment of anabolic activity of soil microorganisms, i.e., the estimation of balance between C bound in soil microbial biomass and humic substances, and CO_2 released from soil would contribute both to the elucidation of global carbon budget. Also, it could be used as an ecological parameter for evaluating soil quality.

Among the major elements required for all forms of life, nitrogen is unique in that it exists in vast amounts in the atmosphere. From that reservoir, N_2 must be bound into a living matter by the activity of free-living and/or symbiotic microorganisms mainly in soils, and also in waters. Although human activities may add as much fixed N_2 to terrestrial ecosystems as do natural sources [42] for the enormous importance of N_2 fixing in the global cycle of nitrogen, the assessment of that process should not fail in the evaluation of soil quality.

Beside the N_2-fixation, also other key processes of the N cycle, such as ammonification of nitrogen bound in organic compounds, an oxidation of NH_4^+-N to NO_3^--N (nitrification), and a reduction of NO_3- /NO_2^- to N_2 (denitrification), represent microbially mediated soil processes which could be of use in a system of ecologically based parameters of soil quality. This is because an anthropogenical alteration of primary processes in the cycle of nitrogen may have multiple consequences, such as (i) an increased concentration of the greenhouse gas N_2O and/or other NO_X gases globaly, (ii) losses in soil nutrient, (iii) acidification of soils, and (iv) increased transfer of N from soil to aquatic environments [42]. For the emission of NO_2 soils have been identified as being by far the largest source [43].

Not only the cycles of carbon and nitrogen may became strongly affected by human activities. Similar is true also for the S, P and other elemental cycles. However, the key roles of C and N in the biosphere, the important role of microbially mediated processes in the transformation of these elements, and also a variety of analytical methods available, may account for the priority involvement of C and N transformations in the biological assessment of soil quality.

An anthropogenic pressure on soil organisms and their activities may cause a threat to biodiversity in soil environments. Measuring of phospholipid fatty acid profiles (PLFA), DNA hybridization or DNA/RNA fingerprinting techniques give the opportunity to estimate microbial diversity. However, there is no sufficient knowledge available yet on the dependence of the most of soil processes on a degree of biodiversity in a soil environment. Also, no standardized monitoring methods exist to control this parameter in soil habitats. Nevertheless, in general the preservation of biodiversity should be attempted with respect to soil quality conservation. A higher biodiversity would mean longer food-chains, more cases of symbiosis and greater possibilities for negative feedback control, which could reduce oscillations and hence increase stability of a soil ecosystem [44]. In a broadly based soil monitoring the response of the microbial community to stresses could be measured with a large variety of existing microbiological techniques. Although these methods have severe limitations, an estimation of total microbial biomass and colony forming units of physiologically different microorganisms could be important characteristics of soil quality.

6.5. EFFECTS OF POLLUTANTS ON SOIL MICROBIAL ACTIVITIES

Soil comprises a structurally complex mosaic of heterogeneous microsites colonized by different organisms. Even in a biologically rather less active podzolic soil, e.g., up to 9 t ha^{-1} (d.w) of bacteria and about the same amount of fungal biomass exists in a plough layer [45]. From an average size of a bacterial cell, a total surface of about 500 ha bacteria per ha of arable soil was postulated [46]. These numbers and the capability of soil microorganisms to multiply even under undesirable environmental conditions relatively quickly, signalize the high degree of susceptibility of soil microorganisms against either positive or negative effects, e.g., those caused by pollutants. Different ways exist on how anthropogenic pollutants may target soil microorganisms (Fig. 16). In general, either directly or via different degradation or transformation products (metabolites), pollutants can affect the total microbial biomass, individual groups of microorganisms, and also different microbially mediated processes [47]. Because of their ecological importance, all those individual targets should be considers as potentially useful parameters in assessing soil quality.

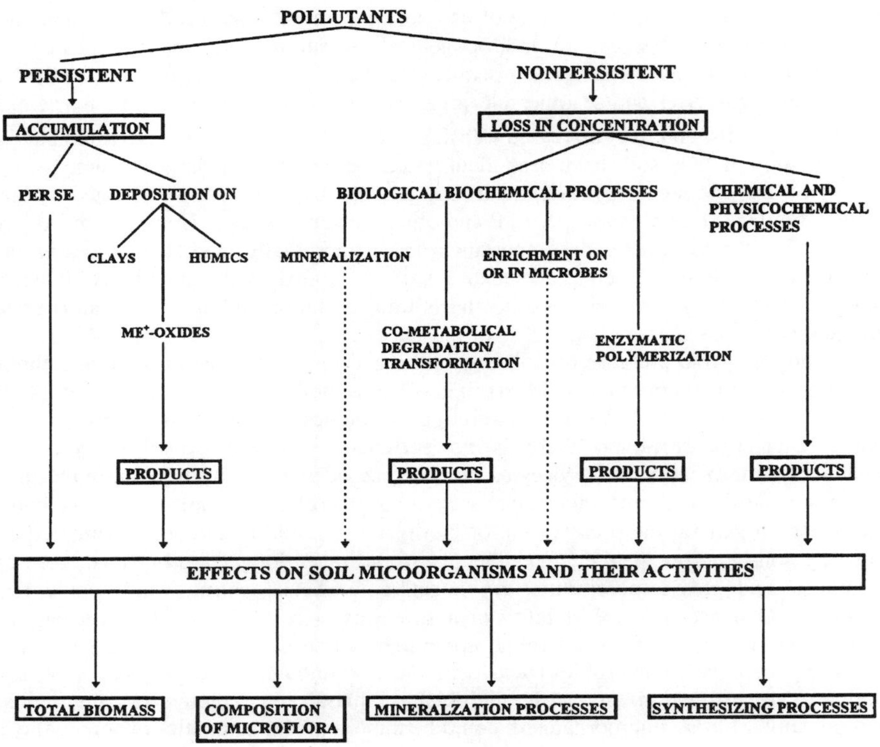

Figure 16. Contaminant Interaction Pathways with Soil Microbial Populations

6.4. CONCERNS AND LEGAL SOLUTIONS OF SOIL PROTECTION IN GERMANY

The main issues of concern in soil protection policies in Germany has been identified as follows: (i) Land consumption; (ii) Hazardous substances in soil; (iii) Soil erosion and compaction; (iv) Waste application on soil [48]. There is no ranking for the individual issues, for their importance may vary in dependence on soil use in the region of concern.

In March 1999 the Federal Soil Protection Act entered into force in Germany. It has the main purpose to protect or restore soil functions on a sustainable basis. This aim should be achieved by preventing harmful soil changes, rehabilitation of contaminated soil sites and affected water resources, and by general precautions against negative impacts on soil.

According to the Federal Soil Protection Act, soil is expected to fulfill the following functions:

A) Natural functions
   - as a basis for life and a habitat for people, animals, plants, and soil organisms
   - as a part of natural systems, especially by means of water and nutrient cycles

- as a medium for matter decomposition, balance and restoration with respect to soil filtering, buffering and substance-converting properties, and also concerning groundwater protection

B) Functions as an archive of natural and cultural history

C) Functions useful to man
- as a medium that holds deposits of raw materials
- as land for settlement and recreation
- as land for use in agriculture and forestry
- as land for other economic public uses such as transport, supply, provision and disposal purposes.

The Federal Soil Protection Act indicates harmful impacts on soil functions that may bring about hazard, considerable disadvantages or nuisances for either individuals or general public. In this respect it differs between *Suspicious soil sites*, i.e., those suspected to carry harmful soil changes, and *Contaminated sites*. Principles of a *Good Agricultural Practice* are listed in the Act. The implementation of the Federal Soil Protection Act should not interfere with already existing legal regulations as far as these may have impact on soil.

In July 1999 the German Federal Government adopted a Federal Soil Protection and Contaminated Sites Ordinance which consists of concrete requirements and provisions for the implementation of the Federal Soil Protection Act. In four Annexes the Ordinance indicates *Precautionary, Trigger,* and *Action values* for different inorganic and organic soil pollutants, and the effective pathways such as *"Soil – Human Being"*, *"Soil – Useful Plant"* and *"Soil – Groundwater"*. Requirements in respect of sampling, analytical procedures and quality assurance during (i) exploratory, and (ii) detailed investigations of a soil material are also listed in the Ordinance. The Ordinance shall be actualized in accordance with scientific progress achieved and practical requirements. At present the Ordinance is lacking recommendations on the assessment and evaluation of biological indicators of soil quality.

6.5. INDICATORS FOR BIOLOGICALLY-BASED SOIL QUALITY

In their comprehensive work on applied ecotoxicology, Römbke and Moltmann [49] summarized the major effect tests actually used for terrestrial media. Those are mainly tests with higher vertebrates (birds, mammals), soil saprophagous invertebrates, plants and (for pesticide side effects) pollinators, such as honey bees. The authors pointed on the existing lack in soil tests with microorganisms. Graefe [50] presented some soil biological diagnostics based purely on analysis, classification and evaluation of soil invertebrates. He diversified among organisms capable of indicating soil freshness, wetness, acidity, alkalinity, etc. Yakovlev [51] recommended the abundance of selected protozoa (*Amoebae*, *Colpodida*) and algae (*Heterotrix* sp.) in soil samples or their behaviour in a soil suspension for the characterization of virgin and anthropogenically affected soils. Debus and Hund [52] prepared soil extracts and recommended using aquatic test organisms in their strategy for the assessment of soil contaminants, although in their previous work, aquatic tests with *Daphnia magna* and *Scenedesmus subspicatus* appeared rather less sensitive[53]. The same authors also performed tests with natural soil samples, but only in order to target for organisms of different trophic levels such as individual microorganisms, algae, nematodes and higher plants [52]. In our opinion, all those assessment strategies might have some advantage of simplicity and the methods used might be easy to standardize. However, they oversimplify soil both in its biological and abiotic structural complexity and completely disregard soil ecological functions. In an ecologically based approach of soil quality assessment, a population level can be involved but possibly for organisms with recognized importance for some ecologically important functions, such as nitrogen fixation and/or other processes (Fig. 17).

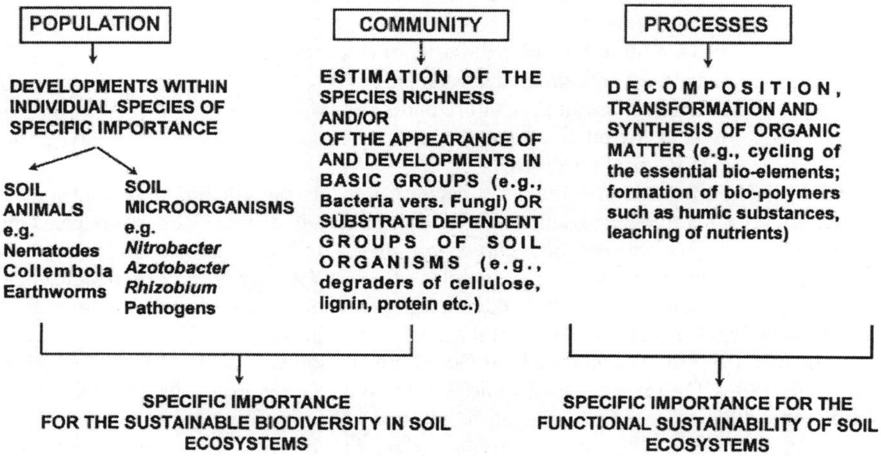

Figure 17. Ecological Indicators for Soil Quality

Contrary to this, there is no doubt that firm linkages exist between microbial communities, their activities, and between ecologically important processes, such as mineralization and transformation of organic matter (Figs. 15, 17). These facts should open several ways in developing assessment and evaluation of soil quality of both natural and anthropogenically stressed soil sites by employing adequate methods.

Based on the above discussion, we preselected prospective soil biological parameters as shown in Fig. 18. These include the estimation of microbial biomass, composition of microbial communities, mineralization processes and synthesizing processes occurring in soil. Simultaneously, basic physical-chemical soil characteristics have to be estimated. To make that system feasible for application, internationally adopted and standardized methods should be applied for soil analyses as far as possible. The system remains open to suitable novel methods or other improvements.

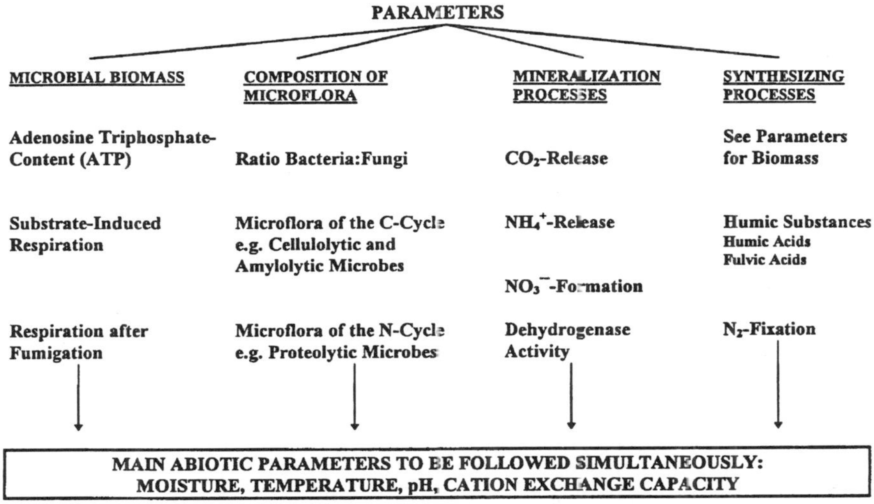

Figure 18. Prospective Soil Biological Parameters for Quality Assessment

6.6. EXPERIENCE FROM AN INTERNATIONAL PROJECT ON SOIL QUALITY TESTING

Between 1995 and 1997 the above mentioned approach to assessing soil quality by biological and biochemical methods was tested in a project co-ordinated from Germany and performed by five groups of soil biologists from the Czech Republic (headed by J. Kubat, Prague, and M. Tesarova, Brno), Hungary (headed by T. Szili-Kovacs, Budapest), Russia (headed by D.G. Zvyagintsev and M.M. Umarov, Moscow), and Slovakia (headed by P. Bielek, Bratislava). Soil samples from 49 sites representing different European soil types, and which were differently anthropogenically affected, were collected several times and analyzed according to the scheme shown in Fig. 18. All analyses were performed in laboratory because under field conditions natural stresses such as variations in temperature and moisture might mask especially low effects of anthropogenically caused alterations in soil quality. All research teams used the same analytical methods which were based either on the available DIN/ISO standards or the ASA/SSSA methods of soil analyses [54].

Fig. 19 shows a quite high sensitivity of parameters such as N_2-fixing bacteria, dehydrogenase activity, and respiration activity against a long-term pollution of a soddy podzolic soil by lead. In Table 7, an overview of relative sensitivities obtained for different parameters used in the international investigations is given. Again, the nitrogen fixing bacteria, soil enzymatic activity (dehydrogenase), respiration and nitrification/denitrification activity appeared among the most sensitive ones. Since the same parameters are closely linked with ecologically important soil processes of the C and/or N cycles, their alteration may effectively indicate an alteration of soil quality.

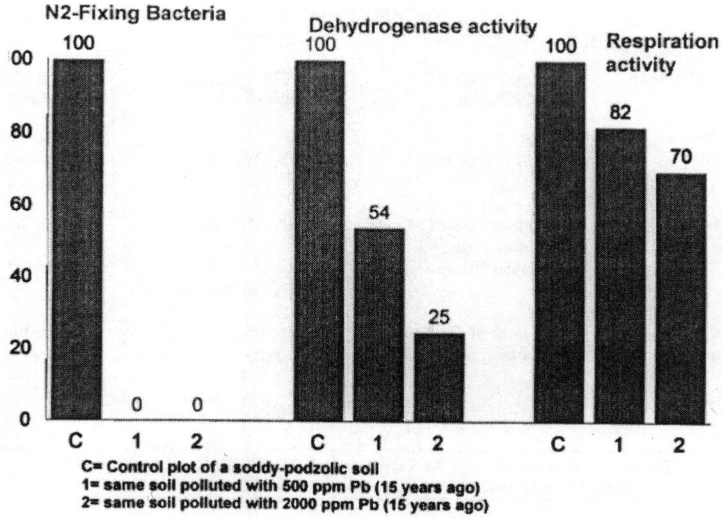

Figure 19. Results from German Soil Quality Analysis

Until now, an actual need exists in the determination of limiting values for the individual parameters under different soil conditions. Dahlin et al. [55] demonstrated this problem on soils with a low level of metal contamination. Problems may be also linked with a remarkable degree of spatial variability of data which can be observed even for pedogenically homogenous sites [56]. Solutions based on statistics have been suggested [57,58] but also alternative methodological developments should be further forced. The latter concerns especially modern molecular-biological and physico-chemical methods in biology, as soon as their practicability will be confirmed. In this way, the search of biological indicators for soil quality has been currently continued and different approaches experimentally tested [59-61].

In the conceptual approach presented here, an attempt was made to characterize process-related biological and biochemical parameters. This is because the entire biosphere strongly depends on organic matter biotransformations particularly those occurring in soil. If implemented using internationally adopted and standardized methods, this approach could fulfil requirements and expectations of ecologists, practitioners, and also policy makers interested in the assessment and evaluation of soil quality.

TABLE 7: Relative sensitivity of the selected microbiological and biochemical parameters for the assessment of soil quality. (Evaluation based on long-term soil analyses from 49 differently polluted soil sites).

Parameter	Relative sensitivity
Microbial Biomass	*/**
<u>Composition of Microbial Communities</u>	
Copiotrophic Bacteria (Colony forming units)	*/**
Oligotrophic Bacteria	**
Actinomycetes	**
Microscopic Fungi	**
Proteolytic Spore Forming Bacteria	-/*
Cellulose Decomposer	*/**
N_2-Fixing Bacteria	****
Pseudomonads	-/*
<u>Biochemical Process-Linked Activities</u>	
Respiration (CO_2 Release)	***
Ammonification (NH_4^+ Release)	**
Nitrification/Denitrification	**/***
Dehydrogenase Activity	***/****
Humification Activity	**

Sensitivity (relative to control) : - = Not detected ; * = Low ; **** = Maximum

7. Summary and Conclusions

This chapter represented an attempt to summarize the state-of-the-art of biodegradation and bioremediation of a wide range of organic contaminants (petroleum hydrocarbons, chlorinated solvents, polycyclic aromatic hydrocarbons), as well as current and proposed methodology of soil quality assessment. It is clear from this treatise that, whereas the scientific knowledge base regarding process understanding is maturing, the transitioning of this information to effectively remediate contaminated sites at best has accumulated a spotty record. Well-documented success stories in the field are rare, and the required design parameters that would allow engineers to effectively use the proposed technologies are generally not available. Moreover, there is no general agreement on the metrics of success (chemical and biological analysis, soil quality indicators) of bioremediation technologies, leaving informed decision making by regulators a difficult task which can at best be accomplished on a site-specific basis. Finally, there is an urgent need to incorporate uncertainty analysis to bridge site assessment and laboratory-based data with the purpose of developing a risk-based decision framework for technology implementation.

8. Acknowledgements

Support from the Michigan Department of Environmental Quality and the Environmental Protection Agency Great Lakes and Mid-Atlantic Center for Hazardous Substance Research are acknowledged.

9. References

1. Lendvay, J.M., Adriaens, P. , Barcelona, M.J., Major, Jr., C.L., Tiedje, J.M., Dollhopf, M., Löffler, F., Fathepure, B.Z., Petrovskis, E.A., Gebhard, M.,Daniels, G., Hickey, R., Heine, R., and Shi, J. (2001). Preventing Contaminant Discharge to Surface Waters: Plume Control with Bioaugmentation. Proceed. Battelle Conference, San Diego, CA.
2. Lendvay, J.M., Barcelona, M.J., Daniels, G., Dollhopf, M., Fathepure, B.Z., Gebhard, M., Heine, R., Hickey, R., Krajmalnik-Brown, R, Löffler, F.E., Major, Jr., C.L., Petrovskis, E., Shi, J., Tiedje, J.M., and Adriaens, P. (2002). Bioreactive Barriers: Bioaugmentation and Biostimulation for Chlorinated Solvent Remediation. Environ. Sci. Technol., *Submitted.*
3. Keith, L.H. and Telliard, W.A. (1979) Priority pollutants. I. A perspective view. *Environ. Sci. Technol.* **13**, 416 – 423.
4. Wick, L. Y., Springael, D., Harms, H. (2001) Bacterial strategies to improve the bioavailability of hydrophobic organic pollutants. *Environ. Sci. Technol.* Submitted.
5. Bastiaens, L., Springael, D., Wattiau, P., Harms, H., De Wachter, R., Verachtert, H., Diels, L. (2000) Isolation of adherent polycyclic aromatic hydrocarbon (PAH)-degrading bacteria using PAH-sorbing carriers. *Appl. Environ. Microbiol.* **66**, 1834 – 1843. Cerniglia, C. E. (1992) biodegradation of polycyclic aromatic hydrocarbons. *Biodegradation* **3**, 351 – 368.
6. Komatsu, T. Omori, T., Kodama, T. (1993) Microbial degradation of polycyclic aromatic hydrocarbons acenapthene and acenapthylene by a pure bacterial culture. *Biosci. Biotech. Biochem.* **57**, 864 – 865.
7. Grifoll, M., Selifonov, S. A., Gatlin, C. V., Chapman, P. J. (1995) Actions of a versatile fluorene-degrading bacterial isolate on polycyclic aromatic compounds. *Appl. Environ. Microbiol.* **61**, 3711 – 3723.
8. Kästner, M., Breuer-Jammali, M., Mahro, B. (1994) Enumeration and characterization of the soil microflora from hydrocarbon-contaminated soil sites able to mineralize polycyclic aromatic hydrocarbons. *Appl. Microbiol. Biotechnol.* **41**, 267 – 273.
9. Mueller, J. G., Chapman, P. J., Blattmann, B. O., Pritchard, P. H. (1990) Isolation and characterization of a fluoranthene-utilizing strain of *Pseudomonas paucimobilis. Appl. Environ. Microbiol.* **56**, 1079 – 1086.
10. Thibault, S. L., Anderson, M., Frankenberger, J. R. (1996) Influence of surfactants on pyrene desorption and degradation in soils. *Appl. Environ. Microbiol.* **62**, 283 – 287.
11. Walter, U., Beyer, M., Klein, J., Rehm, H. J. (1991) Degradation of pyrene by *Rhodococcus sp.* UW1. *Appl. Microbiol biotechnol.* **34**, 671 – 676.
12. Dunn, N. W., Gunsalus, I.C. (1973) Transmissible plasmid coding early enzymes of naphtalene oxidation in *Pseudomonas putida. J. Bacteriol.* **114**, 974 – 979.
13. Churchill, S. A., Harper, J. P., Churchill, P. F. (1999) Isolation and characterization of a *Mycobacterium* species capable of degrading three- and four-ring aromatic and aliphatic hydrocarbons. *Appl. Environ. Microbiol.* **65**, 549 – 552.

14. Sayler, A.A., Whitt, D.D. (1994) Tuberculosis, in *Bacterial pathogenesis: A molecular approach. Chapter 26*. ASM Press, Washington, D.C, pp. 307 – 320.
15. Yabuuchi, E., Yano, I., Oyaizu, H., Hashimoto, Y., Ezaki, T., Yamamoto, H. (1990) Proposals of *Sphingomonas paucimobilis* gen. Nov. And comb. nov., *Sphingomonas parapaucimobilis sp.* nov., *Sphinomonas yanoikuyae sp.* nov., *Sphingomonas adhaesiva sp.* nov., *Sphingomonas capsulata* comb. nov. and two genospecies of the genus *Sphingomonas. Microbiol. Immunol.* **34**, 99 – 119.
16. Vitousek, P.M.,Mooney, H.A., Lubchenco, J., and Melillo, J.M. (1997). Human domination of Earth's ecosystems. Science **277**, 494-499.
17. Matson, P.A., Parton, W.J., Power, A.G. and Sift, M.J. (1997). Agricultural intensification and ecosystem properties. *Science* **277**, 504-509.
18. Filip, Z. (1973). Healthy soil - fundament of a healthy environment. *Vesmir* **52**, 291-293 (in Czech).
19. Kovda, V.A. (1975). *Biogeochemical cycles in nature and their disturbance caused by humans.* Moscow, Nauka, 72 pp (in Russian).
20. Möller, F. (1983). Modellierung des natürlichen Selbstreinigungsvermögens des Bodens. *Zbl. Mikrobiol.* **138**, 595-604.
21. Barth, H., L'Heremite P. (eds.) (1987). *Scientific basis for soil protection in the European Community*, London, Elsevier.
22. Howard, P.J.A. (1993). Soil protection and soil quality assessment in the EC. *Sci. Tot.Environ*, **129**, 219-239.
23. Thormann A. (1984). Bodenschutz - Teil einer vorsorgerden Umweltpolitik. *Z. Kulturtechnik und Flurbereinigung* **25**, 195-202.
24. Arshad., M.A., Coen, G.M. (1992) Characterization of soil quality. Physical and chemical criteria. *Am. J. Alter. Agric.* **7**, 25-31.
25. Karlen, D.L., Stott, D.E. (1994). A framework for evaluating physical and chemical indicators of soil quality, in J.W. Doran, D.C. Colleman, D.F. Bezdicek, and B.A. Steward (eds.), *Soil Quality for a Sustainable Environment*, Madison, Wisconsin, Soil Sci. Soc. Amer. Inc., pp. 53-72.
26. Chaussod, R. (1996) La qualité biologique des sols: évaluation et implications. *Etude et Gestion des Sols* **3**, 261-277.
27. Linden, D.R., Hendrix, P.F., Coleman, D.C., Van Vliet, P.C.J. (1994) Faunal indicators of soil quality, in J.W. Doran, D.C. Colleman, D.F. Bezdicek, and B.A. Steward (eds.), *Soil Quality for a Sustainable Environment*, Madison, Wisconsin, Soil Sci. Soc. Amer. Inc., pp. 91-106.
28. Torstenson, L. (1997) Microbial assays in soil, in J Taradelles et al. (eds.), *Soil Toxicity*, CRC Lewis Publ., pp. 207-234.
29. Turco, R.F., Kennedy, A.C., Jawson, M.D. (1994) Microbial indicators of soil quality, in J.W. Doran et al. (eds.), *Soil Quality for a Sustainable Environment*, Madison, Wisconsin, Soil Sci. Soc. Amer. Inc., pp. 73-90.
30. Visser, S., Parkinson, D. (1992). Soil biological criteria as indicators of soil quality: Soil microorganisms. *Am. J. Alter.Agric.* **7**, 33-37.
31. Vortragstagung 1996. Neue Konzepte in der Bodenbiologie. *Mitt. Dtsch. Bodenkd. Ges.* **81**,Oldenburg, DBG, 384 pp.
32. Pankhurst, C.E. (1997) Biodiversity of soil organisms as a indicator of soil health, in Pankhurst, C.E. et al. (eds.), *Biological Indicators of Soil health*, CAB Int., Wallingford, 297-324.
33. Torsvik, V., Sorheim, R., Goksoyr, J. (1996) Total bacterial diversity in soil and sediment communities: a review. *J. Ind. Microbiol. Biotechnol.* **17**, 170-178.

34. Dobson, A.P., Bradshaw, A.D., Baker, A.J.M. (1997). Hopes for the future : Restoration ecology and conservation biology. *Science* **277**, 1997: 515-521.
35. Bertram, H.G. (1991) Carbon dioxide and climate: climatic impact of soil-borne CO_2 , in J. Berthelin (ed.), *Diversity of Environmental Biogeochemistry* , Amsterdam, Elsevier, 391-395.
36. Titljanova,A.A., Tesarova, M. (1991) Natural systems of biological cycles. Novosibirsk, Nauka, 148 pp (in Russian).
37. Paustian, K., Andren, O., Cearholm, M., hansson, A.C., Johannson, G., Lagerlof, J., Lindberg, T., Peterson, R., Soblenium, B. 1990. Carbon and nitrogen budgets of four agro-ecosystems with annual and perrenial crops with and without N fertilization. *J. Appl. Ecol.* **27**, 60-84.
38. Curry, J.P., Good, J.A. (1992) Soil faunal degradation and restoration, in R. Lal and B.A. Steward (eds.), *Advances in Soil Science,* Vol. 17 , New York, Springer Verlag, 171-215.
39. Norby, Z. (1997) Inside the black box. *Nature* **388**, 522-523.
40. Vitousek, P.M., Aber, D.J., Howard, R.W., Likens, G.E., Matson, P.A., Schindler, D.W., Schlesinger, W.H., Tilman, D.G. (1997) Human alteration of the global nitrogen cycle: sources and consequences. *Ecol. Applications* **7**, 737-750.
41. Banin, A.: Global budget of N_2O (1997) The role of soils and their changes. *Sci. Tot. Environ.* **55**, 27-38.
42. Bianchi, A., Bianchi, M. (1995) Bacterial diversity and ecosystem maintenance, in D. Allsop., R.R. Colwell., and D.L. Hawksworth (eds.), *Microbial Diversity and Ecosystem Function*, CAB International, 185-198.
43. Nikitin, D.I., Kunc, F. (1988) Structure of microbial soil associations and some mechanisms of their autoregulation, in V. Vancura and F. Kunc (eds.), *Soil Microbial Associations*, Amsterdam, Elsevier, 157-190.
44. Kas, V. (1966) *Mikroorganismen im Boden*. Wittengerg Lutherstadt, A. Ziemsen Verlag, 208 pp.
45. Filip, Z. (1995) Einfluss chemischer Kontaminanten (insbesondere Schwermetalle) auf die Bodenorganismen und ihre ökologisch bedeutenden Aktivitäten. *UWSF-Z. Umweltchem. Ökotox.* **7**, 92-102.
46. Franzius, V., Bachmann, G. (1999) Soil protection policy in Germany, in R.A. Kraemer et al. (eds.), *Soil Protection Policies within the European Union*. Fed. Ministry for the Environment, Nature Conservation and Nuclear Safety, Bonn, 185-194.
47. Römbke, J., Moltmann, J.F. (1996) *Applied Ecotoxicology*. Boca Raton, CRC Lewis Publ., 282 pp.
48. Graefe, U. (1997) Von der Spezies zum Okosystem: Der Bewertungsschritt bei der bodenbiologischen Diagnose. *Abh. Ber. Naturkundemus-Görlitz* **69**, 45-53.
49. Yakovlev, A.S. (1997) Biological diagnostics of natural and anthropogenically affected soils, Sc.D. Thesis, Moscow State Univ., Moscow, 56 pp (in Russian).
50. Debus, R., Hund, K. (1997) Ecotoxicological tests for effect assessment of bioavailable portion of soil, in M. Herrchen, R. Debus, and R. Pramanik-Strehlow (eds.), *Bioavailability as a Key Property in Terrestrial Ecotoxicity Assessment and Evaluation* , Stuttgart, Fraunhofer IRB Verlag, 97-107.
51. Hund, K. (1994) *Entwicklung von biologischen Testsystemen zur Kennzeichnung der Bodenqualität.* Texte 45/94, Berlin, Umweltbundesamt, 88 pp.
52. Page, A.L., Miller, R.M., Kenny, D.R. (eds.) 1982. *Methods of Soil Analysis. part 2*, Madison, Wisconsin Amer. Soc. Agron., 1159 pp.

53. Dahlin, S., Witter, E., Mortensson, A., Turner, A., Baath, E. (1997) Where is the limit ? Changes in the microbiological properties of agricultural soils at low levels of metal contamination. *Soil Biol. Biochem.* **29**, 1405-1415.
54. Robertson, G.P., Klingensmith, K.M.. Klug, M.J., Faul, E.A., Crum, J.R., Ellis, B.G. (1997) Soil resources, microbial activity and primary production across an agricultural ecosystem. *Ecol. Applications* **7**, 158-170.
55. Halvorson, J.J., Smith, J.L., Papendick, R.I. (1996) Integration of multiple soil parameters to evaluate soil quality : a field example. *Biol. Fertil. Soils* **21**, 207-214.
56. Smith, J.L., Halvorson, J.J., Papendick, R.I. (1993) Using multiple-variable indicator kriging for evaluating soil quality. *Soil Sci. Soc. Am. J.* **57**, 743-749.
57. Trazar-cepeda, C., Lerio, C., Gil-Sotres, F., Scoane, S. (1998) Towards a biochemical quality index for soils: An expression relating several biological and biochemical properties. *Biol. Fertil. Soils* **26**, 100-106.
58. Bruggen van, A.H.C., Semenov, A.M. (2000) In search of biological indicators for soil health and disease suppression. *Appl. Soil Ecol.* **15**,13-24.
59. Brohon, B., Delolme, C., Gourdon, R. (2001) Complementarity of bioassays and microbial activity measurements for the evaluation of hydrocarbon-contaminated soils quality. *Soil Biol.Biochem.* **33**, 883-891.

ADVANCES IN PHYTOREMEDIATION: PHYTOTRANSFORMATION

CLAUDIA BOCK[1], MARIT KOLB[1], MARIA BOKERN[1], HANS HARMS[1], MARTINA MACKOVA[2], LUDMILA CHROMA[2], THOMAS MACEK[3], JOSEPH HUGHES[4], CRAIG JUST[5] and JERALD SCHNOOR[5]

[1]*Institute of Plant Nutrition and Soil Science, Federal Agricultural Research Center (FAL), Bundesallee 50, D-38116 Braunschweig, Germany*

2Dept. of Biochemistry and Microbiology, Fac. of Food and Biochemical Technology, ICT Prague, Technicka 3, 16628 Prague, Czech Republic

[3]*Inst. of Organic Chemistry and Biochemistry, CAS, Flemingovo n. 2, 16610 Prague, Czech Republic*

[4]*Environmental Science & Engineering, Rice University, 6100 Main Street, MS 317,Houston, TX 77005-1892 USA*

[5] *Department of Civil and Environmental Engineering, 4105 Seamans Center University of Iowa, Iowa City, IA 52242 USA*

1. Introduction

Phytoremediation is the use of vegetation for in-situ treatment of contaminated soils, sediments, and water. It is applicable at sites containing organic, nutrient, or metal pollutants that can be accessed by the roots of plants and sequestered, degraded, immobilized, or metabolized in place. In the last few years, a greater understanding has been achieved regarding the uptake and metabolism of organic xenobiotic chemicals by plants, especially chlorinated solvents, petrochemicals, some pesticides, and explosives [1-8]. In addition, inorganics (nutrients, selenium and arsenic) and metals (lead, cadmium, nickel and zinc) have been successfully remediated using plants. These are toxic chemicals that contaminate a large number of hazardous waste sites. This chapter focuses on advancements in phytoremediation, especially with respect to phytotransformation, the enzymatic conversion of organic chemicals within plant tissues following plant uptake.

Phytoremediation is popular because of its cost effectiveness, aesthetic advantages, and long-term applicability [1]. Applications include hazardous waste sites where other methods of treatment are too expensive or impractical, low-level contaminated sites where only "polishing treatment" is required over long periods of time, and sites where phytoremediation is used in conjunction with other technologies as a final cap and closure. Limitations of the technology include the potential for introducing the contaminant or its metabolites into the food chain, long times required for clean-up to below action levels, and toxicity encountered in establishing and maintaining vegetation at waste sites.

Plants have shown the capacity to withstand relatively high concentrations of metals or organic chemicals without toxicity [5, 9]. Also, they can uptake and transform organic chemicals to less toxic metabolites in some cases [3, 10-13]. Plants can stimulate the degradation of organic chemicals in the rhizosphere by the release of root exudates, enzymes, and the build-up of organic carbon in the soil [14, 15].

For metal contaminants, phytoextraction has been utilized (i.e., uptake and recovery of metals into above-ground biomass at waste sites). Filtering metals from water onto root

D. Reible and K. Demnerova (eds.), Innovative Approaches to the On-Site Assessment and Remediation of Contaminated Sites, 115–140.

systems has also been successful [16], and stabilizing wastes by hydraulic control (phytostabilization) is a widely recognized strategy [16-18].

Publication of research articles on phytoremediation has been enormous in recent years. Today, there are hundreds of waste sites where phytoremediation has been applied, usually in conjunction with other technologies. Much information is available on the Internet, and Table 1 provides a summary of phytoremediation applications, contaminants, and vegetation that has been utilized.

2. Uptake and Transformation of Organics

Organic chemicals may undergo root sorption, uptake, translocation, metabolic transformation, and/or volatilization. Figure 1 is a schematic of a typical contaminant treated using phytoremediation, trichloroethylene (TCE). The Figure shows how chemicals can be transported from soil to roots, translocated by the plant through the roots, stem and leaves (xylem or sap flow), and ultimately transformed by plant tissues or volatilized to the atmosphere. The first step is sorption to roots. When chemical contaminants in soil water or groundwater come into contact with roots, they may sorb or bind to the root structure and cell walls. Hemi-cellulose in the cell wall and the lipid bilayer of plant membranes can bind hydrophobic organic chemicals strongly. Such sorption may be reversible, and it can be measured using standard sorption isotherms. If the sorption is not reversible, the contaminant has undergone a chemical or biological reaction at the root surface, a type of phytotransformation which may lead to bound residue formation.

TABLE 1. Applications of phytoremediation.

Application	Description	Contaminants	Types of Plants
Soils			
Phyto-transformation	Sorption, uptake, and transformation of contaminants	Organics, including nitroaromatics and chlorinated aliphatics	Trees and grasses
Rhizosphere Biodegradation	Microbial biodegradation in the rhizosphere stimulated by plants	Organics; e.g., PAHs, petroleum hydrocarbons, TNT, pesticides	Grasses, alfalfa, many other species including trees
Phytostabilization	Stabilization of contaminants by binding, holding soils, and/or decreased leaching	Metals, organics	Various plants with deep or fibrous root systems
Phytoextraction	Uptake of contaminants from soil into roots or harvestable shoots	Metals, inorganics, radionuclides	Variety of natural and selected hyperaccumulators, e.g., *Thalaspi, Alyssum, Brassica*
Water/Groundwater			
Rhizofiltration	Sorption of contaminants from aqueous solutions onto or into roots	Metals, radionuclides, hydrophobic organics	Aquatic plants, (e.g., duckweed, pennywort), also *Brassica,* sunflower
Hydraulic Control Plume Capture/ Phytotransp./Phyto-volatilization	Removal of large volumes of water from aquifers by trees	Inorganics, nutrients, chlorinated solvents	Poplar, willow trees
Vegetative Caps	Use of plants to retard leaching of hazardous compounds from landfills	Organics, inorganics, wastewater, landfill leachate	Trees such as poplar, plants (e.g., alfalfa) and grasses
Constructed Wetlands	Use of plants as part of a constructed ecosystem to remediate contaminants from aqueous wastestreams	Metals, acid mine drainage, industrial and municipal wastewater	Free-floating, emergent, or submergent vegetation; reeds, cattails, bamboo

2.1. UPTAKE AND TRANSLOCATION

Organic chemicals can be uptaken by roots and translocated to the shoots and leaves through the xylem. Direct uptake of organic chemicals by plants is important for shallow contaminated sites with hydrophilic and moderately hydrophobic organic chemicals (octanol-water partition coefficients, log K_{ow} = 1.0-3.5). This includes most BTEX chemicals, chlorinated solvents, and short-chain aliphatic chemicals. Hydrophobic chemicals (log $K_{ow} > 3.5$) are bound so strongly to soils and roots that they cannot be easily translocated within the plant. Such chemicals are candidates for phytostabilization or rhizosphere bioremediation. Chemicals that are quite water soluble (log $K_{ow} < 1.0$) may not be sufficiently uptaken by roots nor actively transported through plant membranes [9].

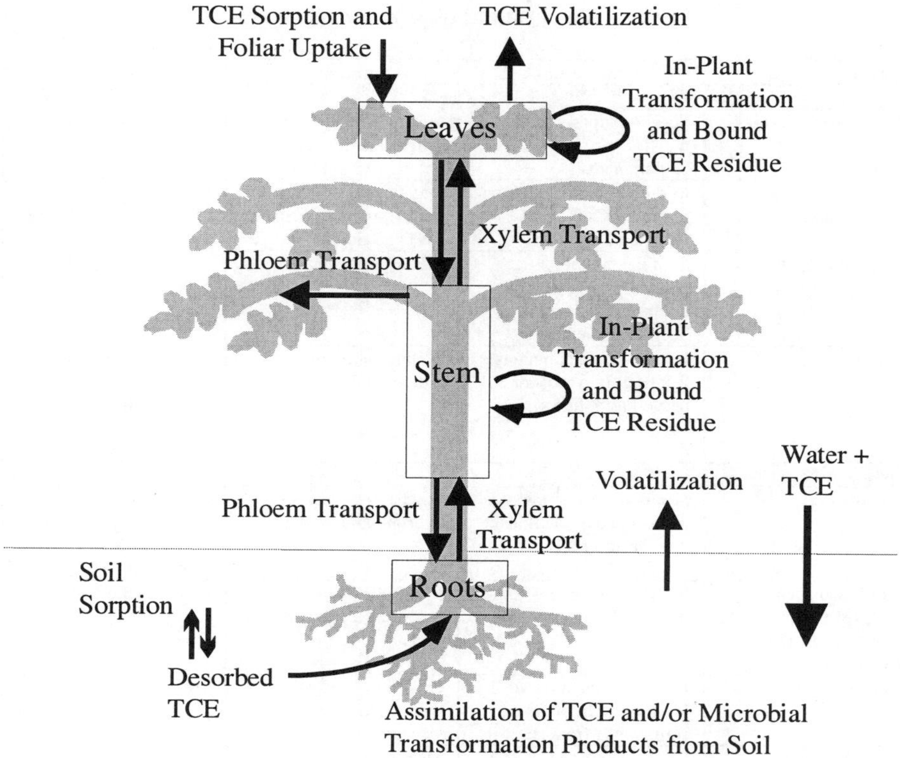

Figure 1. Potential uptake and transformation pathways of trichloroethylene in a plant-soil system [19].

The direct uptake of a chemical into a rooted vascular plant depends on the uptake efficiency, transpiration rate, and the concentration of chemical in soil water as shown in Equation 1 [5].

$$U = (TSCF)\,(T)\,(C) \tag{1}$$

Where
U = the rate of chemical uptake by plant, mg/day
TSCF = efficiency of uptake, dimensionless
T = transpiration rate, L/day
C = soil water concentration of chemical, mg/L

Uptake efficiency for rooted vascular plants is termed the Transpiration Stream Concentration Factor (TSCF). TSCF is the dimensionless ratio between the concentration of chemical in the transpiration stream of the plant to the concentration in soil water. Uptake efficiency or TSCF depends on physical-chemical properties of the contaminant, chemical speciation, and the nature of the plant itself. Some measured values for TSCF appear in Table 2. TSCF can vary from zero (no uptake) to complete

TABLE 2. Measured Transpiration Stream Concentration Factor (TSCF) and Root Concentration Factor (RCF) for some typical contaminants and physical-chemical properties [5, 6, 15].

Chemical	log K_{ow}	Solubility $-\log C_w^{sat}$ @25°C, (mol/L)	Henry's Constant K_H @25°C (dimensionless)	Vapor Pressure $-\log P^o$ @25°C (atm)	Transpiration Stream Conc. Factor (TSCF)	Root Conc. Factor, (RCF) (L/kg)
Benzene	2.13	1.64	0.2250	0.90	0.82	1
Toluene	2.69	2.25	0.2760	1.42	0.81	3
Ethylbenzene	3.15	2.80	0.3240	1.90	0.80	2
m-Xylene	3.20	2.77	0.2520	1.98	0.78	11
TCE	2.33	2.04	0.4370	1.01	0.75	3
Aniline#	0.90	0.41	2.2×10^{-5}	2.89	0.32	420
Nitrobenzene	1.83	1.77	0.0025	3.68	0.82	3
Phenol*	1.45	0.20	$>1.0 \times 10^{-5}$	3.59	0.48	12
Pentachlorophenol	5.04	4.27	1.5×10^{-4}	6.75	0.04	30
Atrazine	2.69	3.81	1×10^{-7}	9.40	0.57	8
1,2,4-Trichlorobenzene	4.25	3.65	0.1130	3.21	0.04	19
1,4-Dioxane	-0.27	Miscible	2.0×10^{-4}	0.05	0.72	<1
Methyl-*tert*-butyl ether	1.1	0.36	0.56	0.49	0.65	<1
TNT	1.90	3.36	-	-	0.46	49
RDX	0.87	3.57	-	-	0.16	1.3
HMX	0.19	4.77	-	-	0.21	5.6
	Physical chemical properties [19]				Measured data from hydroponic studies with hybrid poplars [5, 19]	

pKa = 4.87, test conducted at pH 6.8
* pKa = 9.99, test conducted at pH 5.8

uptake, 1.0 (i.e., the chemical is translocated at the same concentration present in soil water). Chemicals which undergo biochemical reaction at the root-water interface do not follow the above relationship because, in this case, uptake is determined by site binding and biochemistry, and not by the rate of passage through membranes into the transpiration stream. Transpiration rate (T) is a key variable that determines the rate of chemical uptake. It depends on the plant type, leaf area, nutrient availability, soil moisture, temperature, wind conditions, and relative humidity.

Transpiration Stream Concentration Factors have been measured for herbicide related chemicals (such as substituted phenylureas and o-methylcarbamoyloximes) with crop species (barley), and for xenobiotic chemicals typically found at waste sites with hybrid poplar cuttings [5, 9]. Equations 2 and 3 [5, 9] predict a maximum uptake for chemicals in the moderately hydrophobic range (log K_{ow} = 1-3.5).

$$\log TSCF = 0.756 \exp\{-(\log K_{ow} - 2.50)^2 / 2.58\} \tag{2}$$
$$\log TSCF = 0.784 \exp\{-(\log K_{ow} - 1.78)^2 / 2.44\} \tag{3}$$

Recent developments have indicated that neutral, water soluble chemicals (log K_{ow} < 1.0) can sometimes be taken-up by rooted vascular plants. Chemicals such as 1,4-dioxane, methyl-tert-butyl ether (MTBE), and ethylene glycol can be taken-up by rooted vascular plants, perhaps via hydrogen bonding with water during transpiration.

2.2. ENZYMATIC PHYTOTRANSFORMATION

Phytotransformation refers to the uptake of organic contaminants from soil and groundwater and the subsequent metabolism or transformation by plants. Although certain organic compounds are stable in the biochemical environment of the plant, most organic pollutants will undergo biotransformation in plant tissues [20] as a part of a three stage metabolic strategy comparable to that which occurs animals [21, 22]. Once an organic chemical is taken-up and translocated, it undergoes one or more phases of transformation [11].

Phase I – Conversion: oxidations, reductions, hydrolysis
Phase II – Conjugation: with glutathione, sugars, amino acids
Phase III—Compartmentation: conjugates from phase II are converted to other conjugates and deposited in plant vacuoles or bound to cell wall material (hemi-cellulose or lignin).

Phase III conjugates are sometimes termed "bound residues" because of the ineffectiveness of extraction by chemical methods. These conjugates are likely covalently bound to stable tissues in the plant. But one concern is whether under different conditions, such as the gut of a worm or snail or butterfly, there could exist lignases or other enzymes capable of cleaving covalent bonds and liberating the parent compound or toxic conjugate from the plant tissue.

Plant metabolism of organic pollutants often varies markedly from metabolic processes carried out be bacteria and fungi. For example, aromatic hydrocarbon degradation in microorganisms commonly results in mineralization, while aromatic moieties are very stable in plants [23], generally becoming irreversibly bound within plant tissue over time. This is referred to as "plant detoxification" in agronomic studies [24]. The "Green Liver" model has been studied in certain plant species with a range of organic compounds – although most information has focused on plant species and chemicals used in agriculture. The fate of many important environmental contaminants in the wide range of terrestrial and aquatic plant species that exist in nature is largely unexplored. However, the existing body of information from agronomic studies is an excellent resource that can be used as a basis for understanding metabolic processes in plants that may influence the fate of organic pollutants in plants.

2.2.1. *Conversion*

Conversion and transformation reactions constitute the first phase of plant xenobiotic metabolism and include oxidation, reduction, and hydrolysis reactions. Oxidation is the most often observed [20], while certain organics, such as nitroaromatics, tend to favor reduction processes [4]. Initial transformation reactions within plants are most often enzymatic attacks on common functional groups (nitro, carboxyl, alkyl or halogens). Enzymes that catalyze initial transformations are often thought to be constitutive with broad specificity [23], although recent research indicates that several of the enzymes involved in xenobiotic metabolism are extremely substrate-specific [22]. Transformation reactions often result in less toxic compounds, with increased polarity, that are amenable to subsequent conjugation reactions in Phase II [20].

Oxidation is the primary reaction of transformation and is most often mediated by mixed functional oxidases and peroxidases. Mixed functional oxidases (MFO's) are usually induced enzyme systems [20] that catalyze substrate oxidation with molecular oxygen. The system is mediated by members of the cytochrome P-450 family that have been widely investigated in mammalian liver metabolism [22]. The most significant reactions mediated by plant MFO's include aliphatic hydroxylation, deamination, N-dealkylation, N-dealkoxylation, ring hydroxylation, and N- and S-oxidation [23]. Peroxidase enzymes, commonly found in plants, catalyze two types of reactions, each resulting in hydroxylated products. One oxidative reaction of peroxidases results from the formation of hydroxyl radicals by cleavage of hydrogen peroxide. The other involves molecular oxygen similar to MFO's. Plants high in peroxidase have been investigated in the removal of chlorinated phenols, such as 2,4-dichlorophenol, from wastewater [25, 26].

Hydrolytic reactions have been documented for some classes of herbicides. Ester, amide and nitrile groups are the most common functional groups subject to hydrolytic attack. Studies of the metabolism of 3',4'-dichloropropionanilide (DPA) in rice plants revealed high levels of a hydrolytic enzyme in resistant plants [27]. In rice plants, DPA is hydrolyzed to 3,4-dichloroaniline (with the production of propionic acid and lactic acid) that is then more slowly complexed with lignin in the plant tissues. Also the hydrolysis of chloro-s-triazine herbicides is thought to be the mechanism of deactivation of atrazine in corn [20].

Reduction reactions are relatively rare compared with oxidation and hydrolysis, and the enzyme systems are less well characterized; however, nitroaromatics are known to be reduced to a corresponding amine [28]. For example, Höhl and Barz [29] report an "unknown reduction process to the primary amine" as part of the metabolic pathway of the insecticide phoxim in soybean cell suspension cultures, and aryl nitro reduction is known to occur in nitro-substituted diphenyl ether pesticides [20]. Interestingly, in this second case, in vitro studies with isolated enzymes required anaerobic conditions for the reaction to proceed.

2.2.2. *Conjugation*

Following transformation, the second phase of xenobiotic metabolism occurs. This process, known as conjugation, involves the covalent linkage of the transformed xenobiotic and an endogenous plant molecule; usually a sugar, an amino acids, glutathione (or other peptide), or a cellular macromolecule (e.g., lignin, hemicellulose, protein, and pectin). Substrate conjugation with sugars, amino acids, and glutathione generally form polar and water soluble products. In contrast, reactions with macromolecules generally result in insoluble biopolymers often classified as "bound residues" [21]. The formation of insoluble, non-extractable bound components is difficult to distinguish from processes that occur in Phase III compartmentalization reactions, and often are considered together. Most conjugates display significantly different chemical properties from those of the parent molecule and for this reason can be difficult to isolate, characterize and identify.

Investigations into a variety of conjugation reactions have been carried out using heterotrophic plant-cell suspension cultures. Several reviews of these studies [21, 28, 30] along with standardized procedures [30-33] to evaluate this process have been published. In particular, cell-suspensions of wheat and soybean species have been widely used to analyze the production of bound residues and soluble metabolites of several pesticides. Many of the specific pathways of xenobiotic metabolism as well as enzyme and terminal residue identification have been elucidated using such systems. These systems present many advantages in the study of metabolic products as cell cultures provide a bacteria-free environment where the biochemistry of plant cells can be examined in isolation. In addition, cell cultures do not exhibit the degree of Phase III compartmentalization found in intact plants [29].

Unlike secondary metabolism in the mammalian liver, which favors formation of D-glucuronyl conjugates, plants tend to form glucose conjugates [22] and several uridine diphosphate (UDP)-glucose dependent glucosyltransferases have been isolated from a variety of plants [22]. Herbicides with phenolic, N-arylamine or carboxylic acid functional groups (either inherent or arising from Phase I metabolism) are frequently conjugated as β-glycosides.

These compounds include O-glucosides, N-glucosides, glucose esters and polysaccharides [23].

Following reduction of aryl nitro groups to the corresponding amine, N-glucosylation is frequently observed. The products being of N-glucosylation are relatively non-toxic and stable. In several cases, herbicide resistance is correlated with the formation of N-glucosides of the parent compound [28]. In the late 1960's, Frear isolated a soluble enzyme system from soybean that catalyzed the formation of an N-glucosylarylamine from the corresponding arylamine [34]. Other investigations have succeeded in isolating and characterizing N-glucosyl transferases that specifically catalyze the conjugation of 3,4-dichloroaniline in cell culture [35, 36], excised plant tissues [37], and intact plants [35].

Malonic acid conjugation reactions in plant systems often compete with glycoside conjugations and may be favored in the case of aromatic amine groups [28]. In plants, where D-amino acids are not naturally occurring, N-malonyl conjugates are possibly formed from aromatic amines, presumably as a defense mechanism [28]. Malonic acid conjugates of xenobiotic compounds have been observed as a dominant mechanism for detoxification of 3,4-dichloroaniline [29, 35, 36-39]. Several N-malonyltransferase enzymes that utilize malonyl CoA have been isolated and characterized from a variety of sources [36, 39].

Frequently, N-glucosylation and N-malonylation are competing reactions in the metabolism of anilines. Bockers et al. [38] report the presence of both glucose and malonyl conjugates of 3,4-dichloroaniline in exposed soybean and wheat plants. The ratio of these metabolites varies according to species, tissue type, and time exposed. The researchers also reported the presence of mal-O(6')Glc-β-3,4-dichloroaniline in several tissues, indicating a malonyl conjugation subsequent to glucosylation.

Another class of conjugation involves reaction with the tripeptide glutathione or simple amino acids. Of the two, glutathione conjugation is a far more common process [23, 28]. Glutathione is an ubiquitous molecule in most living systems and has the structure of γ-Glu-Cys-Gly (homoglutathione (γ-Glu-Cys-Ala) is the dominant thiol molecule in some legumes and functions in much the same manner [20]. In conjugation reactions, the sulfhydryl group of cysteine is the reactive site. Although conjugation can proceed spontaneously, xenobiotic conjugation is catalyzed by glutathione-S-transferase (GST) enzymes [23]. Classes of chemicals prone to glutathione conjugation include those with a hydrophobic region and an electrophilic center prone to nucleophilic attack. The electrophilic region can be formed by modification of functional groups on the parent compound during Phase I metabolism. Studies indicate that glutathione transferase enzymes can be induced in certain plant species increasing plant resistance to certain herbicides [28]. Specific examples of compounds subject to glutathione conjugation include atrazine and other triazines, diphenyl ethers, 1-chloro-2,4-dinitrobenzene, carbomathioate sulfoxides and chloroacetamides [20, 28]. Amino acid conjugation is not as common as glutathione conjugation and the enzyme systems are less well characterized, although research reveals reactions such as the conjugation of 2,4-dichlorophenoxyacetic acid (2,4-D) with aspartate [28].

A large number of xenobiotics are metabolized to a group of entities collectively known as bound or terminal residues. This nomenclature arises from the observation that complete recovery of applied ^{14}C-radiolabeled chemicals is almost never seen in intact plants. Unextractable ^{14}C-label is frequently associated with plant structural tissues, covalently bound to lignins, pectins and other high molecular weight biopolymers. These residues are distinguished from other conjugates in that they are less polar and more insoluble in aqueous solution. Parent compounds that form bound residues have aromatic or heterocyclic ring structures which may be incorporated into plant tissue immediately or after Phase I and Phase II reactions [23]. Almost uniformly, the percent of unextractable parent compound in plant tissues increases with time [27, 37, 38, 40, 41]. Isolation of conjugated biopolymers can be difficult, but researchers have been successful at isolating lignin conjugates for further characterization [42]. Bound residues are generally thought to be terminal products of xenobiotic metabolism and there has been no indication of bioavailability or residual ecotoxicity upon animal consumption of these products [23]. It is important to note that structural tissues can be recycled and reutilized by the plant under certain circumstances, leading to the question of liberation of the parent compound by catabolic enzymes.

2.2.3. *Compartmentation*

Plants do not break down conjugation products to extract energy or carbon for metabolism, instead they store and compartmentalize metabolized xenobiotics [24]. This process removes the transformed xenobiotic from the area of active cell metabolism, such that they do not interfere with ongoing plant processes (hence the term "plant detoxification"). Excretion refers to the transport of materials out of the plant; compartmentalization includes storage in vacuoles or binding to lignin. Because plants lack a well-defined excretory system, the excretion of metabolites becomes problematic to anticipate/measure, but may be an important fate of certain hazardous organic compounds, including the volatilization of certain pollutants from plant leaves. The purpose of conjugation and excretion is to transform the compound and remove it from the plant in a manner where it is not readily reassimilated. Thus, excreted conjugates may not be subject to subsequent re-uptake. Alternatively, bound residues (from Phase II or Phase III) "irreversibly" incorporate metabolites into the plant structure.

3. Rhizosphere Biodegradation

In the soil-rhizosphere, plants can facilitate microbial biodegradation by providing a stable, nutrient rich environment for microorganisms by supplying growth substrates, oxygen, and cycling nutrients with high efficiency [2, 14, 43]. A similar cooperation may exist in aquatic systems, but very little is known about the contributions of microbial transformation of hazardous organics in aquatic plant systems. The considerable study of terrestrial plant/rhizosphere interactions has shown that the contribution of microorganisms can be an important factor in contaminant fate [44-46]. For example, significant evidence exists indicating indigenous bacteria in the rhizosphere degrade a wide variety of organic compounds including aliphatic, aromatic and heterocyclic contaminants [2]. Experiments with atrazine [47] and 2,4-D [48] indicated that the microbial biodegradation of these compounds is a significant fate process in the soil rhizosphere. More recently, the unique contribution of rhizosphere microorganisms to the degradation of non-agricultural xenobiotics has been investigated. Rhizosphere microbes have been found to enhance degradation of trichloroethene [49] and pentachlorophenol [50], and to a lesser extent, polycyclic aromatic hydrocarbons [51]. Similar evidence exists for the role of microrizae fungi in the degradation of xenobiotics. Fungal mediated transformation processes can result in a covalently and irreversibly bound residue in plant/soil systems [26, 52]. Fungi have also been investigated in the degradation of complex lignin conjugates formed from xenobiotic metabolism in plants [42].

While studies of the degradation of xenobiotics by aquatic plant-associated microbes are not found in the literature, their potential participation must be considered. Presumably the processes of uptake competes with microbial degradation, yielding multiple pathways of contaminant metabolism (i.e., microbial- and plant-derived). Depending upon the nature of the contaminant and the plant species present, the extent of either process could vary markedly.

4. Modeling Plant-Contaminant Interactions

Investigations of the transport of organic chemicals within terrestrial plants has resulted in several mathematical models to predict the fate of organic compounds in plant tissues [9, 53-55]. Since the active uptake of any anthropogenic chemical into plants has yet to be documented, uptake is described by contact of the contaminant with plant tissue followed by diffusion and/or partitioning (where the plant mimics an organic phase) into the plant. Summaries of mathematical expressions developed to interpret the uptake of various xenobiotics by plants are presented by Ryan et al. [56] and Trapp [24]. Subsequent transport within the plant is evaluated by the advective water flow in the transpiration stream coupled with retardation coefficients. With this approach, the distribution of chemicals within the plant has been predicted by the physiochemical properties of the compound, in particular its hydrophobic partitioning properties [24, 53, 57]. The sorption of contaminants to plant roots has been explained by a linear isotherm often called the Root Concentration Factor (RCF). The transport from roots to apical sections is defined as the Transpiration Stream Concentration Factor (TSCF) [24, 58, 59].

Briggs et al. [9] proposed models to estimate RCF and TSCF in barley plants (Hordeum vulgare) based on the octanol/water partition coefficient of organic compounds, and also defined a Stem Concentration Factor (SCF) to describe the translocation of compounds within the stem. From these initial descriptions, they model a plant as a series of interconnected compartments into and through which chemicals pass. The validity of Briggs' models as generalized estimates for plant contamination is reviewed in Polder et al. [60]. In addition, Trapp [55, 61, 62] expanded and modified Briggs' model to evaluate the transport of chemicals into and within the plant. The generalized plant model [62] describes the flux and mass transfer for compartments which include exchange between soil and roots, advective transport into roots, stems and leaves in the transpiration stream, diffusion into stem tissues and movement into fruits. The model also considered loss of volatile chemicals through leaf stomata and cuticle and dilution of the chemical due to biomass increase. Trapp includes a first order reaction rate constant to describe metabolism. The model was tested against model systems of carbamates [61] and bromacil [62] with encouraging results. Lindstrom [54], Boersma [63], and McFarlane [64] have developed a similar model to describe the soil/plant system as a series of compartments through which the compound is transported.

5. Cell Suspension Cultures for Understanding Metabolism of PCB, PAH, and Nonylphenol by Plants

In order to be certain that plants can actually metabolize xenobiotic organic chemicals, standardized test methods have been developed. Besides tests with intact plants under septic and aseptic conditions, in vitro techniques, such as plant cell suspension cultures or differentiated root cultures, have been used for studying the metabolic fate and persistence of various xenobiotic compounds like polychlorinated biphenyls (PCBs), polycyclic aromatic hydrocarbons (PAHs), and the surfactant degradation product 4-nonylphenol (4-NP).

5.1. UPTAKE AND EXTRACTION FROM PLANTS

A special extraction scheme (Figure 2) was used to investigate the metabolism and to separate applied compounds and metabolites. Generally, the plant material is incubated with the ^{14}C-labeled compound to be investigated. After the incubation, the media is separated from the cells by filtration. The cells are then destroyed by ultrasonic probe or with a mixer and extracted with CH_2Cl_2/MeOH using the Bligh-Dyer-Extraction (BD-extraction). After filtration and separation, non-extractable residues and a one-phase extract are received. In a second step of the BD-extraction, CH_2Cl_2/H_2O is added to the extract to get a two-phase-system with a more polar MeOH/H_2O-phase, which contains the metabolite-conjugates, and a less polar CH_2Cl_2-phase, which contains the applied compound. The metabolism rate is estimated by the amount of radioactivity in the MeOH/H_2O-phase. The same separation is possible with the medium if it contains a high amount of radioactivity.

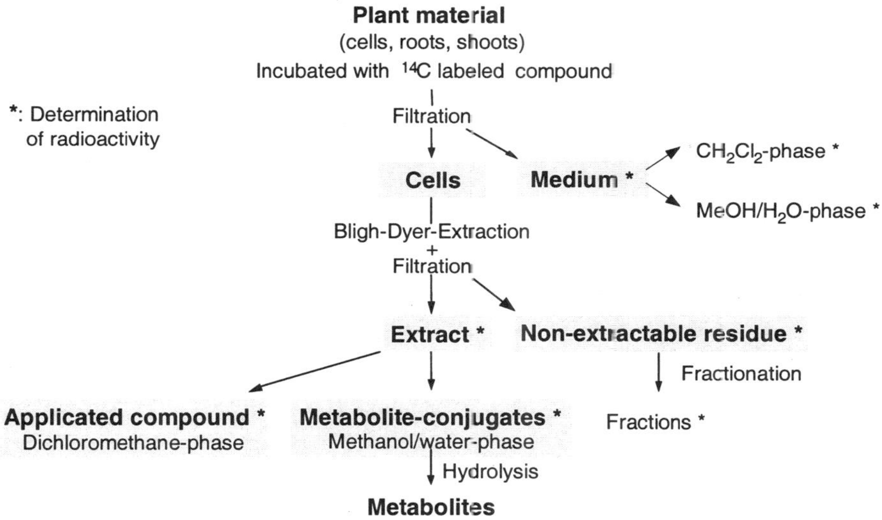

Figure 2: Extraction scheme for plant material

Metabolism in plants is usually a two-step reaction. An enzymatic or acid-catalyzed hydrolysis will break the linkage of the metabolite-conjugate, resulting in identifiable free hydroxy-metabolites which can be identified. The non-extractable residue could be subject to a special fractionation as well.

5.2 POLYCHLORINATED BIPHENYLS

Polychlorinated biphenyls (PCBs) are industrial compounds, with good dielectric properties, making them useful as an insulating oil in capacitors and transformers. Since they have shown toxic effects in animals and humans (for example Yusho disease) the production of PCBs has stopped in most industrialized countries. But, PCBs remain important pollutants in the environment and are detected in almost every compartment of the global ecosystem, due to their persistence and lipophilic nature.

In a screening test, the metabolism of a mixture of 10 different PCB congeners in cell culture systems of 12 different plant species was investigated [65]. The same mixture was added to dead cell cultures, which aren't able to metabolize the PCBs. The recovery rate after 4 days incubation time was calculated in comparison to dead cells. The decrease in PCBs which are found after incubation can be attributed to metabolism while the PCBs found in dead cells can only be attributed to uptake. The results are shown in Table 3.

TABLE 3. Metabolism of different PCBs in various cell cultures.

Group Free Positions	**PCB** (Ballschmiter no.)		**Water Solubility** (10^{-6} M) 22°C	**log K_{ow}**	**Metabolisn**	
					no. of cultures (total 12)[a]	amount[b]
I Free o/m and m/p	2,2'	(4)	5.4	5.0	4	++
	2,4'	(8)	4.7	5.3	5	+++
	2,2',5	(18)	1.7	5.6	4	+++
	2,4',5	(31)	-	5.8	3	++
II Free o/m	4,4'	(15)	0.3	5.3	0	--
	2,4,4'	(28)	0.7	5.7	2	++
	2,2',4,4'	(47)	0.2	6.4	1	+
III Free m/p	2,2'5,5'	(52)	0.1	6.3	2	++
	2,2'4,4',5	(101)	0.0	6.9	0	--
IV No free o/m or m/p	2,2'4,4'5,5'	(153)	0.0	7.4	0	--

a: cultures which are able to metabolize the PCB, b: highest amount of metabolism which is detectable

Those PCBs which possess free o/m and free m/p positions are metabolized best, while those with no free o/m or free m/p positions are metabolized by only some cultures to a less extent. If these positions are chlorinated, no metabolism is detectable. So the penta- and hexachlorinated PCBs, for example PCB 101 and PCB 153, were not metabolized by any of the tested cell cultures. There seems to be a relationship between the structural and physical properties of PCBs. With lower chlorination grade, a higher possibility for metabolism was observed. Generally higher water solubility and lower log K_{ow} are related to better metabolism. PCB 15 is an exception to this rule. It is not metabolized despite its low chlorination grade and its relatively high water solubility. This can be explained by its planar structure which sterically hinders an attack of enzymes.

5.2.1. *3,3',4,4'-Tetrachlorobiphenyl (PCB 77)*

The metabolism of 3,3',4,4'-tetrachlorobiphenyl (PCB 77), one of the more toxic PCBs, was investigated with ^{14}C-labelled PCB 77 in different plant cell suspension cultures. The extraction scheme mentioned above resulted in the following distribution of radioactivity (Table 4).

TABLE 4. Radioactive distribution after incubation of ^{14}C-PCB 77.

	Rosaceae rose (Paul's Scarlet)	**Solanaceae** tomato	potato	**Compositae** sunflower	salad
Medium	4	5	0	1	1
Cell extract					
$MeOH/H_2O$-phase	16	2	1	6	9
CH_2Cl_2-phase	79	91	98	93	89
Bound residue	0	0	0	0	1
Recovery	99	99	99	100	100

Only a small amount of radioactivity was found in the medium and no bound residues were formed. Most of the activity was found in the cell extract. Paul Scarlet rose was able to metabolize 16% of the applied radioactivity. Tomato and potato metabolized 2% and 1% of the parent PCB 77, while sunflower and salad metabolized 6% and 9% respectively. Similarities in plant families are suggested, but such a small number of tested cultures allows no generalization. After hydrolyzation, the PCB 77 metabolites were cleaned up and separated by HPLC. A chromatographic separation showed [66] that several polar compounds were formed. GC/MS analysis revealed that these metabolites were mono- and dihydroxy-compounds of PCB 77 (Figure 3).

Metabolism depends on the different cell cultures. The main metabolites identified are 2-hydroxy-3,3',4,4'-tetrachlorobiphenyl by rose and tomato, 5-hydroxy-3,3',4,4'-tetrachloro-biphenyl by salad, and furthermore a 6-hydroxy-3,3',4,4'-tetrachlorobiphenyl has been identified in cultures of rose, sunflower, tomato, and salad. As a result, a metabolism pathway via a 2,3-arene oxide intermediate is possible. A comparison of the metabolites in tomato cell cultures with metabolites in aseptically grown plants was done by HPLC. The metabolite pattern is qualitatively similar but quantitatively different.

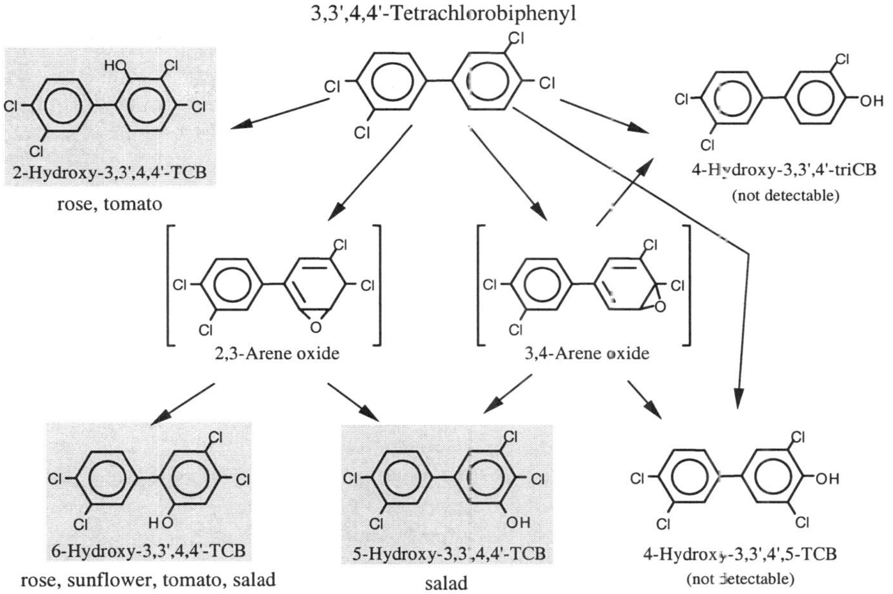

Figure 3. Metabolism of PCB 77 in plants (TCB: Tetrachlorobiphenyl).

5.3. POLYCYCLIC AROMATIC HYDROCARBONS (PAHs)

Polycyclic aromatic hydrocarbons (PAHs) are undoubtedly the most widespread carcinogens in the human environment. Environmental contamination by PAHs can be traced mainly to industrial, household and motor vehicle exhausts or related activities of modern civilization.

5.3.1. *Fluoranthene*

Fluoranthene is one of the most abundant PAHs and has been detected in air, water, soils, sediments and even in biota, including man. For analyzing the metabolism of this hazardous compound, cell suspension cultures of various plant species have been incubated [13]. The results of these experiments are summarized in Table 5.

TABLE 5. Distribution of radioactivity (%) after application of (3-^{14}C)-fluoranthene.

	Rosaceae rose (Paul's Scarlet)	**Solanaceae** tomato	**Poaceae** wheat	barley	**Asteraceae** salad
Medium	27	5	5	4	8
Cell extract					
$MeOH/H_2O$-phase	50	15	9	5	6
CH_2Cl_2-phase	16	72	73	83	67
Bound residue	2	1	2	0	5
Recovery	95	93	89	92	86

Greater amounts of fluoranthene or its metabolites than in case of PCB 77 were found in the medium. Small amounts of radioactivity were found in bound residues, the most in salad with 5%. The highest amounts of radioactivity were found in the cell extracts and significant metabolism amaounts were detected. The cultures of tomato, wheat, barley, salad signigicant metobolism, but rose culture showed a remarkable metabolism of 50%. Fluoranthene showed more complete metabolism than PCB 77.

After hydrolysis with hydrochloric acid and clean-up, the metabolites were subjected to further chromatographic analysis. The pattern seemed to be similar for all tested cultures. GC/MS analysis showed a base peak, which could be attributed to monohydroxylated fluoranthene isomers. As the UV/Vis data of the different fluoranthenols are known from literature [66, 67], HPLC-DAD analyses were performed to confirm the formation of fluoranthenols and to identify the position of the OH-group. The comparison of the UV/Vis spectra with those of reference compounds from literature confirmed that the metabolites being formed in tomato cultures correspond to 8-hydroxyfluoranthene, 1-hydroxyfluoranthene and 3-hydroxyfluoranthene. In cultures of Paul's Scarlet rose, only 8-hydroxy-fluoranthene and 1-hydroxyfluoranthene were detected.

In another experiment, the metabolism of fluoranthene was studied with intact plants. Aseptically grown tomato plants were incubated with ^{14}C-fluoranthene and after 14 days incubation time, the plants were harvested, shoots and roots were separated, homogenized and extracted. The distribution of the applied radioactivity was determined. The highest fraction of the radioactivity (62%), was recovered from the roots. Most of this activity (92%) could be extracted with dichloromethane. Some fraction of that might have been adsorbed at the root surface, especially since the fluoranthene concentration was higher than its water solubility. But, 8% of the radioactivity recovered in the polar methanol/water phase was taken up and metabolized by the roots.

A comparison of HPLC traces of fluoranthene metabolites after hydrolysis formed by tomato cell cultures and roots and shoots of intact plants showed that the qualitative peak pattern were identical. Consequently, the conclusion can be drawn that the fluoranthenols identified in cell cultures are also formed in intact plants. Only the quantitative amounts differed. The monohydroxylated fluoranthenes showed higher amounts of the more polar metabolites in the shoots than in the roots. In the roots, the monohydroxylated fluoranthene isomers dominated the pattern of the metabolites similar to the cell culture.

5.4. SURFACTANT DEGRADATION PRODUCT

5.4.1. *4-Nonylphenol*

4-nonylphenol (4-NP), one of the persistent products in the degradation of alkylphenol-polyethoxylates (non-ionic surfactants), is frequently found in sewage sludges and effluents of sludge treatment. An application of sludge to agricultural land may lead to an uptake and metabolism in crop plants. The compound has been of concern to the public since aquatic toxicity was proven in the 1980's. More recently, discussions on bans or plans to introduce environmental quality standards have been triggered by findings that nonylphenols are weakly estrogenic. The example of nonylphenol accumulating during the degradation of nonylphenolpolyethoxylates shows that the disappearance of a parent compound does not necessarily prove complete degradation and more toxic metabolites may accumulate.

The influence of the plant species on the metabolism of 4-NP was tested with 14 different cell cultures form 7 plant families. The cultures were incubated with 4-NP (^{14}C). All cell cultures took up 4-NP and metabolized it. No $^{14}CO_2$ was detected in the gaseous phase, indicating that plant cells do not mineralize 4-NP to CO_2. Highlights of the results are provided in Table 6.

TABLE 6. Distribution of radioactivity (%) after application of 4-NP to cell suspension cultures.

	tomato	wheat	barley	lettuce
Medium				
$MeOH/H_2O$-phase	6	5[a]	4[a]	16
CH_2Cl_2-phase	0			1
Cell extract				
$MeOH/H_2O$-phase	87	61	53	19
CH_2Cl_2-phase	0	2	20	14
Bound residue	7	16	7	38
Recovery	100	84	84	88

a: medium not separated into two phases

The major portion of radioactivity was recovered in the polar metabolite fraction of the cell extract. This indicates that the metabolites were more polar than the parent 4-NP which was extracted into the nonpolar phase (data not shown).

Another experiment showed that increasing concentrations of 4-NP correlated to increasing amounts of 4-NP equivalents in plant cells. With higher concentrations, higher proportions of 4-NP equivalents were detected in non-extractable residues, indicating an important role of this compartment in detoxification. The high amounts of radioactivity in the medium consisted of metabolites. The great extent of non-extractable residues of cell cultures suggest that the capacity to form non-extractable residues may be associated with higher tolerance to 4-NP. This confirms the hypothesis that the bound residue fraction (e.g. plant cell wall) is one of the important detoxification sites in plant cells. The metabolites were purified and their structures elucidated using mass spectrometric techniques. The two major HPLC peaks appearing in the hydrolyzed extract were identified to be monohydroxy- and dihydroxy-4-nonylphenols. The hydroxylations occurred at all C-atoms of the alkyl sidechain, except for C1 and C9.

The metabolism of ^{14}C-4-NP in intact plants was tested with aseptically grown plants. To take into account more complex interactions between soil and plants, soil/plant-transfer studies were undertaken as well. Uptake rates depend on the tested plant species, which has been shown for all test systems. A higher concentration of applied 4-NP leads to a higher uptake rate in cell suspension cultures and soil/plant systems. Aseptically grown plants weren't investigated.

In soil/plant systems and in aseptically grown plants, radioactivity was transported into the shoot, indicating that 4-NP or its metabolites were transported within the plants. But the uptake rates in the tested intact plants from soil were very low. The amount did not exceed 3% of the applied radioactivity.

No mineralization of 4-NP to $^{14}CO_2$ took place in cell cultures or in the tested plants. Due to the experimental technique employed, a photosynthetic refixation of released CO_2 can be excluded. Mineralization amounts in soil/plant systems were greater (30%) than cell suspensions, which can be attributed to microbial activity in the soil.

The greatest amount of radioactivity in the cell suspension cultures was found in the cell extract, with a high fraction in the bound residue. The main fraction of radioactivity of aseptically grown plants was bound residues in the roots.

Soil/plant systems show great amounts of radioactivity in the bound residue of the soil. Quantitative differences were shown, but a chromatographic comparison of the extract with the metabolites indicated a qualitatively identical metabolism.

5.4.2. *Non-extractable (bound) Residues*

Large amounts of radioactivity were often associated with insoluble plant components. These are called the bound, non-extractable residues, because they can not be released from the plant matrix by extraction with organic solvents.

To investigate to which fraction the radioactivity is associated, a special fractionation procedure has to be used. Formation of bound residues as related to plant species was tested using 14 different cell cultures with 4-NP. The soybean and salad species, with the highest tolerance to 4-NP, integrated high proportions of radioactivity to bound residues. The most sensitive species, like tomato, integrate only a low proportion to the bound residue. The data shows that the fractionation of bound residues and the distribution of radioactivity in cell wall fractions is species-specific.

In most of the cultures, lignin was the fraction to which the major part of radioactivity was associated, but in some of the cultures the protein and the hemicellulose fraction were the main fractions containing radioactivity.

The use of plants for animal or human food raises the question of bioavailability, which is generally accepted for the soluble fraction. The binding type and binding site of a compound to residue fractions might be important for the bioavailability of the compound. The different patterns of binding of a chemical and/or its metabolites to various cell wall components suggest that bound residues may differ in their bioavailability. The assignment of chemicals to certain cell wall components might enable an estimate of the ecotoxicological risks of these chemicals.

6. Plant Peroxidase Enzymes and Metabolism of PCBs

Peroxidases are ubiquitous enzymes playing important roles in various chemical reactions in plant cells. During the last several years multiple functions of peroxidases have been described including their function in removal of hydrogen peroxide from plant cells, auxin metabolism, and as a defence mechanism against, insects and microorganisms. Peroxidases are also involved in formation of radicals from reduction substrates, when free radicals can undergo various reactions including those transforming organic contaminants. Most plant monooxygenases containing cytochrom P450 were studied with such reactions in mind, but recently peroxidase reactions involving organic pollutants have been documented [68, 69]. Oxygenated compounds formed from parent compounds are more polar than the aromatics, resulting in their higher mobility and bioavailability. These metabolites in different bioassays have shown that they can be less, equally, or even more toxic than their parent compounds.

In our experiments the efficiency of PCBs transformation was tested using the cells of different plant species cultivated in vitro. Cultures with the best ability to metabolise PCBs were used for studies of the optimization of incubation conditions, determination of the transformation products of individual PCB congeners and evaluation of POX activities and their involvement of PCB transformation.

6.1. MATERIALS AND METHODS

6.1.1. *Plant cell cultures*

In vitro cultures of different plant species were from the Collection of Plant Tissue Cultures of the Institute of Plant Nutrition and Soil Science, FAL, Braunschweig, Germany and the Institute of Organic Chemistry and Biochemistry, CAS, Prague, Czech Republic.

6.1.2. *Polychlorinated biphenyls*

As a model of PCBs, a standard commercial mixture Delor 103 (Chemko Strazske, Slovakia) containing 59 individual congeners substituted with 3 – 5 chlorines per biphenyl molecule was used. PCBs were added in methanol solution (0.5 mL) to give an initial PCB content 5 mg in 100 mL of nutrient medium (50 mg/L) in each flask.

6.1.3. *Cultivation*

The plant cells were aseptically grown as submerged cultures in 250 mL conical flasks containing 100 mL of nutrient medium inoculated with 5g fresh weight biomass. Cells were incubated with PCBs (50 ppm) and PAHs (3 ppm) for 14 days in the dark (after 5 days preincubation) [70].

6.1.4. *PCB and PAH analysis*

After homogenization and sonication, the content of the flasks was extracted by 10 mL of hexane at 20°C on rotary shaker for 2 hours. Following phase separation, the upper, hexane layer was sampled for analysis. PCB samples were analysed as described by Burkhard et al. [71] or Kucerová et al. [72]. Hydroxylated derivatives were detected by GC-MS according to the method described by Kucerová et al. [72].

6.2. RESULTS AND DISCUSSION

Plant cells of various species (Table 7) cultivated in vitro were incubated with a recalcitrant commercial mixture of 59 individual PCB congeners Delor 103 (or individual congeners of mono- and dichlorobiphenyls) in the medium of Murashige and Skoog for 14 days at 26°C. With the exception of several clones of S. aviculare and S. nigrum (hairy root culture obtained after transformation by Agrobacterium rhizogenes C58Ci), the selection contained suspensions derived from amorphous non-differentiated callus strains. The differentiated and transformed plant cell cultures metabolized xenobiotics with higher efficiency than amorphous and non-transformed cultures of the same species. A hairy root clone of Solanum nigrum SNC-9O metabolized 40-60 % of PCBs present in Delor 103 in 30 days under laboratory conditions (Table 7). When comparing different clones of one plant species, it was clear that clones can significantly differ in their ability to metabolize PCBs within one species and thus the ability to transform PCBs should always be compared with metabolism of the normal plants.

The experiments comparing intracellular and extracellular total POX activities proved higher enzyme activities synthesized by the cultures which were able to readily transform PCBs (Table 8) than those with low transformation abilities. The same trend was detected in the case of the activity of RBBR (Remazol Brilliant Blue) oxidase which correlated with total peroxidase activities (Table 8). Oxidation and decolorization of RBBR is used for screening of the activity of peroxidase lignolytic system (ligninase, manganese dependent peroxidase) in case of white rot fungi and correlated with the ability of these organisms to degrade environmental pollutants.

The activity of extracellular lignin POX and manganese dependent POX was also followed in plant cells incubated with PCBs. No correlation between excretion of these enzymes and ability of plant cells to metabolise PCBs was found [73].

TABLE 7. Transformation of the PCB mixture Delor 103 by various cultures of Solanum aviculare and S. nigrum with different morphological characteristics, specification of the cultures and their growth in the presence (+D) and absence (-D) of Delor 103.

Species + Culture	Morphology	FW (+D) (g)	FW (-D) (g)	PCB converted by plant cells (%)
Solanum	aviculare			
AVC-7	hairy root	3,2	5,4	0
KK1N	callus	7	29	12
KS1	callus	5,5	16,2	10
AGS-3	callus	5,9	25,2	25
AGC-1	callus	13,4	7	40
C	callus	10.3	22,4	25
AGR-1	callus	5	20	28
AGC-3	hairy root	8,4	10,9	40
AVR-1	Shooty terratomas	7,5	6,9	30
Solanum	nigrum			
SNA-6	Shooty terratomas	5.8	6	0
SNC-9O	hairy root	28	33	45
SNC-7H	hairy root	20,7	21,3	45
SNT-4	Shooty terratomas	6.7	8.6	25
Armoracia	rusticana			
K54	Shooty terratomas	14.5	15.7	30
K50	Shooty terratomas	5.8	6.3	30
Nicotiana	tabacum			
NI	callus	7.2	12.5	30
Atropa	bella-donna			
R1BD	Callus	6.3	66.4	12
R2	Callus	4.5	35.4	13
Medicago	sativa			
Alf	Callus	17.4	24.9	30
Lycopersicon	esculentum			
To	callus	10.5	15.3	10
Glycine	max			
S (PC-1026)	callus	26.1	32.6	22
Triticum	aestivum			
W (PC-0998)	callus	5.9	11.8	22
Betula	pendula			
Bk22	callus	6.4	13.2	5
Hordeum	vulgare			
Hv (PC-1118)	callus	10.6	20	19

(+D) – with Delor 10,3 present in initial concentration of 50 ppm, inoculum size was 5g fresh weight.

TABLE 8. Extracellular POX activity and activity of RBBR oxidase measured in medium of submerged cultures of the clones exhibiting high ability to metabolize PCBs.

Culture	D 103	PCB removal (%)	Total POX (RLUx10^7 nkat/mg)	Total RBBR oxidase (nkat/mg)
AGC1	-		340	190
AGC1	+	40	345	110
AVR1	-		349	851
AVR1	+	30	409	299
SNC-2	-		14	185
SNC-2	+	20	6	155
SNC-90			217	87
SNC-90	+	50	216	78
SNC-7H	-		77	41
SNC-7H	+	45	64	37
SNT-2	-		523	106
SNT-2	+	30	370	97
SNT-4	-		115	39
SNT-4	+	25	200	26

Recently Koller et al. [69] showed involvement of horseradish POX in transformation of PCB 9 (2,5-dichlorobiphenyl) and PCB 52 (2,5,2´,5´) in vitro. The experiments in which both PCB were transformed, were performed with commercial horse-radish peroxidase. Different products (benzoic acid, dichlorobenzoic acid, diphenylethers) were identified including hydroxychlorobiphenyls. Unfortunately, their structure was not further characterized.

In our studies, various hydroxychlorobiphenyls formed in vivo after incubation of tobacco (Nicotiana tabacum) and black nightshade (Solanum nigrum) with PCB-1 (2-chlorobiphenyl), PCB-2 (3-chlorobiphenyl) and PCB-3 (4_chlorobiphenyl), were analyzed (Table 9). After incubation of alfalfa with PCB-1 and PCB-2, hydroxylated derivatives were not detected. During incubation of alfalfa with 4-chlorobiphenyls only 4´-hydroxy-4-chlorobiphenyl was found. The same product was also identified after in vivo transformation of PCB-3 with plant cells of horseradish.

TABLE 9. Hydroxylated derivatives of monochlorobiphenyls formed from individual monochlorobiphenyls during 14 days incubation with plant cells of tobacco, black-nightshade.

Silylated sample	Number of products	Retention time	Standards
tobacco (PCB 1)	4	11:13 11:18 11:32 11:37	n.i. n.i. n.i. n.i.
tobacco (PCB 2)	3	10:24 12:50 13:38	n.i. n.i. n.i.
tobacco (PCB 3)	3	12:04 14:03 14:13	n.i. 4-Cl-4´-OH n.i. n.i. n.i.
black-nightshade (PCB 1)	5	8:18 10:06 10:28 10:46 12:59	n.i. n.i. 2Cl-5-OH n.i. n.i.
black-nightshade (PCB 2)	4	9:12 9:29 11:54 12:43	3-Cl-6-OH 3-Cl-4-OH n.i. n.i.
black-nightshade (PCB 3)	2	13:03 13:15	n.i. 4-Cl-4´-OH n.i. n.i.

n.i.- exact structure not identified

7. Conclusions

Plants can transform a wide variety of organic xenobiotic chemicals including PCBs, PAHs, TCE, BTEX, and nonylphenol. Axenic cell suspensions and tissue cultures are a good method to study these plant-contaminant interactions, free from the confounding effects of bacteria. Plant enzymes mediate these reactions, and peroxidase is a ubiquitous enzyme that is important in PCB transformations and possibly many other reactions.

The Green Liver model of Sandermann et al. is a useful approach for considering phytotransformation of organic chemicals. Plants have many of the same enzymes as mammals, and they are particularly endowed with peroxidases, monooxygenases, and glutathione. Once taken-up, organic chemicals are subject to conversion, conjugation, and compartmentation -- three phases of phytotransformation. Research is needed to understand the ultimate fate of plant conjugates and compartmentalized residuals (non-extractable bound residues). Ecotoxicology studies are necessary to ensure that residues do not cause harm to the environment.

Phytotransformation is one application of the emerging technology, phytoremediation. The future looks bright for phytoremediation as a cost effective means for cleaning waste sites contaminated with organic chemicals. It is best applied for low levels of contaminants as a polishing step in tandem with other technologies. It is aesthetically pleasing and provides a natural way to treat industrial wastes using nature's cycles.

8. References

1. Schnoor, J.L., Licht, L.A., McCutcheon, S.C., Wolfe, N.L., and Carreira, L.H., 1995, *Environmental Science & Technology*, Phytoremediation of Organic and Nutrient Contaminants. **29**, A318-A323.
2. Anderson, T.A., Guthrie, E.A., and Walton, B.T., 1993, *Environmental Science & Technology*, Bioremediation in the Rhizosphere. **27**, 2630-2636.
3. Newman, L.A., Strand, S.E., Choe, N., Duffy, J., Ekuan, G., Ruszaj, M., Shurtleff, B.B., Wilmoth, J., Heilman, P., and Gordon, M.P., 1997, *Environmental Science & Technology*, Uptake and biotransformation of trichloroethylene by hybrid poplars. **31**, 1062-1067.
4. Hughes, J.B., Shanks, J., Vanderford, M., Lauritzen, J., and Bhadra, R., 1997, *Environmental Science & Technology*, Transformation of TNT by aquatic plants and plant tissue cultures. **31**, 266-271.
5. Burken, J.G. and Schnoor, J.L., 1998, *Environmental Science & Technology*, Predictive relationships for uptake of organic contaminants by hybrid poplar trees. **32**, 3379-3385.
6. Thompson, P.L., Ramer, L.A., and Schnoor, J.L., 1998, *Environmental Science & Technology*, Uptake and transformation of TNT by hybrid poplar trees. **32**, 975-980.
7. Raskin, I., *Phytoremediation of toxic metals: using plants to clean up the environment*, ed. B.D. Ensley. 2000, New York: Wiley. 352.
8. Terry, N., *Phytoremediation of Contaminated Soil and Water*, ed. N. Terry and G. Banuelos. 2000, Boca Raton: Lewis. 408.
9. Briggs, G.G., Bromilow, R.H., and Evans, A.A., 1982, *Pesticide Science*, Relationships between Lipophilicity and Root Uptake and Translocation of Non-Ionized Chemicals by Barley. **13**, 495-504.
10. Schalk, M., Pierrel, M.A., Zimmerlin, A., Batard, Y., Durst, F., and WerckReichhart, D., 1997, *Environmental Science and Pollution Research*, Xenobiotics: Substrates and inhibitors of the plant cytochrome P450. **4**, 229-234.
11. Ohkawa, H., Imaishi, H., Shiota, N., Yamada, T., and Inui, H., *Pesticide Chemistry and Bioscience - The Food Environment Challenge, Special Publication 233*, G.T. Brooks and T.R. Roberts, Editors. 1999, The Royal Society of Chemistry: Cambridge, U.K. p. 259-264.
12. Werck-Reichhart, D., Hehn, A., and Didierjean, L., 2000, *Trends in Plant Science*, Cytochromes P450 for engineering herbicide tolerance. **5**, 116-123.
13. Kolb, M. and Harms, H., 2000, *Environmental Toxicology and Chemistry*, Metabolism of fluoranthene in different plant cell cultures and intact plants. **19**, 1304-1310.
14. Shimp, J.F., Tracy, J.C., Davis, L.C., Lee, E., Huang, W., Erickson, L.E., and Schnoor, J.L., 1993, *Critical Reviews in Environmental Science and Technology*, Beneficial-Effects of Plants in the Remediation of Soil and Groundwater Contaminated with Organic Materials. **23**, 41-77.
15. Burken, J.G. and Schnoor, J.L., 1996, *Journal of Environmental Engineering-Asce*, Phytoremediation: Plant uptake of atrazine and role of root exudates. **122**, 958-963.
16. Dushenkov, V., Kumar, P., Motto, H., and Raskin, I., 1995, *Environmental Science & Technology*, Rhizofiltration - the Use of Plants to Remove Heavy-Metals from Aqueous Streams. **29**, 1239-1245.
17. Salt, D.E., Prince, R.C., Pickering, I.J., and Raskin, I., 1995, *Plant Physiology*, Mechanisms of Cadmium Mobility and Accumulation in Indian Mustard. **109**, 1427-1433.

18. Salt, D.E., Pickering, I.J., Prince, R.C., Gleba, D., Dushenkov, S., Smith, R.D., and Raskin, I., 1997, *Environmental Science & Technology*, Metal accumulation by aquacultured seedlings of Indian mustard. **31**, 1636-1644.
19. Dietz, A.C. and Schnoor, J.L., 2001, *Environmental Health Perspectives*, Advances in phytoremediation. **109**, 163-168.
20. Hatzios, D.D., *Environmental Chemistry of Herbicides*, ed. R. Grover and A. Cessna. Vol. II. 1991, Boca Raton: CRC Press.
21. Komoβa, D., Langebartels, C., and Sandermann, H., in *Plant Contamination: Modeling and Simulation of Organic Chemical Processes*, S. Trapp and C. McFarlane, Editors. 1995, Lewis: Boca Raton.
22. Sandermann, H., 1994, *Pharmacogenetics*, Higher-Plant Metabolism of Xenobiotics - the Green Liver Concept. **4**, 225-241.
23. Hatzios, K. and Penner, D., *Metabolism of Herbicides in Higher Plants*. 1982, Minneapolis, MN: Burgess.
24. Trapp, S. and McFarlane, C., *Plant Contamination: Modeling and Simulation of Organic Chemical Processes*. 1995, Boca Raton: Lewis.
25. Adler, P.R., Arora, R., Elghaouth, A., Glenn, D.M., and Solar, J.M., 1994, *Journal of Environmental Quality*, Bioremediation of Phenolic-Compounds from Water with Plant-Root Surface Peroxidases. **23**, 1113-1117.
26. Dec, J. and Bollag, J.M., in *Bioremediation of Chlorinated Solvents*, R.E. Hinchee, A. Leeson, and L. Semprini, Editors. 1995, Battelle Press: Colombus, OH. p. 307-312.
27. Yih, R., McRae, D.H., and Wilson, H.F., 1968, *Science*, Metabolism of 3',4'-Dichloropropionanilide: 3,4-Dichloroaniline-Lignin Complex in Rice Plants. **161**, 376-377.
28. Lamoureaux, G.L. and Rusness, D.G., in *Xenobiotic Conjugation Chemistry*, G.D. Paulson, Editor. 1986, American Chemical Society: Washington, D.C. p. 62-105.
29. Höhl, H.U. and Barz, W., 1995, *Journal of Agricultural and Food Chemistry*, Metabolism of the Insecticide Phoxim in Plants and Cell- Suspension Cultures of Soybean. **43**, 1052-1056.
30. Schuphan, I., Haque, A., and Ebing, W., 1984, *Chemosphere*, Ecochemical Assessment of Environmental Chemicals .1. Standard Screening-Procedure to Evaluate Chemicals in Plant-Cell Cultures. **13**, 301-313.
31. Haque, A., Ebing, W., and Schuphan, I., 1984, *Chemosphere*, Ecochemical Assessment of Environmental Chemicals .2. Standard Supplementary Screening-Procedure to Evaluate More Volatile and Less Persistent Chemicals in Plant-Cell Cultures. **13**, 315-320.
32. Ebing, W., Haque, A., Schuphan, I., Harms, H., Langebartels, C., Scheel, D., Vondertrenck, K.T., and Sandermann, H., 1984, *Chemosphere*, Ecochemical Assessment of Environmental Chemicals - Draft Guideline of the Test Procedure to Evaluate Metabolism and Degradation of Chemicals by Plant-Cell Cultures. **13**, 947-957.
33. Harms, H. and Langebartels, C., 1986, *Plant Science*, Standardized Plant-Cell Suspension Test Systems for an Ecotoxicologic Evaluation of the Metabolic-Fate of Xenobiotics. **45**, 157-165.
34. Frear, D.S., 1968, *Phytochemistry*, Herbicide metabolism in plants. I. Purification and properties of UDP-glucose:arylamine N-glucosyl-transferase from soybean. **7**, 381-390.
35. Winkler, R. and Sandermann, H., 1989, *Pesticide Biochemistry and Physiology*, Plant-Metabolism of Chlorinated Anilines - Isolation and Identification of N-Glucosyl and N-Malonyl Conjugates. **33**, 239-248.

36. Sandermann, H., Schmitt, R., Eckey, H., and Bauknecht, T., 1991, *Archives of Biochemistry and Biophysics*, Plant Biochemistry of Xenobiotics - Isolation and Properties of Soybean O-Glucosyl and N-Glucosyl and O-Malonyltransferases and N-Malonyltransferases for Chlorinated Phenols and Anilines. **287**, 341-350.
37. Gareis, C., Rivero, C., Schuphan, I., and Schmidt, B., 1992, *Zeitschrift Fur Naturforschung C-a Journal of Biosciences*, Plant-Metabolism of Xenobiotics - Comparison of the Metabolism of 3,4-Dichloroaniline in Soybean Excised Leaves and Soybean Cell-Suspension Cultures. **47**, 823-829.
38. Bockers, M., Rivero, C., Thiede, B., Jankowski, T., and Schmidt, B., 1994, *Zeitschrift Fur Naturforschung C-a Journal of Biosciences*, Uptake, Translocation, and Metabolism of 3,4-Dichloroaniline in Soybean and Wheat Plants. **49**, 719-726.
39. Matern, U., Feser, C., and Heller, W., 1984, *Archives of Biochemistry and Biophysics*, N-Malonyltransferases from Peanut. **235**, 218-227.
40. Harvey, S., Fellows, R., Cataldo, D., and Bean, R., 1991, *Environmental Toxicology and Chemistry*, Fate of the Explosive Hexahydro-1,3,5-trinitro-1,3,5-triazine (RDX) in Soil and Bioaccumulation in Bush Bean Hydroponic Plants. **10**, 845-855.
41. Still, G.G., 1968, *Science*, Metabolism of 3,4-Dichloropropionanilide in Plants: The Metabolic Fate of the 3,4-Dichloroaniline Moiety. **159**, 992-993..
42. Arjmand, M. and Sandermann, H., 1986, *Zeitschrift Fur Naturforschung C-a Journal of Biosciences*, Plant Biochemistry of Xenobiotics - Mineralization of Chloroaniline Lignin Metabolites from Wheat by the White-Rot Fungus, Phanerochaete-Chrysosporium. **41**, 206-214.
43. Mitsch, W. and Gooselink, J., *Wetlands*. 2nd ed. 1993, New York: Van Norstrand Reinhold.
44. Anderson, T.A., Kruger, E.L., and Coats, J.R., 1994, *Chemosphere*, Enhanced Degradation of a Mixture of 3 Herbicides in the Rhizosphere of a Herbicide-Tolerant Plant. **28**, 1551-1557.
45. Hsu, T.S. and Bartha, R., 1979, *Applied and Environmental Microbiology*, Accelerated Mineralization of 2 Organophosphate Insecticides in the Rhizosphere. **37**, 36-41.
46. Reddy, B.R. and Sethunathan, N., 1983, *Applied and Environmental Microbiology*, Mineralization of Parathion in the Rice Rhizosphere. **45**, 826-829.
47. Mandelbaum, R.T., Wackett, L.P., and Allan, D.L., 1993, *Applied and Environmental Microbiology*, Mineralization of the S-Triazine Ring of Atrazine by Stable Bacterial Mixed Cultures. **59**, 1695-1701.
48. Sandmann, E. and Loos, M.A., 1984, *Chemosphere*, Enumeration of 2,4-D-Degrading Microorganisms in Soils and Crop Plant Rhizospheres Using Indicator Media - High Populations Associated with Sugarcane (Saccharum-Officinarum). **13**, 1073-1084.
49. Walton, B.T. and Anderson, T.A., 1990, *Applied and Environmental Microbiology*, Microbial-Degradation of Trichloroethylene in the Rhizosphere - Potential Application to Biological Remediation of Waste Sites. **56**, 1012-1016.
50. Ferro, A.M., Sims, R.C., and Bugbee, B., 1994, *Journal of Environmental Quality*, Hycrest Crested Wheatgrass Accelerates the Degradation of Pentachlorophenol in Soil. **23**, 272-279.
51. Aprill, W. and Sims, R.C., 1990, *Chemosphere*, Evaluation of the Use of Prairie Grasses for Stimulating Polycyclic Aromatic Hydrocarbon Treatment in Soil. **20**, 253-265.
52. Bollag, J.M., 1992, *Environmental Science & Technology*, Decontaminating Soil with Enzymes. **26**, 1876-1881.
53. Shone, M.G.T., Bartlette, B.O., and Wood, A.V., 1974, *Journal of Experimental Botany*. **25**, 401-409.

54. Lindstrom, F.T., Boersma, L., and McFarlane, C., 1991, *Journal of Environmental Quality*, Mathematical-Model of Plant Uptake and Translocation of Organic-Chemicals - Development of the Model. **20**, 129-136.
55. Trapp, S., Matthies, M., Scheunert, I., and Topp, E.M., 1990, *Environmental Science & Technology*, Modeling the Bioconcentration of Organic-Chemicals in Plants. **24**, 1246-1252.
56. Ryan, J.A., Bell, R.M., Davidson, J.M., and Oconnor, G.A., 1988, *Chemosphere*, Plant Uptake of Non-Ionic Organic-Chemicals from Soils. **17**. 2299-2323.
57. Reiderer, M., in *Plant Contamination: Modeling and Simulation of Organic Chemical Processes*, S. Trapp and C. McFarlane, Editors. 1995, Lewis: Boca Raton.
58. Russell, R.S. and Shorrocks, V.M., 1959, *Journal of Experimental Botany*. **10**, 301-316.
59. Shone, M.G.T. and Wood, A.V., 1974, *Journal of Experimental Botany*. **25**, 390-400.
60. Polder, M.D., Hulzebos, E.M., and Jager, D.T., 1995, *Environmental Toxicology and Chemistry*, Validation of Models on Uptake of Organic-Chemicals by Plant- Roots. **14**, 1615-1623.
61. Trapp, S. and Pussemier, L., 1991, *Chemosphere*, Model-Calculations and Measurements of Uptake and Translocation of Carbamates by Bean-Plants. **22**, 327-339.
62. Trapp, S., McFarlane, C., and Matthies, M., 1994, *Environmental Toxicology and Chemistry*, Model for Uptake of Xenobiotics into Plants - Validation with Bromacil Experiments. **13**, 413-422.
63. Boersma, L., McFarlane, C., and Lindstrom, F.T., 1991, *Journal of Environmental Quality*, Mathematical-Model of Plant Uptake and Translocations of Organic-Chemicals - Application to Experiments. **20**, 137-146.
64. McFarlane, C., Pfleeger, T., and Fletcher, J., 1990, *Environmental Toxicology and Chemistry*, Effect, Uptake and Disposition of Nitrobenzene in Several Terrestrial Plants. **9**, 513-520.
65. Wilken, A., Bock, C., Bokern, M., and Harms, H., 1995, *Environmental Toxicology and Chemistry*, Metabolism of Different Pcb Congeners in Plant-Cell Cultures. **14**, 2017-2022.
66. Babson, J.R., Russorodriguez, S.E., Wattley, R.V., Bergstein, P.L., Rastetter, W.H., Liber, H.L., Andon, B.M., Thilly, W.G., and Wogan, G.N., 1986, *Toxicology and Applied Pharmacology*, Microsomal Activation of Fluoranthene to Mutagenic Metabolites. **85**, 355-366.
67. Rice, J.E., Lavoie, E.J., and Hoffmann, D., 1983, *Journal of Organic Chemistry*, Synthesis of the Isomeric Phenols and the Trans-2,3-Dihydrodiol of Fluoranthene. **48**, 2360-2363.
68. Stiborova, M., Schmeiser, H.H., and Frei, E., 2000, *Phytochemistry*, Oxidation of xenobiotics by plant microsomes, a reconstituted cytochrome P450 system and peroxidase: a comparative study. **54**, 353-362.
69. Koller, G., Moder, M., and Czihal, K., 2000, *Chemosphere*, Peroxidative degradation of selected PCB: a mechanistic study. **41**, 1827-1834.
70. Kucerová, P., Mackova, M., Polachova, L., Burkhard, J., Demnerova, K., Pazlarova, J., and Macek, T., 1999, *Collection of Czechoslovak Chemical Communications*, Correlation of PCB transformation by plant tissue cultures with their morphology and peroxidase activity changes. **64**, 1497-1509.
71. Burkhard, J., Mackova, M., Macek, T., Kucerova, P., and Demnerova, K., 1997, *Analytical Communications*, Analytical procedure for the estimation of polychlorinated biphenyl transformation by plant tissue cultures. **34**, 287-290.

72. Kucerová, P., Mackova, M., Chroma, L., Burkhard, J., Triska, J., Demnerova, K., and Macek, T., 2000, *Plant and Soil*, Metabolism of polychlorinated biphenyls by Solanum nigrum hairy root clone SNC-9O and analysis of transformation products. **225**, 109-115.
73. Ollikka, P., Alhonmaki, K., Leppanen, V.M., Glumoff, T., Raijola, T., and Suominen, I., 1993, *Applied and Environmental Microbiology*, Decolorization of Azo, Triphenyl Methane, Heterocyclic, and Polymeric Dyes by Lignin Peroxidase Isoenzymes from Phanerochaete-Chrysosporium. **59**, 4010-4016.

NATURAL ATTENUATION

PERRY L. MCCARTY[1], DAVID E. ELLIS[2]
1Department of Civil and Environmental Engineering, Stanford University, Stanford, CA 94305-4020
[2] *Dupont Engineering, P.O. Box 8002, Wilmington, DE 19880-0027*

1. Introduction

The fate of contaminants in the subsurface is controlled by a combination of interacting processes. These processes can be broadly classified as physical, chemical, or biological. The physical processes control the rate and direction of travel as contaminants move through soil and the subsurface away from their source. The chemical and biological processes determine the extent to which the initial compounds will be transformed in the subsurface. In some cases the most important transformations occur abiotically, while in other cases they are mediated by subsurface microorganisms. Both the nature of the initial compounds and the chemical and biological environment of the subsurface are important in determining which reactions will occur. The task of assessing whether natural attenuation is a useful approach for removing pollutants from contaminated sites requires an understanding of all the interacting processes. Given in the following is an overview of the common classes of subsurface contaminants, their geologic and hydrologic settings, and processes that influence their fate and transport. Although the primary focus here is on processes that influence natural attenuation of groundwater contaminants, information presented also applies to natural attenuation of soils and sediments.

Natural attenuation, as used in the context of hazardous chemical contamination of soil, groundwater, surface waters, and sediments, refers to the ability of natural processes over time to bring about the reduction in toxicity to the point where harm to human health and the environment is adequately reduced or eliminated. In cases where natural attenuation is effective, costly engineered processes for remediation or long-term institutional controls of a site may not be necessary. Natural attenuation cannot be assumed to be operational at a given site, detailed investigation and continued monitoring is generally necessary to insure that the natural attenuation processes believed to be operational are in fact occurring and at a pace that is acceptable. Because of the additional expense involved, use of natural attenuation could even be more costly than other approaches for a given site, and may take longer than is acceptable because of desires for more immediate use of a site(many are already in constructive use). Thus, natural attenuation is not always the most desired approach, even when it might be effective in the long run for a site.

Natural attenuation is often not the sole remedy used at a site. Its is often coupled with source removal and control, to the extent possible, to reduce contaminant mass and the longevity of need for natural attenuation monitoring. Also, engineered solutions often do not result in sufficient contaminant removal to be fully protective to human health and the environment. In such cases, which are common, natural attenuation may need to be relied upon as a the polishing step for the contamination. Thus, natural attenuation often becomes part of an overall remedy package, whether planned for or otherwise.

There is no general statement about natural attenuation that is applicable to all chemical contaminants and sites. Each hazardous chemical and each site has its own characteristics that need to be evaluated. Some organic chemicals are readily biodegraded by common

D. Reible and K. Demnerova (eds.), Innovative Approaches to the On-Site Assessment and Remediation of Contaminated Sites, 141–181.

organisms either in the presence or absence of oxygen. Investigative studies for such chemicals need not be as extensive as for other organic chemicals that are biodegradable only under certain conditions or where degrading microorganisms are not ubiquitous in the environment. Another difficulty is that some organic chemicals may be transformed to form hazardous intermediates so that observations of a chemicals disappearance by itself is not sufficient to insure protection. Hazardous inorganic chemicals such as heavy metals cannot be degraded like organic chemicals. Natural attenuation for these materials depends upon the ability of natural processes to stabilize or sequester the hazardous chemicals so that they will remain in place and out of contact with humans and other life forms needing protection. The manner and rate at which a given chemical moves between different media (air, water, land) in the environment is also chemical and site specific, and this needs to be considered in evaluating natural attenuation.

It becomes obvious that in order to take advantage of natural attenuation for site remediation, a good knowledge of natural attenuation processes, detailed site characterization, modeling of chemical fate processes, and long-term monitoring may all be necessary. The purposes of this chapter are to provide a more formal definition of natural attenuation, provide a scientific basis for natural attenuation, discuss how natural attenuation might be evaluated at a given site, briefly present some formalized protocols that might be used for evaluation of natural attenuation, and finally, indicate the importance of community involvement when natural attenuation is contemplated. A full discussion of these issues can be found in a publication of the National Research Council [1]. Much of what is presented here is taken directly from that source.

2. Definitions

Natural attenuation has been evaluated by a number of agencies and organizations in the United States. Among these are the U.S. Environmental Protection Agency [2], the U.S. Air Force [3], the U.S. Army [4] and the American Society for Testing Materials [5]. In general, the different organizations use a similar definition. That issued by the U.S. EPA in a directive [2] is the most generally accepted. It defines what it terms "monitored" natural attenuation, using the word monitored to mean that sites should be observed until it is clear that the contaminants are fully treated:

> Monitored natural attenuation is the "reliance on natural attenuation processes (within the context of a carefully controlled and monitored site cleanup approach) to achieve site-specific remediation objectives within a time frame that is reasonable compared to that offered by other more active methods. The 'natural attenuation processes' that are at work in such a remediation approach include a variety of physical, chemical, or biological processes that, under favorable conditions, act without human intervention to reduce the mass, toxicity, mobility, volume, or concentration of contaminants in soil or groundwater. These in-situ processes include biodegradation; dispersion; dilution; sorption; volatilization; radioactive decay; and chemical or biological stabilization, transformation, or destruction of contaminants."

While many processes result in natural attenuation, biodegradation of organics and stabilization of inorganics are the favored processes because they provide more permanent remedies that are more likely to be protective [1]. Thus, disappearance of contaminant mass is not the only criteria that needs to be evaluated, but the mechanism by which this occurs is also significant in judging the suitability of natural attenuation as a remedy. In order to determine the important natural attenuation processes for a given site, a scientific understanding of these processes is needed to guide site investigation in order not only to obtain sound evidence that the hazard presented is actually being reduced, but the mechanisms involved.

3. Contaminants and Geologic Setting

3.1. CONTAMINANTS

Approximately eight million synthetic and naturally-occurring organic compounds have been widely disseminated in the 19th and 20th centuries [6, 7]. As many as 100,000 compounds have been manufactured and used in commerce. Many of these compounds play a vital role in a variety of industrial processes including production of fuels, solvents, pesticides, and food additives. Furthermore, a broad array of inorganic wastes have been generated by commercial, industrial, and military activities. Sources of inorganic contaminants include materials used in mining operations, wastes produced during metal refining activities, and by-products of nuclear energy and weapons operations.

The location and physical/chemical state of contaminant compounds are of crucial importance for understanding their behavior and fate. Organic compounds are often released as non-aqueous phase liquids (NAPLs). Depending on the density of the NAPLs and their hydrogeological context, they may remain in the soil, be spread throughout the vadose zone, be perched on the water table, and/or penetrate or congeal in pools on impermeable boundaries in the saturated zone. In each of these locations, NAPLs act as contaminant sources that constantly equilibrate with their gaseous and/or aqueous phase surroundings.

3.2. GEOLOGIC SETTINGS

Figure 1 illustrates a typical soil profile and two different liquid contaminant types. At the top is the surface soil. This is followed by the vadose zone, which contains a mixture of air and water. Below that is the groundwater. The vadose zone and the groundwater are separated by a capillary fringe. Liquid contaminants such as acetone and ethanol may be highly soluble in water, while others, such as chlorinated solvents and gasoline, are not. The latter represent non-aqueous phase liquids (NAPLs), which are divided into two types. LNAPLs, represented by gasoline, are lighter than water and tend not to penetrate below the water table surface. DNAPLs, such as chlorinated solvents, are denser than water and can penetrate far below the groundwater table, creating some of the most difficult to solve groundwater contamination problems. NAPLs are sparingly soluble in water. As groundwater passes by a NAPL, contaminants dissolve into the groundwater to a limited extent, which causes a groundwater contaminant plume to form that may extend kilometers beyond the source. The NAPL is generally referred to as the "source" of the groundwater contamination. Because groundwater movement is generally slow, it may take decades to centuries for a NAPL source to completely dissolve in passing groundwater. Therefore, the usual approach to remediation when a NAPL source is involved is to remove as much of the NAPL as possible through NAPL extraction or physical means NAPLs are very difficult to locate and remove completely, and thus physical or hydraulic containment of the source is then necessary. Following action to reduce the influence of the NAPL source on the contaminated plume, the plume itself can be addressed. Here, either an engineered approach or natural attenuation may become the best options. The attention here is on natural attenuation of the contaminated plume, although natural attenuation can also apply to natural processes that either directly degrade the NAPL source itself, or speed up the rate of its dissolution. Many contaminants, such as many pesticides, polycyclic aromatic hydrocarbons (PAHs), and polychlorinated biphenols (PCBs) are not mobile and thus are not often groundwater contaminants. They tend to remain in surface soils or in stream and river sediments. Natural attenuation can apply to the decontamination of these materials as well.

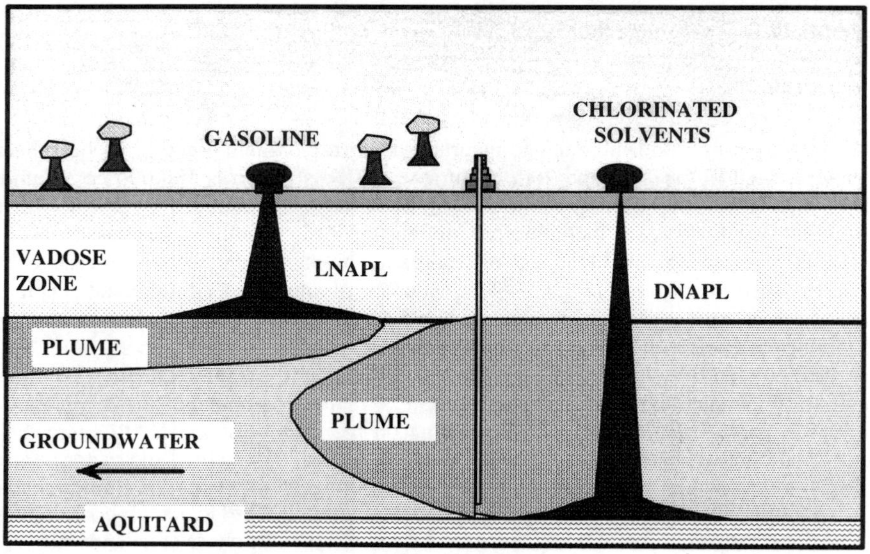

Figure 1. Groundwater contamination by NAPLs.

Groundwater may discharge to a surface water body and may result in surface water contamination. The bottom sediments of surface water bodies are often chemically and biologically active. As contaminants migrate through bottom sediments, rapid microbial transformations may occur and may contribute to natural attenuation. Groundwater contaminants may also be drawn downward into an intermediate flow system due to ground water pumping from a lower aquifer. In some cases contaminants enter deeper systems directly via injection or migration down an open or unfinished borehole, causing a greater problem for remediation. The potential for contaminant migration between environmental compartments needs to be considered when evaluating the potential for natural attenuation.

Processes of importance in the movement and fate of groundwater contaminants include advection, dispersion, sorption, volatilization, and transformation. Figure 2 illustrates the importance of these processes (except volatilization). Here we consider that the relative time of groundwater travel from a point of initial contamination to a location down gradient from the contaminant source to be 1. If advection were the only operating process, then the contaminant would move with the water as plug flow and the concentration at the down gradient location would suddenly increase from zero to that of the contamination source. Dispersion is a mixing process that results from different flow paths around the soil material that causes different flow times for each path. This results in the spreading of the contaminant such that some arrives before the average flow time of one, and some arrives later as illustrated in Figure 2. Sorption represents a partitioning of the contaminant between the water and the soil matrix. This partitioning results in a slowing of contaminant movement relative to water. The ratio of the speed of water movement relative to contaminant movement is called the retardation factor. A retardation factor of 1 indicates the contaminant and the water move together. This is the case for hydrophilic compounds with little sorption to the soil. A retardation factor of 2 means the contaminant moves with half of the speed of water. A mass balance analysis will indicate that for this case, the mass of contaminant sorbed to the soil equals the mass in solution. Hydrophobic compounds tend to have higher retardation factors, which may be 10 or more with some compounds. Finally, compound transformation will result in the loss of the compound mass with distance from the source, the amount depending upon the relative rate of the transformation. Volatilization would act similar to transformation in that mass would decrease with distance from the source. Sorption and dispersion are factors affected by soil properties as well as by contaminant properties. The speed of water movement is a function of site hydrogeology. Thus, the ability to predict contaminant movement and fate in the subsurface environment requires a detailed understanding of both the contaminants and of site hydrogeology.

For modeling, the fate processes are brought together. For a given flow domain the above effects are combined to give the total increase of mass within the system:

Mass increase of chemical in solution of interest	=	Net flow in by advection	+	Net flow in by dispersion	+	Net additions to sorbed phase	-	Net loss by reactions

This is a statement of conservation of mass . When applied over a small time interval and for a small volume, the above statement of mass conservation becomes a partial differential equation relating the concentration C to time and position. In order to find C as a function of position, it is necessary to define conditions at the flow boundary and for time zero as well as to determine appropriate transfer coefficients and the velocity field. More complex formulations can include coupled partial differential equations containing major chemical species and multiple phases.

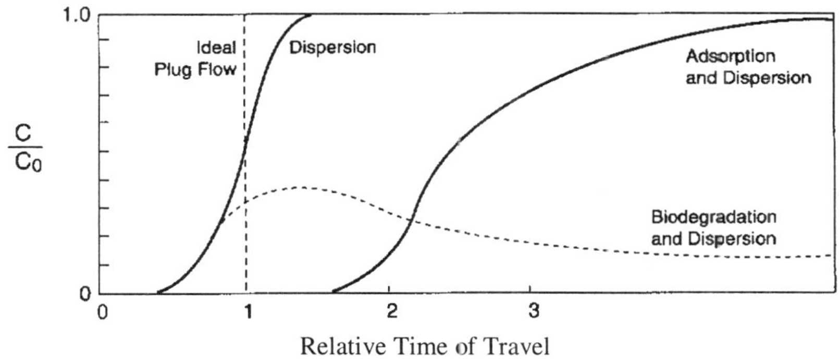

Figure 2. Relative concentration of a contaminant at a monitoring location versus relative travel time from a source as a function of the transport and fate processes involved (After Roberts et al., 103). Reprinted with permission of the Water Environment Federation.

3.3. EXAMPLE OF AQUEOUS PHASE TRANSPORT AND FLOW

In August 1982 approximately 12 m^3 of solution was injected into a shallow sand aquifer at the Canadian Air Forces Base in Borden, Ontario. The injected solution included chloride and bromide ions as conservative tracers along with 5 organic solutes: bromoform, carbon tetrachloride (CTET), tetrachloroethene (PCE), 1,2-dichlorobenzene and hexachloroethane. A dense, three-dimensional array consisting of over 5000 sampling points was used to collect nearly 20,000 samples over a 3-year period in order to identify the resulting distribution of solutes in time and space. Transport of the solutes has been described in detail [8-12].

Profiles of the chloride plume over a two-year period are presented in Figure 3. The plume advanced generally from SW to NE. The chloride movement was not affected by sorptive or reactive processes. From the distance the chloride moved the advective velocity was estimated to be 0.091 m/d [8]. Dispersion was manifested as a spreading of the chloride plume as shown in Figure 3. The longest dimension of the plume increased from a few meters at Day 1 to about 30 m on Day 647. Relative elongation in the direction of flow is a consequence of a longitudinal dispersion, which was considerably larger than the transverse dispersion.

The movement of CTET relative to chloride is illustrated by comparing the plumes depicted in Figures 3. The slower movement for the CTET plume is a consequence of retardation due to sorption [8]. The distance traveled by the center of mass for carbon tetrachloride (about 25 m) is much less than for chloride (about 60 m), suggesting a retardation factor of 2.4. The organic solutes with the largest K_{ow} values (1,2-dichlorobenzene and hexachloroethane) progressed very slowly whereas the others were faster, although PCE was considerably slower than bromoform and CTET.

There was strong evidence of reactions during the experiment leading to a loss of mass of bromoform, dichlorobenzene and hexachloroethane, but not of CTET nor PCE. This conclusion is based on the decline of mass with time for the first three chemicals (Figure 4).

This experiment clearly illustrates four basic processes that affect chemical fate in the subsurface, that is advection, dispersion, sorption, and reaction. Instructive also is that natural attenuation here was responsible for destruction of three of the organic chemicals added (hexachloroethane, 1,2-dichlorobenzene, and bromoform), but not two others (PCE and CTET). Subsequent studies [13] indicated that hexachloroethane was not entirely destroyed, but rather was converted into PCE, a hazardous chemical that was also injected in the Borden study. This indicates that loss of a compound by natural attenuation does not necessarily result in destruction of the hazard.

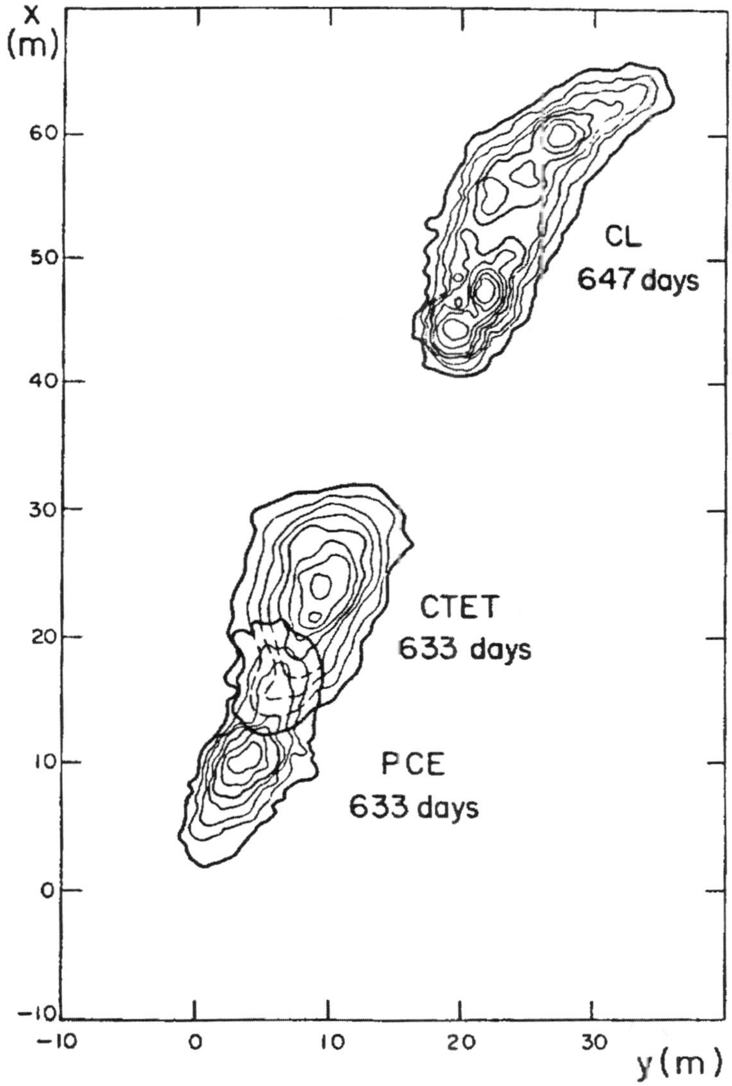

Figure 3. Movement of chloride, carbon tetrachloride, and tetrachloroethene at the Borden site after almost two years following injection at 0,0. (After Roberts, et al., 11).

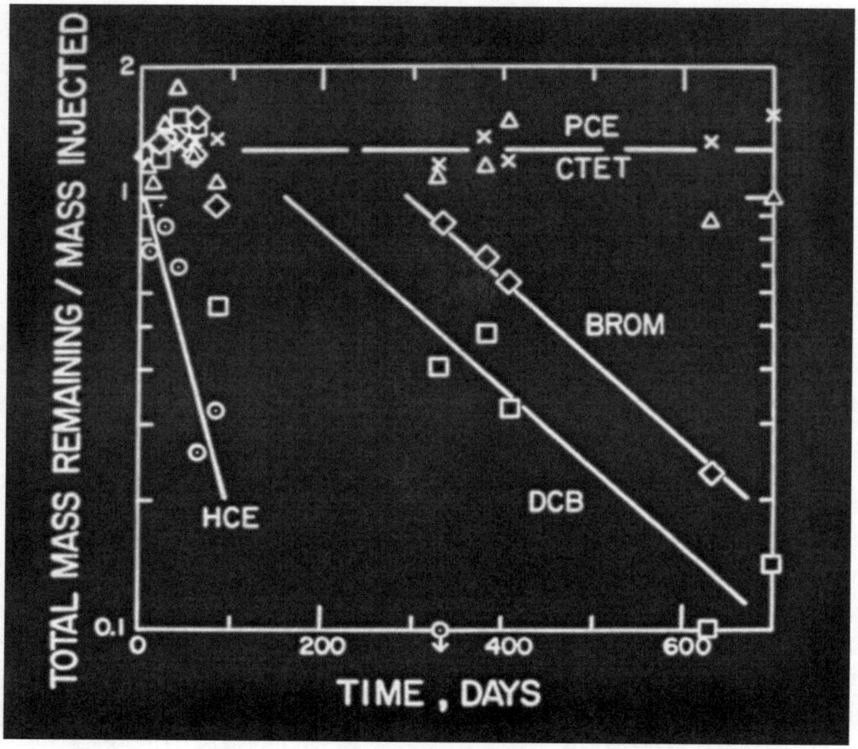

Figure 4. Change in mass with time of groundwater contaminants at Borden. (After Roberts et al., 11)

4. Microbial Transformations

Organic compounds that are most susceptible to microbial transformations are naturally occurring, have a simple molecular structure, are water soluble, exhibit low sorption, are nontoxic, and serve as a growth substrate for aerobic and anaerobic microorganisms. By contrast, compounds that are resistant to microbial metabolism exhibit properties such as complex molecular structure, low water solubility, strong sorptive interactions, toxicity, and/or do not support microorganism growth. Table 1 provides a generalized overview of the categories of contaminants, mechanisms of microbe-contaminant interaction, type of contaminant alteration, and the susceptibility to microbial

Table 1. Overview of biodegradation potential for categories of environmental contaminants (after NRC, 5).

Chemical Classes[a]	Mechanisms of microbe-contaminant interaction[b]	Type(s) of contaminant alteration[c]	usceptibility icrobiological ansformation	
			erobic	Anaerobic
Organic				
Hydrocarbons				
BTEX	Carbon and electron-donor source	Mineralized to CO_2	1	2
low MW, gasoline, fuel oil	Carbon and electron-donor source	Mineralized to CO_2	1	2
high MW, oils, PAHs	Carbon and electron-donor source	Mineralized to CO_2 or partially degraded	1,2	2,4
Creosote	Carbon and electron-donor source	Mineralized to CO_2 or partially degraded	1,2	2,4
Oxygenated hydrocarbons				
low MW alcohols, ketones, esters, ethers	Carbon and electron-donor source	Mineralized to CO_2	1,2	2
MTBE	Cometabolized; not fully used as carbon and electron-donor source	Partially degraded	2-5	4,5
Halogenated Aliphatics				
Highly chlorinated	Electron acceptor under anaerobic conditions; cometabolized	Partially degraded	2-5	2-5
Less chlorinated	Electron acceptor under anaerobic conditions; carbon and electron-donor source, cometabolized	Partially degraded	2-5	2-5
Halogenated Aromatics				
Highly chlorinated	Electron acceptor under anaerobic conditions; carbon and electron-donor source, cometabolized	Partially degraded	2-5	2,3
Less chlorinated	Electron acceptor under anaerobic conditions; carbon and electron-donor source	Partially degraded	1,2	2

Table 1. Continued

Chemical Classes[a]	Mechanisms of microbe-contaminant interaction[b]	Type(s) of contaminant alteration[c]	Susceptibility to microbiological transformation	
			Aerobic	Anaerobic
PCBs				
Highly chlorinate	Electron acceptor under anaerob conditions	Partially degraded		2,3
Less chlorinated	Electron acceptor under anaerobic conditions; carbon and electron-donor source	Mineralized to CO_2 or partially degraded	1,2	2,4
Dioxins	Electron acceptor under anaerobic conditions	Partially degraded	4	4
Nitroaromatics				
(TNT, RTX)	Carbon and electron-donor source, cometabolized	Partially degraded; immobilized by precipitation or polymerization	2	2
Inorganic				
Metals				
Cu, Ni, Zn	Sorbs to extracellular polymers and biomass	Immobilized by sorption	2	2
Cd, Pb	Sorbs to extracellular polymers and biomass	Immobilized by sorption	2	2
Fe, Mn	Electron acceptor under anaerobic conditions; oxidized to form insoluble hydroxides; sorbs to extracellular polymers and biomass	Mobility (solubilization) increased by reduction; immobilized by precipitation and sorption	1	1
Cr	enzymatically oxidized or reduced to promote detoxification; cometabolized; sorbs to extracellular polymers and biomass	Immobilized by precipitation	2	2
Hg	sorbs to extracellular polymers and biomass	Volatilized or immobilized by sorption and precipitation	2	2

Table 1. Continued

Chemical Classes[a]	Mechanisms of microbe-contaminant interaction[b]	Type(s) of contaminant alteration[c]	Susceptibility to microbiological transformation: Aerobic	Susceptibility to microbiological transformation: Anaerobic
Non-metals				
As	sorbs to extracellular polymers and biomass	Volatilized or immobilized by sorption and precipitation	2	2
Se	sorbs to extracellular polymers biomass	Volatilized or immobilized by sorption and precipitation	2	
Oxyanions				
Nitrate	Electron acceptor under anaerobic conditions	Converted to nontoxic nitrogen	4	1
Perchlorate	Electron acceptor under anaerobic conditions	Reduced to nontoxic chloride ion	4	2,5
Radionuclides				
U	Electron acceptor under anaerobic conditions, sorbs to extracellular polymers and biomass	Immobilized by precipitation	4	2
Pu	Cometabolized, sorbs to extracellular polymers and biomass	Mobility increased by reduction to soluble Pu(III); immobilized by precipitation and sorption	4	2
Tc	Cometabolized, sorbs to extracellular polymers and biomass	Immobilized by precipitation	4	2

The numeric entries for each compound class provide a rating of susceptibility to microbial transformation under aerobic conditions (in the presence of oxygen) and anaerobic conditions (when oxygen is absent): 1 = readily mineralized or transformed; 2 = degraded or transformed under a narrow range of conditions, 3 = metabolized partially when second substrate is present (cometabolized); 4 = resistant, 5 = insufficient information.
NOTE: BTEX = benzene, toluene, ethylbenzene, and xylene; MTBE = methyl *tert*-butyl ether; PCB = polychlorinated biphenyl; RDX = royal Dutch explosive; TNT = trinitrotoluene.

transformations [1]. Compounds most susceptible to natural attenuation through biodegradation are those that are readily mineralized or transformed (category 1)by both aerobic and anaerobic processes.

Proof of biotransformation is not in itself sufficient to guarantee protection, the biotransformation must not result in the production of harmful intermediates with long lifetimes. For this reason, biotransformation pathways must also be understood. Additionally, good mass balances are difficult to make. Alternative approaches are often necessary to provide proof that the desired transformation processes are occurring. Evidence of biotransformation is provided by "footprints" of transformation, which can be the presence of biotransformation intermediates or end products, or of the consumption of other chemicals that are associated with biotransformation. Examples are the presence of transformation intermediates or end products, such as vinyl chloride (VC), ethene, and chloride in the case of anaerobic transformations of the chlorinated solvents, trichloroethene (TCE) and PCE. Others are the depletion of electron acceptors, such as oxygen, nitrate, and sulfate, or the formation of reduced end products of biotransformation, such as hydrogen sulfide and methane. Such changes in chemical characteristics of groundwater are generally the result of biological transformations and thus serve as clues to the presence of microbial activity. Thus, an understanding of the likely intermediates and end products of biotransformation of chemicals is equally important for evaluating natural attenuation as is knowledge that a compound may be transformed under given conditions in an aquifer.

4.1. HYDROCARBON BIOTRANSFORMATION

Petroleum hydrocarbons are naturally occurring chemicals featuring a variety of molecular weights and functional groups. Benzene, toluene, ethylbenzenes, and xylenes (BTEX) are components of gasoline that are widely distributed and have been the focus of substantial biodegradation and bioremediation research. BTEX, the low molecular weight fuels, and structurally-simple alcohols and ketones are readily mineralizable aerobically. Hence, these compounds have been successfully removed from contaminated sites via established bioremediation procedures. Recently, many BTEX components, traditionally considered resistant to anaerobic microbial attack, have also been found to be biodegradable under a variety of anaerobic conditions [14-21, 92-97]. The high molecular-weight petroleum components and creosotes are only slowly metabolized – partly as a result of their structural complexity, low solubility, and strong sorptive characteristics. Thus, bioremediation techniques for these latter classes of petroleum hydrocarbons are still emerging [22].

One prominent oxygenated hydrocarbon, notably resistant to biodegradation, is methyl *tert*-butyl ether (MTBE). MTBE is often added to gasoline at up to 15% by volume. It features three traits (a stable ether linkage, a tertiary carbon structure, and reactivity with microbial membrane) which prevent MTBE from readily serving as a carbon and electron source for microbial growth; consequently MTBE is often a threat to water quality [98]. Slow MTBE biodegradation was suggested in one field study [22] and in both aerobic [23] and anaerobic [24] laboratory studies. When microorganisms possess one of a variety of nonspecific oxygenase enzyme systems, oxygen atoms can be fortuitously inserted into MTBE by a process known as cometabolism [25]. Oxygen insertion renders MTBE susceptible to both chemical and enzymatic attack.

Polycylic aromatic hydrocarbons (PAHs) are low-solubility aromatic hydrocarbons that are frequently found in soils. PAHs are neutral, nonpolar organic molecules composed of two or more benzene rings fused together in various configurations [26, 27]. PAHs may also contain alkyl substituents, or become heterocyclic if carbon atoms in the aromatic rings are replaced with nitrogen, oxygen or sulfur atoms. PAHs are a generally hazardous class of materials found in emissions from fossil fuel utilization and conversion processes. PAHs are formed from coking of coal, petroleum refining and coal tar refining industries, and are found as contaminants in many soil, sediment and groundwater systems impacted by industrial and hazardous wastes. The fate of PAHs in subsurface systems is governed largely by their hydrophobicity. Also important is whether the PAHs exist as potentially mobile nonaqueous phase liquids (NAPLs). PAH compounds have low aqueous solubilities, ranging from about 31 mg/L for naphthalene, a two-ring PAH, to several μg/L and less for molecules composed of five rings or more. Adsorption may also influence PAH metabolism via surface phenomena controlling chemical and biological processes.

Physiological, biochemical and genetic aspects of PAH metabolism by bacteria, fungi and algae have been studied using pure cultures and models of terrestrial and aquatic environments [26, 27]. PAHs composed of up to 5 benzene rings can serve as growth substrates (carbon and energy sources) for some microorganisms. Bacteria employ dioxygenase enzymes that initially oxidize aromatic hydrocarbons to cis-dihydrodiols; further oxidation leads to the formation of catechols that are substrates for other microbial reactions causing aromatic ring cleavage. Fungi oxidize PAHs via monooxygenases, yielding either a trans-dihydrodiol or an alcohol. Some algae can also hydroxylate PAHs [26, 27]. Because of oxygen's ability to react with highly stable aromatic rings of PAHs, it is generally accepted that effective microbial degradation of PAHs requires aerobic environments. However, microbial degradation of lower molecular weight PAHs have been reported in the absence of oxygen — under denitrifying [28, 29], and sulfate reducing conditions [30, 31].

4.2. HALOGENATED COMPOUND BIOTRANSFORMATION

Though halogenated organic compounds have been found in nature [32], these are not of commercial significance compared to the synthetic chemicals listed in Table 1. When halogen atoms (chlorine, bromine, and fluorine) are introduced into organic molecules, many properties, such as solubility, volatility, density, hydrophobicity, stability, and toxicity change markedly. These changes confer improvements that are valuable for commercial products, but also have serious implications for microbial metabolism. The susceptibility of the chemicals to both chemical and enzymatic attack is sometimes drastically altered by halogenation and persistent compounds often result.

Halogenated aliphatic compounds are non-aromatic hydrocarbons in which varying numbers of hydrogen atoms have been replaced by halogen atoms. Halogenated aliphatics are effective solvents and degreasers that have been widely used in many manufacturing and service industries. Some highly chlorinated representatives of this class, such as PCE, are generally resistant to aerobic microbial attack, while being susceptible to anaerobic reductive dehalogenation under methanogenic and other anaerobic conditions (Table 2). In fact, from laboratory and field evidence complete reductive dechlorination from PCE to the relatively non-toxic ethene can occur [33, 34, 35, 36, 38].

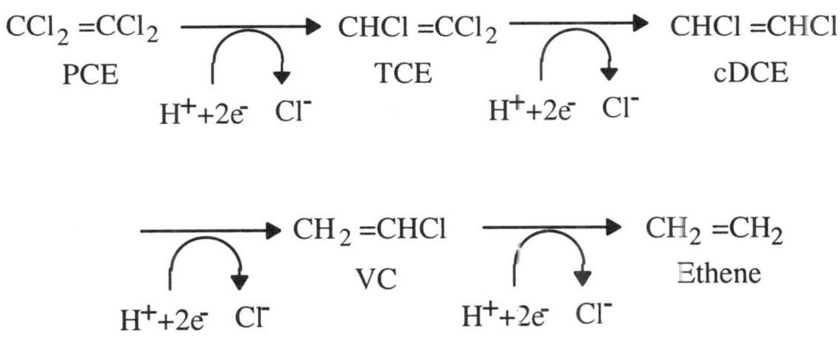

Furthermore, as the degree of halogenation in aliphatics diminishes, susceptibility to other metabolic reactions increases. For instance, VC (a compound created by reductive halogenation of 3 of the 4 chlorines on the tetrachloroethene molecule) can be converted to CO_2 by iron-reducing [39] and aerobic [40, 41] microorganisms. Aerobic cometabolism of partially halogenated aliphatics has been proven for microorganisms supplied with substrates such as methane, toluene, or phenol that activate the oxygenase enzymes. For example, oxidation of methane in groundwater can be accompanied by cometabolism of some chlorinated ethenes:

Table 2. Major chlorinated aliphatic hydrocarbon (CAH) contaminants found in groundwater and biotransformation pathways known to exist (indicated with an X) (After NRC, 5).

			Primary Substrate			Cometabolism	
				Anaerobic			
CAH	Formula	Acronym	Aerobic Donor	Donor	Acceptor	Aerobic	Anaerobic
Methanes:							
Carbon Tetrachloride	CCl_4	CTC					X
Chloroform	$CHCl_3$	CF				X	X
Dichloromethane	CH_2Cl_2	DCM	X			X	X
Chloromethane	CH_3Cl	CM	X			X	X
Ethanes:							
1,1,1-Trichloroethane	CH_3CCl_3	TCA				X	X
1,1,2-Trichloroethane	$CH_2ClCHCl_2$	1,1,2-TCA				X	X
1,1-Dichloroethane	CH_3CHCl_2	1,1-DCA				X	X
1,2-Dichloroethane	CH_2ClCH_2Cl	1,2-DCA	X	X		X	X
Chloroethane	$CH3CH_2Cl$	CA	X			X	X
Ethenes:							
Tetrachloroethene	$CCl_2{=}CCl_2$	PCE			X		X
Trichloroethene	$CHCl{=}CCl_2$	TCE			X	X	X
cis-1,2-Dichloroethene	$CHCl{=}CHCl$	c-DCE		X	X	X	X
trans-1,2-Dichloroethene	$CHCl{=}CHCl$	t-DCE		X		X	X
1,1-Dichloroethene	$CH_2{=}CCl_2$	1,1-DCE				X	X
Vinyl Chloridee	$CH_2{=}CHCl$	VC	X	X	X	X	X

Methane Oxidation

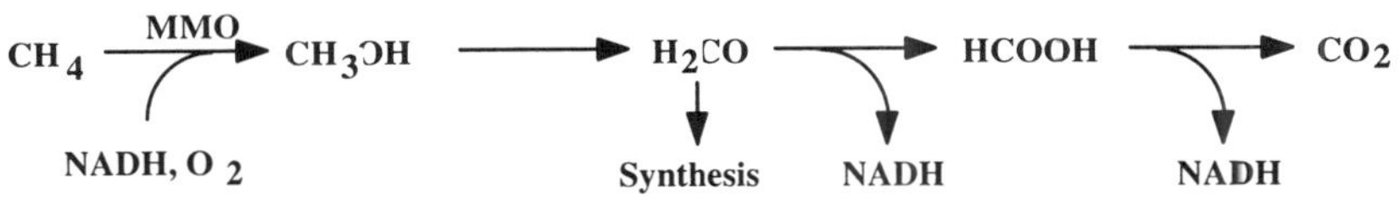

(1)

TCE Epoxidation

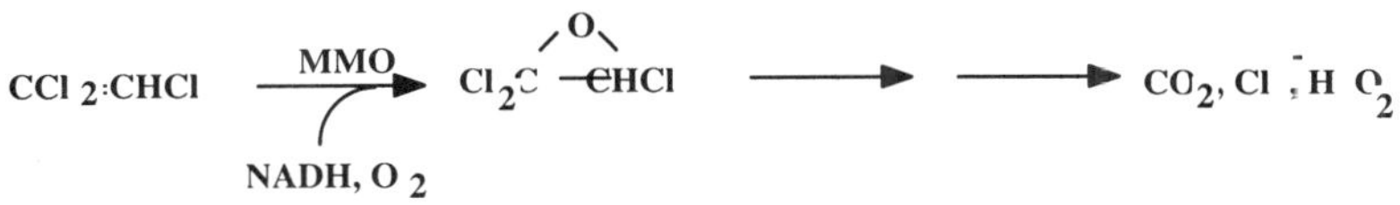

(2)

This radically alters both the toxicity of biodegradability of the halogenated ethenes. One structurally simple chlorinated aliphatic compound is Dichloromethane (DCM). This has been shown to support growth of microorganisms by serving as a carbon and energy source under aerobic, denitrifying, methanogenic, and fermenting conditions [42-44]. Thus, DCM is readily amenable to natural attenuation.

Halogenated aromatics, such as phenoxy acetic acid pesticides, pentachlorophenol, polychlorinated biphenyls (PCBs) and others (Table 1) consist of one or more benzene rings that bear halogens as well as other chemical functional groups (i.e., hydroxyls, carboxyls, etc.). The aromatic benzene nucleus is susceptible to both aerobic and anaerobic metabolism, though the latter occurs relatively slowly. Overall however, the presence of halogen atoms on the aromatic ring, their position, and their interaction with functional groups is what governs biodegradability [94,95]. A high degree of halogenation may prevent aromatic compounds from being oxidatively (aerobically) metabolized, as is the case for PCBs. However, as discussed above for the aliphatic compounds, the highly halogenated aromatics are subject to anaerobic reductive dehalogenation. As hydrogens replace the halogen atoms, the molecules become susceptible to aerobic attack. Thus a common bioremediation scenario for treating soils, sediments, or water contaminated with halogenated aromatic chemicals is anaerobic dehalogenation followed by aerobic mineralization of the residual compounds. It should be noted, however, that when the proper substituent group accompanies the halogens on the aromatic ring, aerobic metabolism might proceed rapidly, as is the case for pentachlorophenol. Other prominent chlorinated aromatic-contaminants are the dioxins (e.g., tetrachlorodibenzo-p-dioxin, TCDD) and polychlorinated dibenzofurans (e.g., TCDF). TCDD and TCDF are potent teratogens and carcinogens whose complex chemical structure and strong sorptive properties render them nearly nonbiodegradable — although dechlorination reactions have been reported [45].

4.3. NITROAROMATIC COMPOUND BIOTRANSFORMATION

Nitroaromatic organic contaminants associated uniquely with military activities include the explosives, TNT, RDX, and HMX (octahydro-1,3,5,7,-tetranitro-1,3,5,7-tetrazocene). Manufacturing and decommissioning operations have generated large quantities of explosives waste products, often deposited in soils and unlined lagoons where leaching can deliver these materials to groundwater. Despite a preponderance of sites contaminated with explosives, rigorous field studies documenting the transport, fate, and influence of microbial activity on explosives are rare. One of the most complete field studies examining TNT was reported by

USGS personnel [46]. These investigators used hydrogeological and ground water chemistry data from a site near Hawthorn, Nevada to calculate transport parameters, including linear velocity per year of TNT and associated compounds in groundwater impacted by an ammunition plant. Based on field mobilities alone, Van Denburgh et al. [46] suggested that microbial transformation of TNT to inorganic nitrogen contributed to depletion of TNT. However, they failed to provide key information, such as mass balances and field-derived geochemical footprints mechanistically linking TNT metabolism to depletion of final electron acceptors or the production of expected metabolites. Other TNT contaminated sites have been examined [e.g., 47. 48] but the data tying field observations to the physiological processes of microorganisms has been incomplete. Major causes for the absence of unequivocal evidence of natural attenuation of explosives is that these chemicals are resistant to microbial metabolism, are toxic, and undergo complex chemical reactions. From laboratory studies, the potential for microorganisms to metabolize nitroaromatic compounds has been reported [49-52]. However, despite decades of study, the explosives, TNT and RDX, have not been shown to serve as readily utilizable carbon or energy sources for microbial growth. Under both aerobic and anaerobic conditions, the nitrate moieties on explosives are reduced to amino groups that can increase toxicity, undergo polymerization reactions, and/or strongly sorb onto soil solids [49, 53]. Recent reports have shown that aerobically grown bacteria can use TNT as a nutritional nitrogen source [55]. Nonetheless, the known microbial transformations of nitroaromatics are limited in extent.

4.4. INORGANIC COMPOUND BIOTRANSFORMATION

Microbiological processes also affect inorganic environmental contaminants [19, 56-69]. Unlike organic compounds, the majority of inorganic contaminant compounds are subject only to changes in speciation that may influence contaminant mobility. Microbial interactions may cause precipitation, volatilization, sorption, and solubilization of inorganic compounds. Mechanistically, these reactions can be the result of direct enzymatic processes such as oxidation, reduction, methylation, or uptake. Reaction mechanisms can also be indirect (nonenzymatic) — resulting from microbiological production of metabolites or biomass that can strongly influence the behavior of inorganic contaminants via redox, acid/base, coprecipitation, sorption, and other geochemical means.

One nearly universal means by which microorganisms have been shown to reduce concentrations of inorganic contaminants in water (e.g., Cu, Ni, Zn, Cd, and Pb) is by immobilizing aqueous-phase inorganics in microbial biomass and/or microbial exopolymers [70, 72]. Many inorganic contaminants, especially metals, become relatively soluble hence mobile at low pH. The highly abundant nontoxic iron and manganese exist in reduced and oxidized states. The oxidized states [Fe (III), Mn (IV)] react chemically to form oxyhydroxide precipitates that serve as physiological electron acceptors for anaerobic microbial food chains [64]. The end products of Fe- and Mn-reduction [Fe (II) and Mn (II)] are relatively soluble and may migrate to aerobic habitats, where reoxidation and precipitation can also be catalyzed by microorganisms [72]. The behavior of many of the toxic metals discussed below is intimately tied to the microbially-mediated cycling of Fe and Mn. The toxic metals may be immobilized (through coprecipitation and sorptive reactions with many Fe and Mn oxides) or solubilized (by being reduced via chemical reactions with Fe (II) and Mn (II)). Thus, many inorganic compounds undergo immobilization reactions via sorption and precipitation.

Chromium is a metal whose key oxidation states are (VI) and (III). In aqueous environments, Chromium (VI) predominates as the mobile and highly toxic anion, chromate (CrO_4^{2-}) and dichromate ($CR_2O_7^{2-}$). Reduced chromium (III) is less toxic and less mobile because of precipitation reactions as oxides and hydroxides at pH 5 and above. A variety of both aerobic and anaerobic microorganisms have been shown to enzymatically reduce Cr (VI) to Cr (III).

Mercury is a toxic metal whose predominant forms include mercuric ion [Hg (II)], elemental mercury [Hg (0)], and the biomagnification-prone organic mercury compounds, monomethyl- and dimethyl-mercury. All transformations of mercury by microorganisms are considered detoxification reactions that are intended to mobilize mercury away from microbial cells. Most reactions are enzymatic, carried out by both aerobes and anaerobes, and involve uptake of Hg (II) followed by its reduction to volatile forms (elemental Hg (0), methyl-, and dimethyl-mercury). Mercuric ion [Hg (II)] also forms highly insoluble precipitates with sulfide.

In addition to mercury, microorganisms are capable of methylating other metals (Cd, Pb, Sn, Te, Se [73]. Additional methylation reactions may occur as a result of nonbiological transmethylation by microbially-produced methylated donor compounds such as trimethyl tin [73]. These donors may react with ionic forms of Pd, Th, Pt, and Au, but the resultant reduced metals may not be chemically stable.

Arsenic is a toxic element capable of existing in five different valence states [As (-III), (0), (II), (III), and (V)]. Forms of arsenic range from sulfide minerals (e.g., AS_2S_3) to elemental As to arsenic acid to arsenite (AsO_2^-) to arsenate (AsO_4^{3-}) to various organic forms that include methylated arsenates and trimethyl arsine. Clearly, the chemical and microbiological reactions of As are complex [73, 74). Both anionic forms (arsenite and arsenate) are highly soluble and highly toxic — interfering with various enzyme functions and oxidative phosphorylation, respectively. However, no form of As is nontoxic. Microorganisms transform arsenic for several fundamental physiological reasons: (1) under anaerobic conditions, arsenate [As (V)] can be used as a final electron acceptor; (2) under aerobic conditions, reduced As (e.g., arsenite) oxidation had been shown to generate energy (ATP); and (3) under both anaerobic and aerobic conditions. As can be mobilized away from microbial cells by methylation, oxidation, or reduction mechanisms.

Although selenium is an important and beneficial micronutrient for plants, animals, humans, and some microorganisms (largely because of its role in some key amino acids), this element can be toxic at greater than trace concentrations. In natural environments, selenium has four predominant inorganic species: Se (VI) (selenate, SeO_4^{2}), Se (IV) (selenite, SeO_3^{2-}), Se (0) (elemental selenium), and Se (-II) (selenide) [19, 73, 74]. Like arsenic, selenium also has many volatile organic forms; these include: dimethyl selenide, dimethyl diselenide, methane selanone, methane selenol, and dimethyl selenyl sulfide. Each of these compounds exhibits its own chemical and biochemical behavior, mobility, and toxicity. The various forms of selenium are transformed by microorganisms according to their physiological needs and the ambient thermodynamic conditions. Reduced inorganic selenium compounds can be oxidized under aerobic conditions, though not linked to microbial growth. Oxidized selenium (selenate) can serve as a final electron acceptor for anaerobic microorganisms – resulting in production of selenide and/or elemental Se. Methylation of the various Se compounds is thought to be a protective detoxification mechanism that mobilizes Se away from microbial cells. Thus like As, the environmental fate of Se is governed by complex interactions between chemical and physiological processes. For instance, anaerobic microbial reduction of selenate and selenite to insoluble elemental selenium represents an important mechanism for immobilizing and removing Se from aqueous solution.

Oxyanions are water soluble, negatively charged chemical species in which a central atom is surrounded by oxygen. Nitrate (NO_3^-) is a naturally-occurring oxyanion commonly found at low concentrations in soils, sediments, and both surface and groundwaters as a result of the biogeochemical cycling of organic matter. Nitrate is a readily-utilizable form of nitrogen that can be assimilated into amino acids by plants and microorganisms. Although serving to supply nitrogen, an essential nutrient, nitrate is also a serious health concern for at least two reasons: (1) it can be a chemical or microbiological precursor for nitrite, which spontaneously combines with amino compounds to form highly carcinogenic nitrosamines; and (2) nitrate can be reduced to nitrite in stomachs of infants, which can cause the respiratory stress diseases, methemaglobinemia. Nitrate is produced from ammonia by nitrifying microorganisms under aerobic conditions. The major microbial process that destroys nitrate is dissimilatory reduction to dinitrogen gas (denitrification). Genetic, biochemical, physiological, ecological, environmental, and engineering aspects of denitrification [75, 76] have been examined for decades. As shown in Table 1, nitrate is used as a physiological electron acceptor under oxygen-free (anaerobic conditions). The denitrification process is widespread among microorganisms, and occurs reliably in every anaerobic habitat with abundant carbon and electron sources. Denitrification is a well-established bioremediation process and is used routinely in sewage treatment plants to curb eutrophication.

The oxyanions chlorate (ClO_3^-) and perchlorate (ClO_4^-) or their precursors (chlorine dioxide, hypochlorite, and chlorite) are produced by a variety of paper manufacturing, water disinfection, and both aerospace and defense industries. Although not naturally occurring, these highly oxidized forms of chlorine are energetically very favorable physiological electron acceptors for microorganisms. Compared to denitrification, knowledge of chlorate and perchlorate biodegradation reactions is quite limited. However, laboratory studies using both pure bacterial cultures and environmental samples (soil, freshwater sediments, and sewage) have shown that in the presence of common electron donors (carbohydrates, carboxylic acids, amino acids, even H_2 and H_2S), microorganisms can grow at the expense of perchlorate and chlorate; thus reducing them to the nontoxic chloride ion [77-79].

Uranium is a radioactive element that releases alpha, beta, and gamma radiation that can be toxic. Uranium can exist in the oxidation states of (0), (III), (IV), (V), and (VI), though in nature insoluble U (IV) O_2 and soluble U $(VI)O_2^{2+}$ predominate. Recently, U (VI) has been shown to serve as a terminal electron acceptor for anaerobic microorganisms; thus, in anaerobic habitats growth-linked reduction (hence immobilization) of U should be a reliable process [64]. Although the reverse process, microbial oxidation of U (IV) to U (VI) under aerobic conditions has been demonstrated, this process has not been shown to be physiologically beneficial to the responsible microorganisms. Direct chemical oxidation of U (IV) by molecular oxygen [creating U (VI)] may also influence the robustness of U bioremediation strategies. Another radioactive element, plutonium (Pu) has also been shown to be susceptible to microbial transformation. Iron reducing microorganisms were found to reduce insoluble PU^{4+} to the more soluble PU^{3+}; thus, rendering the soluble form more susceptible to mobilization.

Technetium (99Tc) is a beta-emitting radionuclide produced by the nuclear industry. Tc can exist in oxidation states that range from (-I) to (VII); the latter highly-stable state is typified by the very soluble and mobile pertechnetate anion, which predominates in aerobic soils. A variety of anaerobic microorganisms, pure and uncharacterized soil cultures, have been shown in laboratory studies to convert Tc (VII) to less oxidized, insoluble forms that can be immobilized onto microbial cells and/or other solids [80].

5. Likelihood of Success

As becomes evident from the above discussion, some organic compounds are readily biodegraded under all redox conditions and by organisms that are ubiquitous in the environment. The likelihood that natural attenuation will occur and will be sufficient to normally provide adequate protection to human health and the environment is high. Probably most organic contaminants discharged in the environment, including natural products such as proteins, carbohydrates, and fats, fall under this category. The environment has a natural capacity to rid itself of such materials. Biological transformations of many inorganic species of concern also can occur, changing their physical, chemical, and biological properties in ways that may be protective or may be harmful. Some organic contaminants, both manmade and natural (lignin for example), are difficult to degrade and may do so only under certain conditions and in the presence of organisms that may not be widespread in the environment. When such compounds constitute a hazard to humans or the environment, whether natural products (such as BTEX) or manmade (such as chlorinated solvents and PCBs), then justifiable environmental concerns about their presence results. The more difficult a compound is to degrade, the fewer environmental conditions under which it degrades, the more poorly distributed are the organisms that can effect complete biodegradation, and the higher the potential for formation of hazardous daughter products, then the greater the uncertainty of whether or not natural attenuation is likely to a successful remedial approach for the contamination. Uncertainty also increases in cases where we have insufficient knowledge of fate processes than when our scientific knowledge about the processes is extensive, well documented, and well accepted. The "likelihood of success" is thus a function of both compound characteristics and the adequacy of existing scientific knowledge.

The U.S. National Research Council estimated the likelihood of success for natural attenuation to be a successful remediation alternative for many contaminants of concern based upon environmental circumstances and level of scientific understanding of the natural attenuation processes for that contaminant. A summary of the collective judgement of the NRC committee is presented in Table 3 [1]. The definitions they provided for the various levels (low, moderate, and high) are provided in footnotes to the Table. Only a few of the contaminants are rated as "high." The level of effort required to demonstrate that natural attenuation is occurring and is protective is a function of the likelihood of success. More effort is required for those with low likelihood to provide the proof required by regulatory agencies and the public that the natural processes are indeed occurring, and will continue to occur over time.

6. Evaluating Natural Attenuation

6.1. CREATING A CONCEPTUAL MODEL

Evidence that a compound is disappearing at a site or that a plume appears to be shrinking with time is insufficient in itself to confirm that natural attenuation is occurring. Changes in site hydrogeology may cause a plume to shift in direction so that changes in concentration at a given monitoring location may not indicate that real changes in mass have actually occurred. Because of the large expense of installing monitoring wells, the number of wells established is frequently insufficient to be certain where the centerline of a plume lies and the exact direction of its movement. For this reason, monitoring locations thought to be along the center line may tend to suggest the plume is not extending, when indeed it is extending, but in a direction different from that thought. Many contaminants can be removed biologically or chemically, but through the formation of daughter products, that may be more hazardous than the parent compounds. Thus, information in addition to simple observation that a compound appears to be disappearing is required to confirm natural attenuation is occurring, and indeed that it is occurring in sufficient time and with completion so that it is truly protective of human health and the environment. In order to clearly establish that natural attenuation is occurring and is protective, a weight-of-evidence approach is required. Several types of field data combined with conceptual models of the subsurface and an understanding of fate processes for the contaminants of concern generally are needed to clearly link observations of disappearance with underlying mechanisms that may be responsible for that disappearance. The amount of evidence required generally depends upon the knowledge of fate processes for a given compound and the relative ease of transformation of the compounds to harmless end products. Gasoline hydrocarbons, such as BTEX compounds, are readily biodegraded by microorganisms that are ubiquitous in the environment. Thus, disappearance of BTEX compounds in groundwater can generally be expected. The weight-of-evidence required to ascertain adequate natural attenuation here is much less than in the case of chlorinated solvents, which are more difficult to degrade, and their degradation often leads to the formation of other compounds which are also hazardous. If a compound is being transformed, generally footprints are left which can be used to support the case that disappearance is truly associated with degradation. Examples of footprints and their interpretation for the important case of chlorinated solvents are presented in the following. Other examples are provided in the NRC publication upon which much of this chapter is based [1].

6.2. NATURAL ATTENUATION FOOTPRINTS FOR CHLORINATED SOLVENTS

Three fundamental biological processes are responsible for most of the natural biodegradation of chlorinated solvents in the natural environment: reduction, oxidation, and fermentation. Combinations and variations of these processes create the footprints that we look for. Each biological process creates a group of related, sometimes distinctive geochemical patterns. Each of these geochemical signatures is known as a "footprint". A distinctive combination of footprints is called a natural attenuation "pattern". For example, degradation of a chlorinated compound should cause the chloride to be released into ground water as dissolved chloride. These footprints and patterns are discussed in more detail in the NRC report [1] and in the ITRC document "Principles and Practices of Natural Attenuation for Chlorinated Solvents". This following discussion is largely drawn from those documents.

Table 3. Summary of dominant processes affecting attenuation of environmental contaminants, the level of scientific understanding of the processes, and the likelihood of successful remediation through natural attenuation (after NRC, 5).

Chemical Class	Dominant Attenuation Processes	Level of Understanding	Likelihood of Success
ORGANIC			
Hydrocarbons			
BTEX	Biotransformation	High	High
Gasoline, fuel oil	Biotransformation	Moderate	Moderate
MTBE	Biotransformation	Low	Low
PAHs	Biotransformation Immobilization	Moderate	Low
Creosote	Biotransformation Immobilization	Low	Low
Oxygenated Hydrocarbons			
Low MW alcohols, ketones, esters	Biotransformation	High	High
Halogenated Aliphatics			
PCE, TCE, CT	Biotransformation	Moderate	Low
TCA	Biotransformation Abiotic transformation	Moderate	Low
MC	Biotransformation	Moderate	Moderate
Halogenated Aromatics			
Highly Chlorinated			
PCBs, Dioxins, TCDF, PCP	Biotransformation Immobilization	Low	Low
	Biotransformation		
Less chlorinated			
PCBs, Dioxins	Biotransformation	Low	Low
Nitroaromatics			
TNT, RTX	Biotransformation Abiotic transformation Immobilization	Low	Low

Table 3 Continued

Chemical Class	Dominant Attenuation Processes	Level of Understanding	Likelihood of Success
INORGANIC			
Metals			
Ni	Immobilization	Moderate	Moderate to High
Cu, Zn	Immobilization	Moderate	Moderate
Cd	Immobilization	Moderate	Low
Pb	Immobilization	Moderate	Moderate
Fe, Mn	Biotransformation Immobilization	Moderate	Low to Moderate
Cr	Biotransformation Immobilization	Moderate	Low to Moderate
Hg	Biotransformation Immobilization	Moderate	Low
Non-metals			
As	Biotransformation Immobilization	Moderate	Low
Se	Biotransformation Immobilization	Moderate	Low
Oxyanions			
Nitrate	Biotransformation	High	Moderate
Perchlorate	Biotransformation	Low	Low
Radionuclides			
^{60}Co	Immobilization	Moderate	Moderate
^{137}Cs	Immobilization	Moderate	Moderate
^{3}H	Decay	High	Low
^{90}Sr	Immobilization	High	Moderate
99 Tc	Biotransformation Immobilization	Low	Low
$^{238,239,240}Pu$	Immobilization	Low	Low
$^{235,238}U$	Biotransformation Immobilization	Moderate	Low

Three steps are used when evaluating footprints of attenuation. 1) Create a detailed conceptual model of the site. 2) Collect and analyze an appropriate set of site-specific data. Design the data collection so that the data specifically tests the validity of the conceptual model. 3) Establish a long-term monitoring program to document that the natural attenuation remedy processes are operating as anticipated.

When evaluating natural attenuation of chlorinated solvents it is best to find at least several footprints to develop confidence that a proposed degradation process is in fact operating. The level of evidence needed to support natural attenuation as a protective remedy at a chlorinated solvent site correlates directly related to the level of that site's complexity. A complex site should be required to provide a much higher level of proof than a simple site.

There are general similarities between chlorinated solvent footprints. The most useful footprint is the accumulation of dissolved chloride that happens whenever a chlorinated solvent is degraded. Dissolved chloride is not reactive so it is transported as a conservative tracer. Some elements of a footprint may be very hard to observe, in particular the source and usage of low-solubility organic substrates.

Mass balance should always be checked when analyzing natural attenuation. The stoichiometries of common natural attenuation reactions for chlorinated solvents are well known. If the stoichiometries of proposed natural attenuation reactions are incompatible with the site's ground water chemistry, then those reactions cannot be solely responsible for effective natural attenuation. For example, direct oxidation of vinyl chloride (VC) with oxygen requires 1.3 mg/l of oxygen for each 1 mg/l of VC oxidized. If the maximum ground water concentration of oxygen is 8 mg/l then the maximum concentration of VC that can be degraded by this process is 6.2 mg/l. Therefore, a site that has 20 mg/l VC in its ground water cannot look to direct oxidation as a mechanism for complete natural attenuation. As another example, if 100 mg/l of TCE is observed leaving a site's source area and is completely degraded, 83 mg/l dissolved chloride should be seen in ground water down gradient of that destruction. Mass balancing of chlorinated solvent destruction is rarely perfect due to the inherent flaws of the site assessment process.

Natural attenuation footprints cannot be observed if the necessary measurements and chemical analyses are not conducted. A standard list of chemical analyses can allow most patterns to be observed. The recommended analyses are: volatile organic compounds (VOCs) and chlorinated VOCs; speciation of dichloroethene (DCE) isomers; methane; ethane; ethene; propane; dissolved oxygen, redox; pH; conductivity; temperature; BOD, and the inorganic ions Fe^{+2}, Mn^{+2}, Cl^{-}, NO_3, SO_4^{-2} H_2S, and alkalinity. Other key measurements are hydraulic conductivity, the piezometric surface, and aquifer matrix fraction of organic carbon.

6.2.1. *Reductive Dechlorination*

Reductive dechlorination is the most important natural attenuation process for chlorinated solvents because it is the only process that can biodegrade fully chlorinated compounds. The key footprints of this process are the formation of less-chlorinated daughter compounds, and the indicators of highly reduced aquifer geochemistry. For example, when PCE is the parent contaminant it may dechlorinate sequentially to dissolved chloride and TCE, then to cis-DCE, to VC, and to ethene. The VOC pattern for reductive dechlorination of PCE is shown in Figure 5. Carbon tetrachloride may be reductively dechlorinated to CF, and then to DCM which is rapidly fermented. 1,1,1-TCA dechlorinates to form 1,1-DCA, which rapidly biodegrades. Other compounds such as PCB's, chlorinated propanes, hexachlorobutadiene, 1,2-dichloroethane, and FREONS have also been shown to undergo reductive dechlorination. High concentrations of chlorinated solvents suppress acetate-using methanogens, so acetate may sometimes be seen down gradient from high-concentration source areas [81, 83].

The inorganic geochemical footprints and pattern for reductive dechlorination, shown in Figure 6, are based on the consumption of electron donors. Where reductive dechlorination is operating the aquifer geochemistry should be consistent with reductive processes. Dissolved oxygen and nitrate should be absent, redox should be below -100 mv, removal of sulfate by sulfate reduction should be active, and methanogenesis is seen at most sites. The loss of electron donor may be hard to directly observe. Electron donors may be native aquifer carbonaceous materials or may be hydrocarbons, oils, and greases that were dissolved in the solvents before they were released, or co-disposed organic compounds. Oils and greases have been shown to be important substrates, but cannot be observed by groundwater sampling since they are strongly sorbed to the aquifer matrix.

In order for reductive natural attenuation of chlorinated ethenes to proceed to completion certain critical bacteria need to be present. Even if a site's geochemical footprints are completely favorable for natural attenuation, the reactions generally cannot be completed without the bacterium *Dehalococcoides ethenogenes* [35]. This bacterium was shown [79] to be widely distributed in North America and northern Europe. The presence of this bacterium is a key footprint.

6.2.2. *Direct Oxidation*

The chlorinated solvents that can be degraded by direct oxidation are those that can be used as electron donors. These include DCE, VC, DCM, dichloroethane (DCA), and chlorobenzene. The footprints and patterns for oxidation are shown in Figures 7 and 8. The concentration of the solvent begins to decline in the source area at a rate greater than predicted by transport processes alone. Oxygen is partially depleted in the source area. Down gradient it gradually returns to near background levels. The concentration of dissolved chloride increases as the solvent degrades [84].

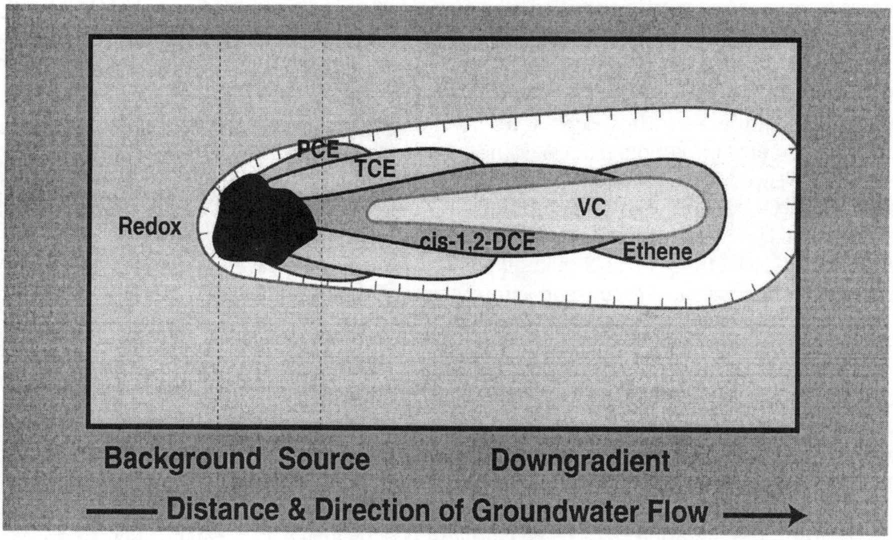

Figure 5. Distribution of PCE and daughter compounds on an idealized reductive dechlorination site. Plan view.

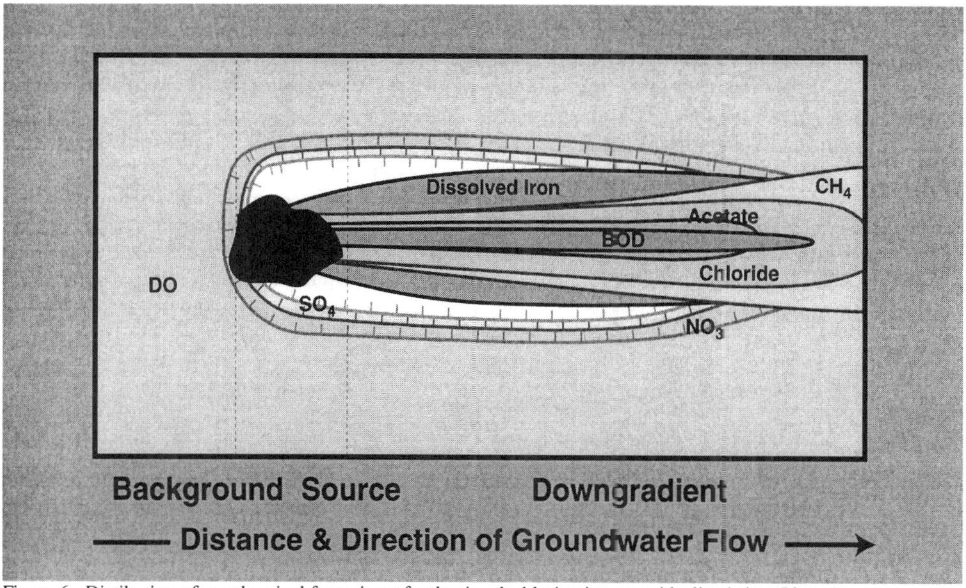

Figure 6. Distibution of geochemical footprints of reductive dechlorination at an idealized site. Plan view.

The stoichiometry of oxidation reactions is very important due to the limited solubility of oxygen in ground water of about 8 mg/l. Eight mg/l of oxygen is enough to fully degrade 21 mg/l of DCM, 6.2 mg/l of VC, 9.9 mg/l of DCA, or 12 mg/l of DCE. If ground water contains less than eight mg/l of oxygen or if it contains higher concentrations of these contaminants natural attenuation via direct oxidation cannot degrade all of the contaminants.

In cases where sites have co-contamination of certain organic, especially toluene, some co-oxidation of TCE may occur. This reaction is relatively inefficient. Ground water containing 8 mg/l of dissolved oxygen is likely to degrade a maximum of about 0.3 mg/l TCE via this reaction. The key footprints of this reaction would be a decline in dissolved oxygen and toluene, the loss of some TCE, and the creation of dissolved chloride.

6.2.3. *Sequential Reduction/Oxidation*

Reductive dechlorination followed by oxidation occurs at sites where the amount of organic substrate co-disposed with the chlorinated solvents is not sufficient to maintain the entire plume in a reduced state. However, before reductive dechlorination is complete these plumes become oxygenated again. This appears to be a common pattern.

The footprints of this pattern are illustrated in Figures 9 and 10. In the up gradient area the footprints and pattern are the same as for reductive dechlorination. The parent VOCs (e.g. PCE) decline or are eliminated in up gradient area; DCE and possibly VC are produced, but may not be fully degraded. VC and ethene disappear down gradient while any remaining TCE or DCE may persist. At times DCE may degrade fully in an oxygenated environment, but this does not always occur.

Oxygen and nitrate are depleted near the source area. Sulfate reduction and methanogenesis are often observed down gradient from the source area. Further down gradient oxygen and nitrate reappear. Sulfate reduction and methanogenesis do not occur down gradient from that point [85, 86].

6.2.4. *Anaerobic Oxidation*

Anaerobic oxidation of certain chlorinated compounds may occur down gradient from a highly reduced source area [87]. In that source area the parent compounds are dechlorinated to ones which can now be used as electron donors. The chlorinated solvents that can be degraded by anaerobic oxidation are the same ones that can be used as electron donors in direct oxidation. These include DCE, VC, DCM, and DCA. In these reactions oxidized iron and manganese compounds in the aquifer matrix act as the electron acceptors. Stoichiometric limits are not calculated for these reactions because aquifers vary widely in the iron and manganese concentrations. The key footprints of these reactions are appearance of dissolved iron and manganese in ground water, and the accumulation of dissolved chloride. Dissolved oxygen should be absent, and the aquifer should not be so reduced that sulfate reduction is active. There are suggestions in field data that sulfate-reducing bacteria can oxidize DCE, but this has not yet been confirmed by rigorous laboratory testing.

When anaerobic oxidation occurs down gradient from an area of reductive dechlorination, the resulting patterns are similar to sequential reduction/oxidation (Figures 9 and 10). The primary difference is the reappearance of dissolved oxygen in the down gradient area.

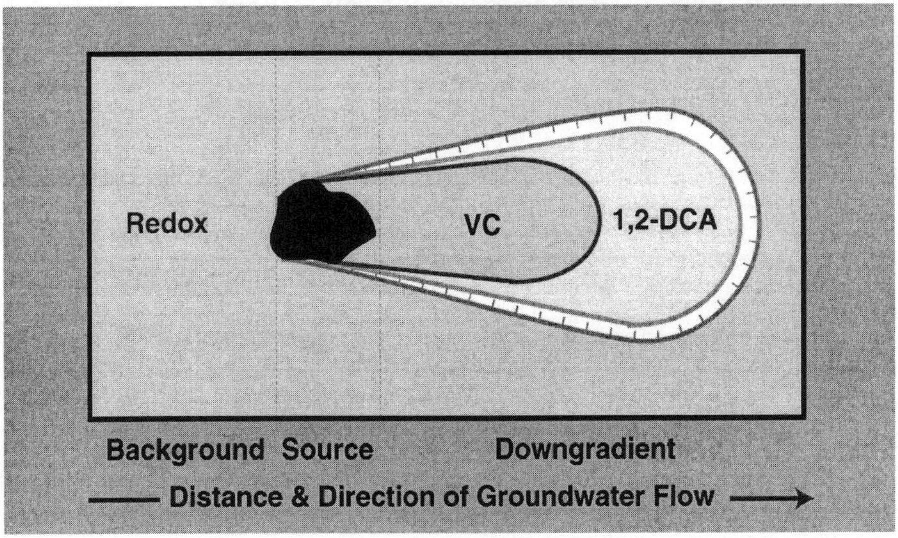

Figure 7. Distribution of compounds on an idealized oxidation site. Map view.

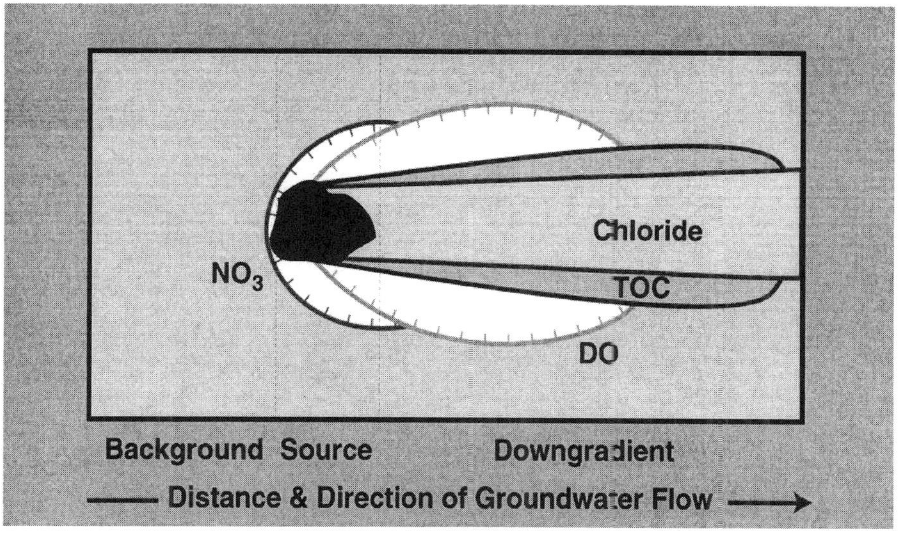

Figure 8. Distribution of geochemical footprints on an idealized oxidation site. Plan view.

6.2.5. *Fermentation*

Fermentation is a very important reaction in natural biodegradation of DCM. DCM is been well documented to ferment to acetate and formate in the presence of sulfate [88]. Although sulfate reducing or methanogenic bacteria do not use DCM directly, the acetate and formate created by the fermentation reaction are substrates that those bacteria readily use. Therefore, DCM fermentation is usually observed simultaneously with sulfate reduction and methanogenesis [42].

DCM is much more soluble than other chlorinated solvents. Its solubility is roughly 40,000 mg/l [89]. The combination of high solubility and high degradability gives a high rate of reaction even under natural circumstances. Field half-life is on the order of 10 days. Due to these factors DCM plumes tend to be very short - generally less than 100 m long [1].

The only organic footprint of natural attenuation for DCM is its disappearance. The inorganic footprints are the same as those for reductive dechlorination illustrated in Figure 7. Dissolved oxygen, nitrate, and sulfate are all depleted as electron acceptors during acetate and formate degradation; a substantial plume of dissolved chloride is created, sulfate is reduced to sulfide, and there is clear methanogenesis. Some times acetate may be observed, but it frequently appears to degrade too rapidly to be observed [90].

6.2.6. *Inorganic Reactions*

Inorganic reactions affecting most chlorinated solvents are very, very slow [89]. However, the inorganic reactions of 1,1,1- trichloroethane (TCA) are fast enough to be important natural attenuation processes. There are two inorganic degradation pathways, hydrolysis and condensation [91]. The hydrolysis reaction yields chloroacetate, which degrades very quickly. The condensation reaction yields 1,1-DCE, which can be reductively dechlorinated or potentially oxidized. Both reactions are temperature dependent. At standard conditions the hydrolysis reaction has a half-life of about six months and the condensation reaction has a half-life of about two years [89].

The acetate generated by the hydrolysis of TCA can support reductive dechlorination. One often observes some 1,1-DCA created or the reduction of other solvents to less chlorinated products. The footprints of these reactions include TCA as a source material, 1,1-DCE down gradient from the source, small amounts of 1,1-DCA near the source area, and a generally reduced geochemical environment.

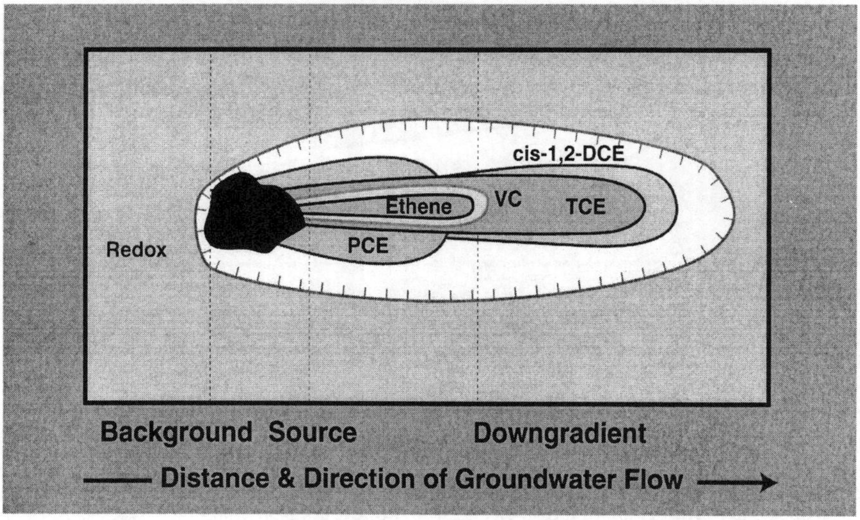

Figure 9. Distribution of parent and daughter compounds on an idealized sequential reduction-oxidation site. Plan view.

6.3. ANALYZING SITE DATA

The steps required to evaluate natural attenuation at a site involve first, the development of a conceptual model of the processes occurring at a site. This includes description of the site hydrogeology, the source and nature of the contamination, and the impact of movement and fate processes for the contaminants present, including advection, dispersion, sorption, volatilization, and transformation. Field data on the distribution of the contaminants are then required, but this is not sufficient in itself. One need also to gather information that can provide footprints of the processes occurring, such as concentrations of chloride, dissolved oxygen, carbon dioxide, methane, nitrate, iron, manganese, sulfate and sulfide. Often then a more detailed mathematical model is needed that brings together information on site hydrogeology; contaminant physics, chemistry and microbiology and the changes in these properties that transformation may cause, and the fate processes expected to be occurring. The modeling effort is necessary to calculate mass balances for chemicals and to make reasonable predictions of contaminant movement and fate into the future. Such models may be quite simple, or they can be very complex. The complexity required depends upon site complexity and the likelihood of success of natural attenuation. For example, with BTEX compounds where the likelihood of success is reasonable high, then the modeling effort need not be so great, for if protective processes are demonstrated to be occurring now, then there is a reasonably good chance that they will continue to be protective in the future. On the other hand, with chlorinated solvents where the likelihood of success is low, more effort is required to demonstrate that the processes are currently occurring, and even more effort is required to be convincing that they are likely to occur into the future. A description of the various modeling efforts required is provided elsewhere [1].

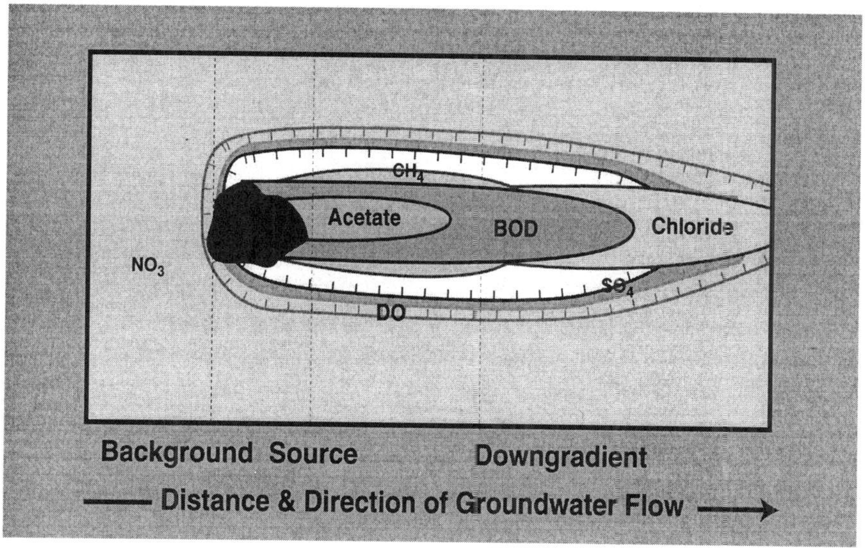

Figure 10. Distribution of inorganic footprints on an idealized sequential reduction/oxidation site. Plan view.

6.4. SITE MONITORING

Once the site investigation has demonstrated that natural attenuation appears to be acting successfully at a site, then a long-term monitoring program needs to be established to insure that model projections into the future do indeed occur, and that natural attenuation continues to be protective. Because of the inability to gather sufficient data from a site and the complexities of site hydrogeology, mathematical models of a site will always be subject to considerable uncertainty. Frequent monitoring over time is thus required to insure that the anticipated protective processes continue to occur, and do so at a sufficient rate. A good monitoring plan is thus an essential component of a good natural-attenuation remediation strategy. As in modeling, the level of monitoring required is dependent upon site complexity and the likelihood of success for the given contaminant or contaminants

6.5. SUMMARY

Successful evaluation of natural attenuation requires the development of a conceptual model for a site, gathering of information on site hydrogeology and contaminant source and distribution, gathering of "footprint" information, development of a mathematical model of movement and fate processes for predicting changes into the future, and the establishment of a good long-term monitoring program. The level of effort required in each of these areas is a function of various factors as summarized in Table 4 [1]. The level 1 effort data gathering and analysis need only be sufficient to establish that flow direction is reasonable constant, that contaminant movement is consistent with the flow direction, and that contaminant concentration decreases sufficiently with distance from the source. At least one footprint that confirms assumed degradation processes are occurring may be sufficient here. A level-2 effort is necessary with more complex hydrogeology or where the likelihood of success is not high. More data is required here, a reasonably good mathematical model may be required, better footprint data is required, and a more extensive long-term monitoring program needs to be established. A level-3 effort, when required, can be much more extensive, and requires extensive information on contaminant sources, site hydrogeology, and information on contaminant spread and daughter product formation. A good mathematical model that includes all relevant fate processes, and is verified to a greater extent through retrospective analysis of field data. The long-term monitoring program must be more extensive and complete to insure that contaminants are following the anticipated course. In some cases, the cost of the analysis and monitoring program for some contaminants with low likelihood of success may be higher than the cost of more engineered solutions to the problem. Such potential costs should be carefully evaluated before beginning efforts to evaluate whether natural attenuation is likely to be a viable alternative for remediation of a given site.

Table 4. Typical level of effort required in site characterization and data evaluation to establish natural attenuation as a viable remedy (After NRC. 5)

	Likelihood of Success of Natural Attenuation of the Contaminant of Concern[a]		
Site Hydrogeology	High	Moderate	Low
Simple flow and uniform geochemistry, and low concentrations	1	2	2
Simple flow, and small-scale physical or chemical heterogeneity, and medium-high concentrations	2	2	2
Strongly transient flow, large-scale physical or chemical heterogeneity, or high concentrations	2	3	3

NOTES: level of effort refers to the number and frequency of samples taken, parameters analyzed in site samples, and type of data analysis required; 1 = low effort; 2 = moderate effort, 3 = high effort.
[a]Likelihood of success refers to judgements in Table 4.

7. Community Involvement

The NRC report on natural attenuation [1] emphasized the importance of active involvement of the community surrounding a contaminated site in developing and evaluating alternatives for remediation, especially where natural attenuation is likely to be one of the remedies evaluated. This community represents the body of people that are more directly impacted by the contamination and are likely to have strong concerns with any remediation approach, and especially with natural attenuation. Many community members view natural attenuation as a do-nothing approach, which it is not, because of the extensive effort at evaluation and monitoring that a successful program requires. If the community is to accept natural attenuation as a viable alternative that is protective to their health, they must have a full understanding of the nature and cost of other alternative remedies, and of the certainty with which natural attenuation will indeed be protective for them into the future. In order to gain their acceptance, it is necessary that the community be brought into the overall process early so that they can feel confident that all viable alternatives are being explored, that their interests are seriously being taken into consideration, and that the process of evaluation is an open one with adequate attention being paid to their concerns. The NRC report provides a thorough discussion of this issue and provides some guidance on how to involve the community in remediation decisions.

8. Protocols for Evaluating Natural Attenuation

Because of a growing interest in the use of natural attenuation as a remedy for contaminated sites, various protocols have been developed by different agencies and organizations to guide efforts for evaluating a site's potential for natural attenuation. The U.S. National Research Council [1] undertook an effort to evaluate the various protocols available to determine the contaminants being addressed, the strengths and weaknesses of existing protocols, and the components that make for good protocols. The following discussion is a condensation of this effort. NRC defined the term protocol broadly. It includes "any policy statement, state regulation, or technical document on how decision making and implementation of natural attenuation should be carried out".

The NRC's Committee on Intrinsic Remediation reviewed fourteen natural attenuation documents. They included: EPA's Monitored Natural Attenuation (MNA) policy directive, EPA ORD's chlorinated solvent MNA protocol, EPA Region 4's draft chlorinated solvent protocol, Dept. of Energy's MNA guidance, US Air Force's fuel protocol, US Air Force's chlorinated solvents protocol, US Navy MNA guidelines, Minnesota's draft guidelines, New Jersey's technical requirements, Chevron's fuels protocol, Chevron's chlorinated solvents protocol, the American Society for Testing and Materials standard, American Petroleum Institute's guidance, and the Remediation Technology Develop Forum's "Principles and Practices".

Several European protocols have been issued recently, for example the UK Environment Agency guidance on natural attenuation [92] and the NOBIS decision document (Sinke et al, 93]. The NRC Committee did not review non-US protocols.

8.1. CHARACTERISTICS OF PROTOCOLS

The NRC Committee on Intrinsic Remediation developed a list of subject areas that high quality natural attenuation protocols should address. The criteria are in three general categories: community concerns, scientific and technical issues, and implementation issues. Using these criteria the Committee reviewed 14 US State and Federal protocols that were available at the time of writing. The great majority of the protocols reviewed deal exclusively with petroleum hydrocarbons or chlorinated solvents. There is a lack of protocols that deal with natural attenuation of metals or other classes of organic compounds.

8.1.1. *Community Concerns*

Community involvement. Protocols should specify points in the evaluation process where community input is especially important. None of the protocols reviewed provide guidance on steps in the evaluation process where public involvement is important. Many of the protocols reviewed may not specifically address public involvement because the developers assume that other guidelines exist. Nevertheless, current and future protocols should specify more clearly the points in the evaluation process at which public input is important.

Institutional controls and long-term monitoring. A protocol should describe criteria for determining when institutional controls are needed, and how the viability of these controls can be ensured. The protocols reviewed provide little or no discussion of the need for institutional controls, or how these should be coordinated with long-term monitoring programs. The protocols are quite varied in their recommendations for monitoring well placement, duration of monitoring, and terminology.

Contingency plans. Protocols should address decision making on contingency plans. Criteria for reevaluating the effectiveness of natural attenuation must be defined, along with whether contingency treatment systems should be pre-positioned. Current protocols provide very limited guidance (if any) on contingency planning.

8.1.2. *Scientific and Technical Issues*

Establish cause-and-effect. A protocol should explain the scientific underpinnings and the evidence used to assess the relationship between field and analytical observations and what is expected. Although several protocols provide detailed guidance on establishing cause-and-effect, many treat issues of documentation in an uneven manner. In particular, the level of detail needed to characterize a contaminant plume is highly variable. Site remedy and closure decisions ultimately depend on the professional judgement of individual regulators. Although investigators and regulators often employ minimalist criteria, these are not an adequate substitute for an experienced professional assessment. Providing guidelines for documenting cause-and-effect is one of the most critical roles for a natural attenuation protocol.

Site condition assessment. Good protocols should describe the level of data required to assess different types of sites and characterize different types of contaminant sources. Except for the US Air Force and US EPA protocols the documents gave surprisingly little attention to characterizing contaminant sources and making decisions whether or not to remove sources.

Sustainability. The protocol should address methods for the long-term viability of natural attenuation. The aquifer or source area must have sufficient electron donors or acceptors to support natural attenuation, complicating factors should not unduly interfere with natural attenuation, and the natural attenuation process must be robust to changes in local conditions. In general, protocols tend to focus on evaluation of present-day rather than future behavior. Better guidance on sustainability and robustness is needed.

Peer review. The protocol should be peer reviewed by individuals representing various disciplines and viewpoints other than those of the organization that wrote it. All protocols would benefit from external peer review to assure that they are based on correct principles and sound methodologies, and that they adequately address public concerns. Peer review is especially important for contaminants where there is limited field experience. Individuals who are not closely associated with the authors' organization should conduct the peer reviews.

8.1.4. *Implementation issues*

Protocols should provide sufficiently detailed explanations so that users can follow them. The qualifications and training of implementers should be discussed. Most protocols are simple for knowledgeable individuals to use. None of the protocols address the training or background needed by implementers of natural attenuation

8.2. DECISION-MAKING TOOLS

Most protocols contain decision-making tools, for example flow charts, check lists, or scoring systems. Such tools have very limited ability to adequately recognize the differences between sites based on size, composition, or geological setting and complexity. Although good tools can help organize thinking about a site, inappropriate or inflexible use of tools could make decision processes rigid rather than fluid.

8.2.1. *Flow Charts*

Flow charts can be very useful for organizing data needs and the decision process. The available flow charts deal with deciding whether to use natural attenuation, combine it with another technology, or use another technology without natural attenuation. Policy directives would benefit from greater use of flow charts.

8.2.2. *Check Lists*

Check lists attempt to list all the potential information needed to make a proper natural attenuation evaluation. They are very useful for maintaining consistency, ensuring thoroughness, and as a reminder of the information to be gathered. A useful checklist will clearly designate which information is mandatory, which information is optional, and where flexibility exists.

8.2.3. *Scoring Systems*

Scoring systems attempt to evaluate potential natural attenuation sites by assigning numerical scores to geochemical and biochemical parameters. Scoring systems are being adopted for uses that the authors never intended. Because the scoring systems are subject to misuse, and because approaches to evaluating natural attenuation have advanced greatly, the NRC Committee recommended the abandonment of scoring systems, instead relying on site-specific conceptual models and footprints.

8.3. TRAINING TO USE NATURAL ATTENUATION PROTOCOLS

Good decisions regarding natural attenuation require a considerable amount of expert judgement. Thus, training of protocol users is essential to ensure that the protocols are implemented properly. The training needed varies depending on the individual's role.

8.3.1. *Responsible Parties*

Responsible parties need to be able to evaluate the work of their remediation consultants and to make decisions based on the recommendations of the consultants. They should have enough technical background and experience to manage their consultants and to negotiate responsibly with their environmental regulators. Those without experience in natural attenuation should attend at least one good short course.

8.3.2. *Consultants*

A consulting firm should have a lead natural attenuation expert who specializes in natural attenuation evaluations. These leaders need the technical skills to understand unique contaminants and situations outside the scope of existing protocols, and the communication skills to effectively communicate their conclusions. High-quality field data are essential to understanding potential natural attenuation processes. Field personnel should be trained to recognize and correct problems as they occur, and to adapt to changeable field conditions.

8.3.3. *Regulators*
Regulators are charged with evaluating the merits of the various remediation proposals they receive and with making responsible judgements on the technical merits of natural attenuation proposals. Regulators should have technical degrees and should attend specialized training before evaluating natural attenuation proposals.

8.3.4. *Community organizations*
Community members may desire technical training to help them understand natural attenuation proposals. The goals of such training are to help community leaders the technical complexities of natural attenuation, its strengths and weaknesses, its effectiveness in treating specific contaminants, its suitability at particular sites, and its short- and long-term risks. Details of training should be worked out in collaboration with community leaders.

8.4. NRC'S RECOMMENDATIONS ON PROTOCOLS

a) A national consensus on guidelines for natural attenuation is needed. EPA should work with other federal agencies, the states, professional organizations, industry groups, and community organizations to create consensus guidelines, and to update them regularly.
b) The national consensus guidelines and all future natural attenuation protocols should be independently peer reviewed.
c) The use of scoring systems should be eliminated in the national consensus guidelines and in all future natural attenuation protocols.
d) Developers of natural attenuation protocols should write easy-to-understand documents to explain the protocols to non-technical audiences.
e) Federal and State agencies, and organizations responsible for contaminated sites should provide additional training on natural attenuation concepts for interested regulators, site owners, remediation consultants, and community or environmental groups.

9. Conclusions

Natural attenuation processes that degrade or transform contaminants can work well in controlling risks from groundwater contamination when the right combination of contaminants and environmental conditions exists. Natural attenuation is most likely to be effective for contaminants that are readily degraded or immobilized under a wide range of environmental conditions. As Table 3 indicates, natural attenuation potential is high for BTEX but moderate or low for most other commonly encountered contaminants. Natural attenuation may work for these other contaminants in some cases under very specific site conditions. For all contaminants, natural attenuation will work best when the geologic system is simple enough for the natural attenuation processes to be effectively monitored.

Documenting natural attenuation requires hard evidence that natural processes at the site are immobilizing or destroying the contamination to an extent that is sufficient to protect public health and the environment. Footprints and patterns of the attenuation reactions should serve as the basis for this evaluation, and protocols are the tools that should be used to format and evaluate this evidence. The public needs to be involved early in the decision-making process at sites where communities are adversely affected by contamination.

10. References

1. National Research Council. 2000. Natural Attenuation for Groundwater Remediation. National Academy Press, Washington, D. C.
2. EPA. 1999. Use of Monitored Natural Attenuation at Superfund, RCRA Corrective Action, and Underground Storage Tank Sites No. Directive 9200.4-17P). EPA, Office of Solid Waste and Emergency Response.
3. Weidemeier, T., et al. 1995. Technical Protocol for Implementing Intrinsic Remediation with Long-Term Monitoring for Natural Attenuation of Fuel Contamination Dissolved in Groundwater No. Vols. 1 and 2). Air Force Center for Environmental Excellence
4. U.S.Army. 1995. Remediation of Contaminated Army Sites: Utility of Natural Attenuation No. Draft Report). U.S. Army Science Advisory Board, Infrastructure and Environmental Group, Department of Army.
5. ASTM. 1997. Standard Guide for Remediation of Ground Water by Natural Attenuation at Petroleum Release Sites No. American Society for Testing and Materials.
6. Swoboda-Colberg, N. G. 1995. Chemical contamination of the environment: Sources, types, and fate of synthetic organic chemicals. In: L. Y. Young, and C. E. Cerniglia (Eds.) Microbial Transformation and Degradation of Toxic Organic Chemicals, pp. 27-7
7. Lenhard, R. J., R. S. Skeen, and T. M. Brouns. 1995. Contaminants at U.S. DOE sites and their susceptibility to bioremediation. In: H. D. Skipper and R. F. Turco (Eds.) Bioremediation Science and Applications, pp. 157-172. SSSA, Special Publication N
8. Mackay, D. M., D. L. Freyberg, P. V. Roberts and J. A. Cherry. 1986. A Natural-Gradient Experiment on Solute Transport in a Sand Aquifer, I. Approach and Overview of Plume Movement. *Water Resources Research*, **22**(13): 2017-2029.
9. Freyberg, D. L. 1986. A natural gradient experiment on solute transport in a sand aquifer. 2. Spatial moments and the advection and dispersion of nonreactive tracers. *Water Resources Research*. **22**(13): 2031-2046.
10. Curtis, G. P., P. V. Roberts and M. Reinhard. 1986. A natural gradient experiment on solute transport in a sand aquifer IV. sorption of organic solutes and its influence on mobility. *Water Resources Research*, **22**(13): 2059-2067.
11. Roberts, P. V., M. N. Goltz and D. M. Mackay. 1986. A natural-gradient experiment on solute transport in a sand aquifer, III. retardation estimates and mass balances for organic solutes. *Water Resources Research*, **22**(13): 2047-2058.)
12. Sudicky, E. A. 1986. A natural gradient experiment on solute transport in a sand aquifer: Spacial variability of hydraulic conductivity and its role in the dispersion process. *Water Resources Research*, **22**(13): 2069-2082.
13. Criddle, C., C. Elliott, P. L. McCarty, and J. F. Barker. 1986. Reduction of Hexachloroethane to Tetrachloroethylene in Groundwater. Journal of Contaminant Hydrology. 1(1/2): 133-143.
14. Anderson, R. T., and D. R. Lovley. 1997. Ecology and biogeochemistry of In situ groundwater bioremediation. Adv. Microb. Ecol. 15:289-350.
15. Beller, H. R. and A. M. Spormann. 1997. Benzylsuccinate formation as a means

of anaerobic toluene activation by sulfate-reducing Strain PRTOL1. Appl. Environ. Microbiol. 63: 3729-3731.

16. Edwards, E. A. and D. Grbic-Galic. 1992. Complete Mineralization of Benzene by Aquifer Microorganisms under strictly anaerobic conditions. Appl. Environ. Microbiol. 58: 2663-2666.
17. Lovley, D. R., and D. J. Lonergan. 1990. Anaerobic oxidation of toluene, phenol, and p-cresol by the dissimilatory iron-reducing organism, GS-15. Appl. Environ. Microbiol. 56:1858-1864.
18. Lovley, D. R. 1997. Potential for anaerobic bioremediation of BTEX in petroleum-contaminated aquifers. J. Ind. Microbiol. Biotechnol. 18:75-81.
19. Lovley, D. R. 1995b. Microbial reduction of iron, manganese, and other metals. Adv. Agron. 54:176-231.
20. Reinhard, M., N. L. Goodman, and J. F. Barker. 1984. Occurrence and distribution of organic chemicals in two Landfill Leachate Plumes. Environ. Sci. Technol. 18: 953-961.
21. Vogel, T. M. and D. Grbic-Galic. 1986. Incorporation of Oxygen from Water into Toluene and Benzene during Anaerobic Fermentative Transformation. App. Environ. Microbiol. 5:200-202.
22. Borden, R. C., R. A. Daniel.L.E. LeBrun, and C.W. Davis. 1997. Intrinsic biodegradation of MTBE and BTEX in a gasoline-contaminated aquifer. Water. Resour. Res. 33: 1105-1115.
22. Mueller, J. G., S. E. Lantz, D. Ross, R. J. Colvin, D. P. Middaugh, and P. H. Pritchard. 1993. Strategy using bioreactors and specially selected microorganisms for bioremediation of groundwater contaminated with creosote and pentachlorophenol. *Environmental Science & Technology*, **27**(4): 691-698.
23. Salanitro, J.P., L.A. Diaz, M.P. Wiliams and H.L. Wisniewski. 1994. Isolation of a bacterial culture that degrades methyl t-butyl ether. Appl. Environ. Microbiol. 60: 2593-2596.
24. Mormile, M.R., S. Liu, and J.M.Suflita. 1994. Anaerobic biodegradation of gasoline oxygenates: Extrapolation of information to multiple sites and redox conditions. Environ. Sci. Technol. 28:1727-1732.
25. Steffan, R. J., K. McClay, S. Vainberg, C. W. Condee, and D. Zhang. 1997 Biodegradation of the gasoline oxygenates methyl tert-butyl ether, ethyl tert-butyl ether, and tert-amyl methyl ether by propane oxidizing bacteria. Appl. Environ. Microbiol. 63
26. Cerniglia, C. E. 1984. Microbial transformation of aromatic hydrocarbons, pp. 99-128. In: R. M. Atlas (Ed.) Petroleum Microbiology, Macmillan Publishing Co., New York, NY.
27. Sutherland, J. B., F. Rafii, A. A. Kahn, and C. E. Cerniglia. 1995. Mechanisms of polycyclic aromatic hydrocarbon degradation, pp., 269-306. In: L.Y. Young and C.E. Cerniglia (eds.) Microbial Transformation and Degradation of Toxic Organic Chemicals.
28. Mihelcic, J. R. and R. G. Luthy. 1988. Microbial-Degradation of Acenaphthene and Naphthalene under Denitrification Conditions in Soil-Water Systems. *Applied and Environmental Microbiology*, **54**(5): 1188-1198.
29. McNally, D. L., J. R. Mihelcic, and D. R. Lueking. 1998. Biodegradation of three- and four-ring polycyclic aromatic hydrocarbons under aerobic and denitrifying conditions. Environ. Sci. Technol. 32:2633-2639.
30. Coates, J. D., J. Woodward, J. Allen, P. Philip, and D. R. Lovley. 1997. Anaerobic degradation of polycyclic aromatic hydrocarbons and alkanes in petroleum-

contaminated marine harbor sediments. Appl. Environ. Microbiol. 63:3589-3593.

31. Zhang, X., and L. Y. Young. 1997. Carboxylation as an initial reaction in the anaerobic metabolism of naphthalene and phenanthrene by sulfidogenic bacteria. Appl. Environ. Microbiol. 63:4759-4764.
32. Gribble, G. W. 1994. The natural production of chlorinated compounds. Environ. Sci. Technol. 38:310A-319A.
33. Freedman, D. L. and J. M. Gossett. 1989. Biological reductive dechlorination of tetrachloroethylene and trichloroethylene to ethylene under methanogenic conditions. Appl. Environ. Microbiol. 55:2144-2151.
34. McCarty, P. L. 1997. Breathing with chlorinated solvents. Science 276:1521-1522.
35. Maymó-Gatell, X., Y.-T. Chien, J. M. Gossett, and S. H. Zinder. 1997. Isolation of a bacterium that reductively dechlorinates tetrachloroethene to ethene. Science 276:1568-1571.
36. Semprini, L. 1995. In situ bioremediation of chlorinated solvents. *Environmental Health Perspectives*, 103 Suppl 5(6): 101-5.
37. Tandoi, V., et al. 1994. Reductive dehalogenation of chlorinated ethenes and halogenated ethanes by a high-rate anaerobic enrichment culture. *Environmental Science & Technology*, **28**(5): 973-979.
38. Bradley, P. M., and F. H. Chapelle, 1996. Anaerobic mineralization of vinyl chloride in Fe(III)-reducing aquifer sediments. Environ. Sci. Technol. 30:2084-2086.
39. Yager, R. M., S. E. Bilotta, C. L. Mann, and E. L. Madsen. 1997. Metabolic adaptation and In situ attenuation of chlorinated ethenes by naturally-occurring microorganisms in a fractured dolomite aquifer near Niagara Falls, NY. Environ. Sci. Technol.
40. Davis, J. W., and C. L. Carpenter. 1990. Aerobic biodegradation of vinyl chloride in groundwater samples. Appl. Environ. Microbiol. 56:3878-3880.
41. Dolan, M. E., and P. L. McCarty. 1995. Small column microcosm for assessing methane-stimulated vinyl-chloride transformation in aquifer samples. Environmental Science and Technology 29(8), 1892-1897.
42. Freedman, D. L. and J. M. Gossett. 1991. Biodegradation of Dichloromethane and its Utilization as a Growth Substrate Under Methanogenic Conditions. Appl. Environ. Microbiol. 57: 2847-2857.
43. Magli A., M. Messmer, T. Leisinger. 1998. Metabolism of dichloromethane by the strict anaerobe Dehalobacterium formicoaceticum. Appl. Environ. Microbiol. 64: 646-650
44. Kohler-Staub, D., S. Frank, and T. Leisinger. 1995. Dichloromethane as the sole carbon source for Hyphomicrobium sp. strain DM2 under denitrification conditions. Biodegradation. 6: 229-235.
45. Barkovski, A. L, and P. Adriaens. 1996. Microbial dechlorination of historically present and freshly spiked chlorinated dioxins and diversity of dioxin-dechlorinating populations. Appl. Environ. Microbiol. 62:4556-4562.
46. Van Denburgh, A.S., D.F. Goerlitz, and E.M. Godsy. 1993. Depletion of nitrogen-bearing explosive wastes in a shallow ground-water plume near Hawthorne, Nevada. USGS Toxic Substances Hydrology Program. Technical Meeting Colorado Springs ,CO Sept. 20-
47. Bradley, P.M., F.H. Chapelle, J.E. Landmeyer, and J.G. Schumacher. 1997. Potential for intrinsic bioremediation of a DNT-contaminated aquifer. Ground Water 35: 12-17.

48. Bradley, P.M., F.H. Chapelle, J.E. Landmeyer, and J.G. Schumacher. 1994. Microbial transformation of nitroaromatics in surface soils and aquifer materials. Appl. Environ. Microbiol. 60: 2170-2175.
49. Funk, S. B., D. L. Crawford, R. L. Crawford, G. Mead, and W. Davis-Hoover. 1995. Full-scale anaerobic bioremediation of trinitrotoluene (TNT) contaminated soil. Appl. Biochem. Biotechnol. 51-52:625-633.
50. Krumholz, L. R., J. Li, W. W. Clarkson, G. G. Wilber, J. M. Suflita. 1997. Transformation of TNT and related amino toluenes in groundwater aquifer slurries under different electron-accepting conditions. J. Ind. Microbiol. Biotechnol. 18:161-169.
51. Lewis, T. A., M. M. Ederer, R. L. Crawford, and D. L. Crawford. 1997. Microbial transformation of 2,4,6-trinitrotoluene. J. Ind. Microbiol. Biotechnol. 18:89-96.
52. Spain, J. C. 1995. Biodegradation of nitroaromatic compounds. Annu. Rev. Microbiol. 49:523-555.
53. Hawari, J., A. Halasz, L. Paquet, E. Zhou, B. Spencer, G. Ampleman, and S. Thioboutot. 1998. Characterization of metabolites in the biotransformation of 2,4,6-trinitrotoluene with anaerobic sludge: role of triaminotoluene. Appl. Environ. Microbiol. 64
54. French, C. E., S. Nicklin, and N. C. Bruce. 1988. Aerobic degradation of 2,4,6-trinitrotoluene by Enterobacter cloacae PB2 and by pentaerythritol tetranitrate reductase. Appl. Environ. Microbiol. 64:2864-2868.
55. Esteve-Nunez, A., and J.L. Ramos 1998. Metabolism of 2,4,6,-trinitrotoluene by Pseudomanas sp JLR11. Environ. Sci. Technol. 32:3802-3808.
56. Babu, G. R. V., J. H Wolfram, and K. D. Chapatwala. 1992. Conversion of sodium cyanide to carbon dioxide and ammonia by immobilized cells of Pseudomonas putida. J. Ind. Microbiol. 9:235-238.
57. Brierley, C. L. 1990. Bioremediation of metal-contaminated surface and groundwaters. Geomicrobiol. 8:201-224.
58. Chapatwala, K. D., G. R. V. Babu, E. R. Armstead, E. M. White, and J. H. Wolfram. 1995. A kinetic study on the bioremediation of sodium cyanide and acetonitrile by free and immobilized cells of Pseudomonas putida. Appl. Biochem. Biotechnol. 51-52:717
59. Hinchee, R. E., J. L. Means, and D. R. Burris (Eds.) 1995a. Bioremediation of Inorganics. Battelle Press, Columbus, OH.
60. Kalin, M., J. Cairns, and R. M. McCready. 1991. Ecological engineering methods for acid mine drainage treatment of coal wastes. Resources Conserv. Recycl. 5:265-276.
61. Lenhard, R. J., R. S. Skeen, and T. M. Brouns. 1995. Contaminants at U.S. DOE sites and their susceptibility to bioremediation. In: H. D. Skipper and R. F. Turco (Eds.) Bioremediation Science and Applications, pp. 157-172. SSSA, Special Publication N
62. Lovley, D. R. 1993. Dissimilatory metal reduction. Annu. Rev. Microbiol. 47:263-290.
63. Lovley, D. R. 1995a. Bioremediation of organic and metal contaminants with dissimilatory metal reduction. J. Ind. Microbiol. 14:85-93.
64. McHale, A. P., and S. McHale. 1994. Microbial biosorption of metals: Potential in the treatment of metal pollution. Biotechnol. Adv. 12:647-652.
65. Saouter, E., M. Gillman, and T. Barkay. 1995. An evaluation of mer-specified reduction of ionic mercury as a remedial tool of a mercury-contaminated freshwater pond. J. Ind. Microbiol. 14:343-348.

66. Summers, A. O. 1992. The hard stuff: metals in bioremediation. Curr. Opinion Biotechnol. 3:271-276.
67. Thompson-Eagle, E. C. and W. T. Frankenberger, Jr. 1992. Bioremediation of soils contaminated with selenium. Adv. Soil Sci. 17:261-310.
68. Videla, H. A. and W. G. Characklis. 1992. Biofouling and microbial influenced corrosion. Biodegradation 29:195-212.
69. Whitlock, J. L. 1990. Biological detoxification of precious metal processing wastewaters. Geomicrobiol. J. 8:241-249.
70. Diels, L. 1997. Heavy metal bioremediation of soil, pp. 283-295. In: Sheehan, J. (ed.). Methods in Biotechnology, 2. Bioremediation protocols. Humana Press, Inc., Totowa, NJ.
71. Ghiorse, W. C. 1994. Iron and manganese oxidation and reduction, pp. 1079-1096. In: Weaver, R. W., S. Angle, and P. Bottomley (eds.). Methods of Soil Analysis, Part 2, Microbiological and Biochemical Properties. ASA-SSSA, Madison, WI.
72. Macaskie, L. E., and G. Basnakova. 1998. Microbially-enhanced chemosorption of heavy metals: A method for the bioremediation of solutions containing long-lived isotopes of neptunium and plutonium. Environ. Sci. Technol. 32:184-187.
73. Ehrlich, H. L. 1996. Geomicrobiology. Third ed. Marcel Dekker, Inc. New York, NY.
74. Frankenberger, W. T., Jr., and M. E. Losi. 1995. Applications of bioremediation in the cleanup of heavy metals and metalloids. In: H. D. Skipper and R. F. Turco (eds.) Bioremediation Science and Applications, pp. 173-210. SSSA, Special Publication Nu
75. Zumft, W. G. 1977. Cell biology and molecular basis of denitrification. Microbiol. Mol. Biol. Rev. 61:533-616.
76. Atlas, R.M., and R. Bartha 1997. Microbial Ecology: Fundamentals and Applications, 4th ed. Benjamin Cummings, Publ. Co., Menlo Park. CA.
77. Malmqvist, A., T. Welander, and L. Gunnarsson. 1991. Anaerobic growth of microorganisms with chlorate as an electron acceptor. Appl. Environ. Microbiol. 57:2229-2232.
78. van Ginkel, C. G., C. M. Plugge, and C. A. Stroo. 1995. Reduction of chlorate with various energy substrates and inocula under anaerobic conditions. Chemosphere 31:4057-4066.
79. Rikkin, G. B., A. G. M. Kroon, and C. G. VanGinkel. 1996. Transformation of (per)chlorate into chloride by a newly isolated bacterium: Reduction and dismutation. Appl. Microbiol. Biotechnol. 45:420-426.
80. Lloyd J.R. and L.E. Macaskie. 1997. Microbially-mediated reduction and removal of technetium from solution. Research in Microbiology. 148: 530-532.
81. Yang, Y. and P. L. McCarty. 2000. Biologically Enhanced Dissolution of Tetrachloroethene DNAPL. Environ. Sci. Technol. 34: 2979-2984.
82. Major, D.W., E.E. Cox, E. Edwards, and P.W. Hare. 1995. Intrinsic dechlorination of trichloroethene to ethene in a bedrock aquifer. In : Intrinsic Bioremediation, R.E. Hinchee, J.T. Wilson, and D.C. Downey (eds). Battelle Press, Columbus, OH. pp 197-
83. Hendrickson, E.R., J.A. Payne, R.M. Young, M.G. Starr, M.P. Perry, S. Fahnestock, D.E. Ellis, and R.C. Ebersole, 2001. Molecular Analysis of Dehalococcoides-Group 16S rDNA from Chloroethene-Contaminated Sites throughout North America and Europe. Appl.Environ. Microbiol., In press.
84. Leethem, J. T., and J. R. Larson. 1997. Intrinsic Bioremediation of Vinyl Chloride in Groundwater at an Industrial Site. In: In Situ and On-Site Bioremediation:Vol.

3.Battelle Press, pp 167-172.

85. Ellis, D. E., E. J. Lutz, G. M. Klecka, D. L. Pardieck, J. J. Salvo, M. A.Heitkamp, D. J.Gannon, C. C. Mikula, C. M. Vogel, G. D. Sayles, D. H. Kampbell, J.T.Wilson and D.T. Maiers. 1996. Remediation Technology Development Forum Intrinsic remediation
86. Cox, E., E. Edwards, L. Lehmicke, and D. Major. 1995. Intrinsic biodegradation of trichloroethene and trichloroethane in a sequential anaerobic-aerobic aquifer. In R. E. Hinchee, J. T. Wilson, and D. C. Cowney, eds., Intrinsic Bioremediation, Batt
87. Bradley, P. M. and F. H. Chapelle. 1997. Kinetics of DCE and VC mineralization under methanogenic and Fe(III) reducing conditions. Environ. Sci. Technol. 31(9):2692-2696.
88. De Best, J.H., J. Ultee, A. Hage, H.J. Doddema, D.B. Janssen, and W. Harder. 2000. Dichloromethane utilization in a packed-bed reactor in the presence of various electron acceptors. Water. Res., vol 34, no, 2, pp. 566-574.
89. McCarty, P. L. 1996. An Overview of Anaerobic Transformation of Chlorinated Solvents. In Symposium on Intrinsic Bioremediation of Ground Water, EPA/540/R-94/515, EPA Office of Research and Development, Washington, D.C., pp. 135-142.
90. Lehmicke, L. G., E. E. Cox, and D. W. Major. 1996. Involvement of dichloromethane in the intrinsic biodegradation of chlorinated ethenes and ethanes. In : Symposium on Natural Attenuation of Chlorinated Organics in Ground Water. EPA/540/R-96/509.
91. Cline, P. V. and J. J. Delfino. 1989. Transformation Kinetics of 1,1,1-Trichloroethane to the Stable Product 1,1-Dichloroethene. In R. A. Larson ed., Biohazards of Drinking Water Treatment, Lewis Publishers, Inc., Chelsa, Michigan, pp. 47-56.
92. Carey, M.A., J.R. Finnamore, M.J. Morrey, and P.A. Marsland. 2000. Guidance on the assessment and monitoring of natural attenuation of contaminants in groundwater. UK Environment Agency, R&D Publication 95.
93. Sinke, A.J.C, T. Heimoiraara, H. Tonnaer, and S. Vam Been. 1998. A decision support system for acceptance of natural attenuation as remediation strategy. NOBIS.
94. Abramowicz, D. A. 1990. Aerobic and Anaerobic Biodegradation of PCBs : a Review. Critical Reviews in Biotechnology, 10(#3): 241-249.
95. Bedard, D. L., and J. F. Quensen, III. 1995. Microbial reductive dechlorination of polychlorinated biphenyls, pp. 127-216. In: L.Y. Young and C.E. Cerniglia (eds.) Microbial Transformation and Degradation of Toxic Organic Chemicals. Wiley-Liss, Inc.,
96. Edwards, E. A., and D. Grbic-Galic. 1994. Anaerobic degradation of toluene and o-xylene by a methanogenic consortium. Appl. Environ. Microbiol. 60:313-322.
97. Edwards, E. A., L. E. Wills, M. Reinhard, and D. Grbic-Galic. 1992. Anaerobic degradation to toluene and xylene by aquifer microorganisms under sulfate-reducing conditions. Appl. Environ. Microbiol. 58:794-800.
98. Evans, P. J., D. T. Mang, and L. Y. Young.1991a. Degradation of toluene and m-xylene and transformation of o-xylene by denitrifying enrichment cultures. Appl. Environ. Microbiol. 57:450-454.
99. Evans, P. J., D. T. Mang, K. S. Kim, and L. Y. Young. 1991b. Anaerobic degradation of toluene by a denitrifying bacterium. Appl. Environ. Microbiol. 57:1139-1145.
100. Grbic-Galic, D., and T. Vogel. 1987. Transformation of toluene and benzene by mixed methanogenic cultures. Appl. Environ. Microbiol. 53:254-260.

101. Lovley, D. R., J. D. Coates, J. C. Woodward, and E. J. P. Phillips. 1995. Benzene oxidation coupled to sulfate reduction. Appl. Environ. Microbiol. 61:953-958.

102. Mackay, D.M., M.D. Einarson, R. D. Wilson and M. Schirmer. Insights from Detailed Field Screening of a 1,700-foot-long MtBE Plume in California. Report by the University of Waterloo, Department of Earth Sciences, to the American Petroleum Institute

103. Roberts, P. V., M. Reinhard, and P. L. McCarty.1980. Organic contaminant behavior during groundwater recharge. *Journal Water Pollution Control Federation*, **52**:161-172.

IN-SITU TREATMENT TECHNOLOGIES

C. H. WARD[1], J.B. HUGHES[1], G.A. POPE[2], M.DELSHAD[2],V. DWARANATH[3], J.SPAIN[4], S. NISHINO[4], J.S.FRUCHTER[5], V.R.VERMEUL[5],M.D.WILLIAMS[5] AND J.E. SZECSODY[5]
[1]*Department of Civil and Environmental Engineering, Rice University, Houston, TX*
[2] *Center for Petroleum and Geosystems Engineering, Univ. of Texas, Austin, TX*
[3]*Duke Engineering Systems, Austin, TX*
[4]*US Air Force Research Laboratory, Tyndall AFB, FL*
[5]*Battelle, Pacific Northwest Laboratory, Richland, WA*

1. Introduction

In the United States, federal legislation and implementation of the Comprehensive Environmental Response, Compensation, and Liability Act (CERCLA/"Superfund") of 1980 and the Resource Conservation and Recovery Act (RCRA) of 1976 led to the discovery of large and wide-spread subsurface contamination. One significant source was from leaking underground storage tanks at gasoline stations and fuel tank farms (depots).

Leaking underground storage tanks in industry and government were also the source of massive groundwater contamination by chlorinated solvents such as trichloroethylene (TCE). Chlorinated solvents have been used in the United States for degreasing operations since the 1940's. Most segments of manufacturing were affected and generally improved by the use of chlorinated solvents. Examples include food and cosmetics production; metal fabrication; manufacture and repair of heat exchangers, electric motors, transformers, and electronics; and in most areas of computer development and manufacture.

Industry and especially government (U.S. Department of Defense (DOD) and Department of Energy (DOE)) were also responsible for localized surface and subsurface contamination by munitions such as 2,4,6-trinitrotoluene (TNT), radionuclides, and heavy metals such as chromium in the hexavalent form (Cr(VI)).

Until recently, subsurface and ground water remediation have primarily employed ground water extraction and above-ground treatment (pump and treat) technology. For the most part, poor results have been obtained. primarily because the physical and chemical nature of the contaminants results in little recovery of product (contaminant) with the water extracted. This is especially true of petroleum products and chlorinated solvents present in the forms of non-aqueous phase liquids (LNAPL and DNAPLs, respectively). Leaching of munitions and heavy metals from surface or subsurface releases also proved to be recalcitrant problems not effectively treated using standard and readily available technologies.

Recognizing that inadequate technology was the primary factor limiting the ability to meet federal and state cleanup goals, three federal agencies (Environmental Protection Agency (EPA), DOE and DOD) launched aggressive programs to develop new, innovative technologies to address the most difficult environmental remediation problems. This chapter describes four research and development activities designed to help solve problems associated with chlorinated solvents, munitions, and chromium contamination. Joseph Hughes describes an innovative microbial approach for degradation of chlorinated solvent source zones. Gary Pope and coworkers review the state-of-the-art of surfactant enhanced aquifer remediation

D. Reible and K. Demnerova (eds.), Innovative Approaches to the On-Site Assessment and Remediation of Contaminated Sites, 183–216.

(SEAR), a technology developed for the removal of non-aqueous phase liquids (NAPLs) by increased solubilization or mobilization. Jim Spain and Shirley Nishino, leading world experts on microbial degradation of nitroaromatic compounds, present a succinct analysis of the mechanisms and technology for bioremediation of explosives. John Fruchter and colleagues, working at a DOE funded laboratory, explain the basis for and application of a highly successful technology for physical-chemical in-situ redox manipulation for stopping the migration of Cr(VI) in a deep aquifer system.

These technologies are illustrative of a set of innovative approaches being developed in the United States, with federal agency support, to help solve historic contamination problems which have proven to be recalcitrant to the application of off-the-shelf technologies. However, other barriers to environmental remediation remain. Most of the new technologies are considered unproven by the regulators and the user communities, who consider the risk of their use and potential failure to be unacceptable. These barriers to environmental restoration have been considered by the National Research Council in a series of thoughtful studies [1, 2, 3] and by others [4].

2. DNAPL Source Zone Bioremediation

Chlorinated solvents are compounds with historic use as industrial solvents, degreasers, paint removers, and dry cleaning agents. Characteristics of these compounds include low water solubilities (ranging from 10^2 to 10^3 mg/L), low interfacial tensions, and high density. When released into the subsurface in sufficient quantities, chlorinated solvents can easily penetrate the water table and form dense non-aqueous phase liquids (DNAPLs). In this state, non-aqueous phase chlorinated solvents are retained for decades – perhaps centuries – and represent a continuous source of groundwater contamination.

Recent breakthroughs in the understanding of anaerobic microbial metabolism have conclusively demonstrated that some chlorinated pollutants as their terminal electron acceptor. Thus, compounds previously thought to be exclusively harmful to living systems are actually required for the existence of these bacteria. Given that these pollutants are the required electron acceptor for energy metabolism by these dechlorinators, it should be quite possible to create conditions in which the microbial colonization of DNAPL globules is favorable. Under conditions where intimate association between dechlorinating bacteria and the NAPL-globule would be favorable, the potential exists to dramaticly increase in DNAPL dissolution rates.

Tetrachloroethene (PCE) and trichloroethene (TCE), are the most frequently detected chlorinated solvents in groundwater. Because of their limited aqueous solubility, and infinite miscibility in other organic solvents, chlorinated ethenes exist as non-aqueous phase liquids (NAPLs) in the subsurface. Depending on the source of contamination, these NAPLs may be comprised of nearly pure chlorinated solvents, or as a chlorinated ethene-hydrocarbon mixture. In either case, the presence of the NAPL greatly complicates the ability to restore contaminated areas. NAPLs represent long-term sources of groundwater contamination, dramatically increasing the time required for natural attenuation processes to restore aquifers. Pump-and-treat is not an effective method for reducing the longevity of NAPLs, and the removal of NAPLs, especially those that are denser than water, is considered to be greatest technical challenge in the field of groundwater engineering".

While the microbial reductive dechlorination of chlorinated ethenes has been well documented [5-11] and anaerobic bioremediation is currently being employed to treat chlorinated ethene contaminated groundwater [12], dechlorination-based source-zone restoration has not been rigorously evaluated. An initial concern in the feasibility of source-zone bioremediation was the potential for toxicity due to high concentrations of contaminants.

Studies have demonstrated, however, that dechlorination can be sustained at chlorinated ethene concentrations similar to those found in source-zones (i.e., several mg/L to levels approaching saturation). Interestingly, source zones may represent an ecological niche for certain halorespiring microorganisms capable of withstanding high PCE concentrations. For these organisms, a PCE-containing NAPL represents a continuous source of terminal electron acceptor required for growth.

In source-zones where anaerobic conditions exist, reductive dechlorination of PCE and TCE is commonly observed. The impact that dechlorination may have on source-zone longevity (i.e., the time required to exhaust the mass of chlorinated ethenes from the NAPL) is poorly understood. In the absence of dechlorination, the longevity of a PCE-containing NAPL will be controlled strictly by the flux of PCE from the NAPL to water flowing into the source zone from upgradient recharge (Figure 1). Dechlorination has the potential to increase PCE removal from the NAPL by depleting PCE from the aqueous phase and increasing the driving force for dissolution (i.e., the difference between the effective solubility and the aqueous phase concentration) – depicted in Figure 2. Under the simplifying conditions described by Seagran [13, 14] where dissolution is described by partitioning, NAPL volume is not changing, biodegradation occurs uniformly throughout the source-zone, and first order kinetics apply, the impact of dechlorination on source-zone longevity can be described with Equation 1.

$$\frac{\partial C^w}{\partial t} + \frac{\theta_o}{\theta_w}\frac{\partial C^o}{\partial t} = D\frac{\partial^2 C^w}{\partial x^2} - v\frac{\partial C^w}{\partial x} - kC^w$$

(1)

Where: C^w and C^o are the concentrations of chlorinated solvent in the water and oil (NAPL) phase, t = time, θ_w and θ_o are the residual saturation of the water and oil phase, D is the dispersion coefficient, x = distance, v is the seepage velocity. and k is a first-order biodegradation rate term. This equation can be written in dimensionless form as is shown in Equation 2.

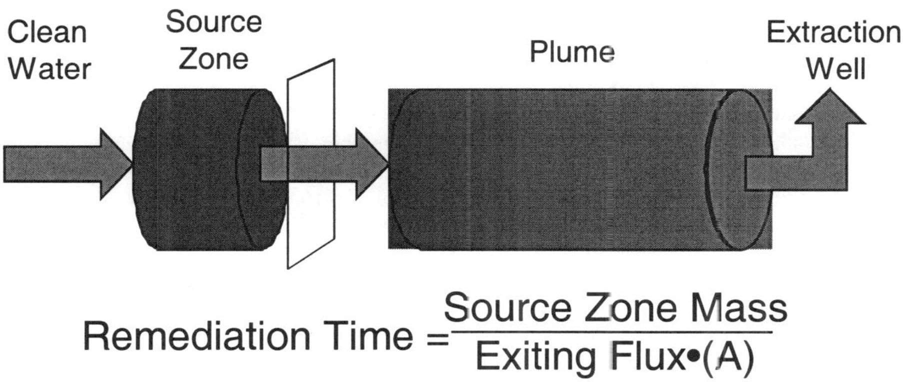

Figure 1. Representation of how flux from a source zone influences the time required to deplete a NAPL

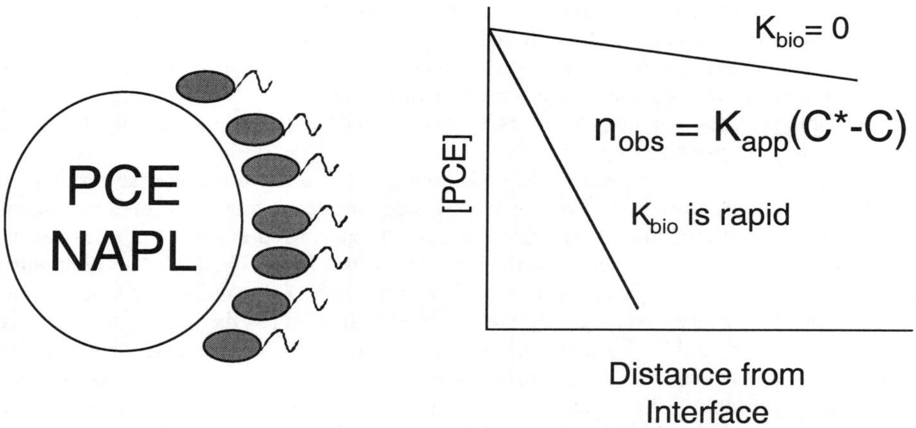

Figure 2. Schematic of a NAPL-globule that is "colonized" by dechlorinating bacteria (left) and the resulting development of a steep PCE concentration gradient with distance (right). In the figure on the right, n_{obs} is the observed flux, K_{app} is the apparent mass transfer coefficient, C* is the saturation coefficient for PCE, C is the concentration of PCE, and K_{bio} is the dechlorination rate coefficient.

$$\frac{\partial C^{w*}}{\partial t^*} = \frac{1}{Pe}\frac{\partial^2 C^{w*}}{\partial x^{*2}} - \frac{\partial C^{w*}}{\partial x^*} - Da_{bio}C^{w*} \tag{2}$$

Where: $C^{w*} = \frac{C^w}{C_o^w}$, $t^* = \frac{vt}{LR}$, $x^* = \frac{x}{L}$, $Pe = \frac{vL}{D}$, $Da_{bio} = \frac{kL}{v}$,

$R = 1 + \frac{\theta_o}{\theta_w}K^{o-w}$, L is the length of the source-zone, and K^{o-w} is the oil-water partition coefficient. The numerical solution of Equation 2 is shown in Figure 3, where the time required to deplete the source of a chlorinated solvent (depicted in the form of when the water phase concentration goes to zero) is presented as a function of the Damköhler number (Da_{bio}). An increase in the Damköhler number is the result of increasing biodegradation rates for a given groundwater velocity and source length.

Implicit in the definition of the Damköhler number is that long sources with slow seepage velocities have the potential for the highest values of Da_{bio}. However, using information presented in Figure 1 for a small source zone (10 feet) with rapid groundwater flow rate (1 foot per day), a biodegradation rate coefficient of only 0.5 per day would be required to reduce NAPL longevity by 80%. Interestingly, rates in excess of 0.5 per day are reported in studies of PCE natural attenuation without intervention.

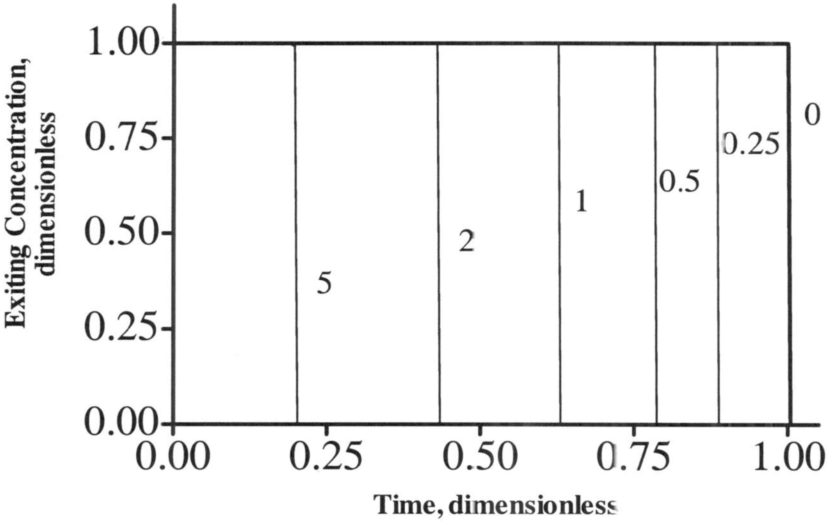

Figure 3. Solution of Equation 2 for a range of Damköhler numbers (as shown on figure). The dimensionless time of 1 equals the time needed for abiotic processes to remove contaminant from the source (i.e., the Damköhler number equals 0). Increasing values of Da_{bio} reflect increases in the rate of dechlorination within the source-zone.

A complicating factor not considered in the previous illustration is the interaction of dechlorination products with the NAPL itself. For example, PCE dechlorination results in the formation of TCE, *cis*-1,2-dichloroethene (*cis*-DCE), and vinyl chloride (VC) before ethene is produced (Figure 4). All chlorinated ethenes must also be removed from source-zones for restoration to be complete, and daughter products of PCE dechlorination will partition into NAPL.

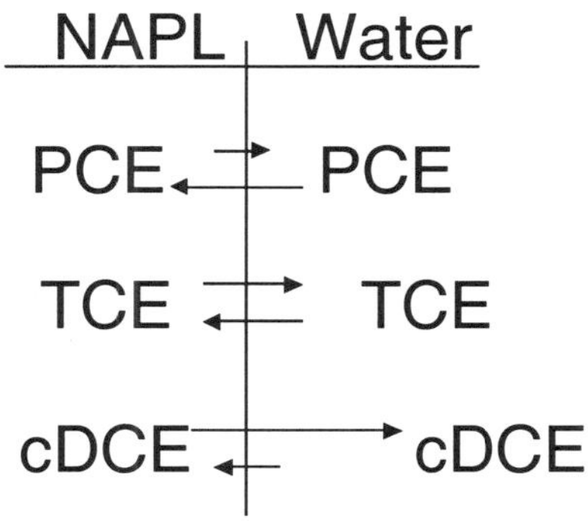

- At equal mole fractions to PCE:

$[TCE]_{aq} = 9 \cdot [PCE]_{aq}$

$[DCE]_{aq} = 33 \cdot [PCE]_{aq}$

$[VC]_{aq} = 90 \cdot [PCE]_{aq}$

Figure 4. The partitioning behavior of dechlorination products relative to PCE

Over time, this results in a changing NAPL composition that impacts source longevity. The partitioning behavior of chlorinated ethenes is such that as the degree of reduction increases, hydrophobicity decreases. For example, the oil/water partition coefficient of *cis*-DCE is 33 times greater than PCE (VC's oil/water partition coefficient is 90 times greater than PCE). Results from both laboratory and modeling studies have demonstrated the importance of dechlorination extent (i.e., the terminal ethene produced by dechlorinating consortia) in the longevity reduction achieved. When dechlorination proceeds beyond the level of *cis*-DCE, truly dramatic reductions in source-zone longevity can occur. This suggests that the microbial ecology present may be closely linked to the extent of source-zone dechlorination achievable. Much less is known about organisms that are capable of dechlorination past *cis*-DCE, or the ability to stimulate their activity *in situ*. Studies conducted at Rice University [15, 16] have commonly observed the rapid loss of ethene production when cultures grown in soluble PCE are transferred to vessels containing PCE in a NAPL (pure PCE or diluted in tridecane). A similar, but slow, loss of VC production is also commonly observed. At this time there is no clear understanding why this occurs.

A second interesting observation from that work is the impact that spatially heterogeneous biological activity has on the effectiveness of rapid source depletion. In column studies using a range of electron donor concentrations (either pyruvate or lactate) the colonization of columns was not homogeneous. Interestingly, an "intermediate" level of electron donor amendment provided for the most "uniform" culture development and most extensive depletion of the PCE and its reduction products (Figure 5). High electron donor concentrations actually resulted in almost complete loss of dechlorination activity, despite the fact that fermentation of the pyruvate to acetate and hydrogen was complete. While we have no clear mechanistic understanding why this effect is observed, it is clear that incorporating heterogeneous aspects of source-zone bioremediation processes will be important to understanding the true potential for reduction in DNAPL longevity.

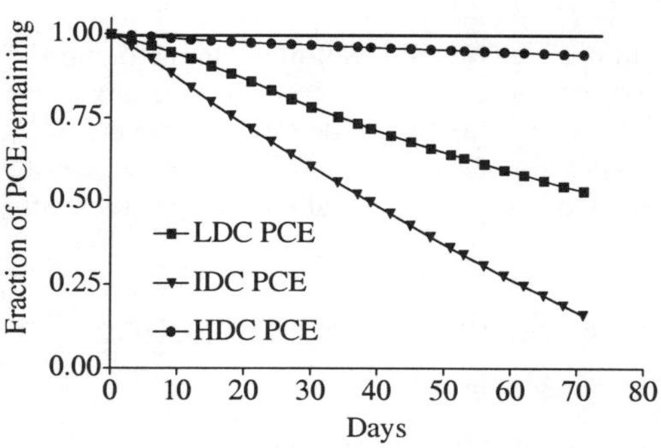

Figure 5. The removal of PCE from a PCE-tridecane NAPL phase in columns fed varying levels of pyruvate as the electron donor: LDC (low donor concentration) = 25 mM pyruvate; IDC (intermediate donor concentration) = 100 mM pyruvate; HDC (high donor concentration) = 250 mM pyruvate. The solid line reflects the rate of loss that would occur due to abioitic washout (approximately 0.3% in 75 days).

In summary, the research that has been conducted to date has demonstrated the potential for source-zone bioremediation to dramatically influence the time that DNAPLs will serve as a source of contamination. Since this is a relatively new area of study, much remains unknown in terms of its true impact at field scales including its influence on the cost of remediation. The first step that is required to address this uncertainty is the development of predictive approaches that can be used at the field scale where the transport, dechlorination, and mass transfer processes operative in source zones are coupled.

3. Surfactant Enhanced Aquifer Remediation (SEAR)

3.1 ENGINEERING DESIGN OVERVIEW

Surfactant enhanced aquifer remediation (SEAR) design has many tightly interlinked facets. For example, a well-characterized site is needed for optimum well placement, yet comprehensive site data cannot be obtained without wells in place. Similarly data about DNAPL composition, the soil and groundwater mineralogy and the existence of an aquitard or lack thereof is needed for optimum surfactant selection, yet, much of this data is often not immediately available. Therefore SEAR design is necessarily an iterative process. While there is a rough sequence to the design process, as some design steps cannot be conducted without earlier data, the reader should keep in mind that in actuality, good SEAR design is performed iteratively. At the same time, it is possible to minimize the number of iterations by developing and following a set of well laid-out design criteria.

The objective of any SEAR operation is to maximize the recovery of NAPL, at minimum cost, while maintaining control over the movement of injected chemicals and NAPL. Since SEAR is aimed towards remediating the subsurface, the most important issue is accurate characterization of the zone being remediated [17]. With accurate characterization, a SEAR test can be designed that ensures efficient sweep of the contaminated subsurface zones, hydraulic control of the injected chemicals and contaminant in the target zone, and optimized pumping rates to minimize the duration of SEAR. In addition, there are usually specific regulatory requirements to be fulfilled to inject surfactant and the other chemicals typically used in the surfactant formulation, such as electrolyte and co-solvent, into the subsurface.

Without careful control, fluids injected as part of the characterization or remediation phases will spread in the aquifer outside the desired treatment zone and perhaps beyond recovery. Uncontrolled spreading in the aquifer is undesirable for two reasons. The first is that the injected compounds will remain in the aquifer, the second is that more chemicals than actually necessary will be required to affect the characterization or remediation. Sheet piling has generally been employed by the environmental industry to control the flow of fluids, however this is generally not a practical solution for anything beyond a shallow demonstration size plot. The control of fluids in the subsurface can also be achieved by the proper balance of injection to extraction rates, proper well location and construction, and the use of hydraulic control wells. Hydraulic control wells are essentially potable water injection wells. Placement of the hydraulic control wells and the rate at which water is injected is such that the injected fluids are constrained to flow to the extraction wells. The use of hydraulic control to contain injected fluids without any other physical control has been successfully demonstrated in numerous PITTs and surfactant floods [18-22].

The objective of SEAR is to remove the maximum possible quantity of contaminant in the minimum possible period of time given the constraints of the site. One key to this process is the selection of a surfactant with optimum attributes. The first attribute a surfactant should display is that it be acceptable for injection into the subsurface. For this reason it is desirable that the surfactant be nontoxic and biodegradable. In terms of logistics and finance, the surfactant should be readily available and low in cost. Raw surfactant cost however can be deceiving, as the true cost of the surfactant will depend on its ability to solubilize the contaminant. Greater efficiency in this regard can offset a higher per pound purchase price. In addition, sorption of the surfactant to the sediments is undesirable, as is the formation of liquid crystals or gels. Either of these problems can reduce the hydraulic conductivity of the formation thus increasing the period of time required for remediation as well as the amount of chemical required. In those cases where it is economically beneficial to recycle the surfactant, ease of separation is also an important selection criteria.

A capillary barrier is usually a formation of low enough hydraulic conductivity (10^{-6} cm/sec or less) to prevent downward migration of DNAPL. DNAPL tends to form pools on top of capillary barriers. The presence or absence of a capillary barrier and the type of capillary barrier at a given site will determine the type of surfactant flooding that can be implemented. At a site with an excellent capillary barrier such as HAFB OU2 (hydraulic conductivity < 10^{-9} cm/sec), a mobilization type surfactant flood is preferable as the mobilized DNAPL will not penetrate formations with such low conductivity. A mobilization flood generally uses a surfactant that has ultra-high contaminant solubilization and ultra-low IFTs. Hence such a surfactant will first mobilize DNAPL and follow up with rapid solubilization of the remaining DNAPL. The combination of ultra-high contaminant solubilization and mobilization of NAPL induced by the ultra-low IFTs requires the use of a very small amount of surfactant, on the order of 1-2 pore volumes. This translates into a lower volume of surfactant, less time in the field to implement the flood and therefore lower costs.

The absence of a capillary barrier at a given site means that downward mobilization of DNAPL and the contaminant-rich microemulsion should be avoided to prevent contamination underlying sediments. Such a situation necessitates the use of a surfactant that will solubilize DNAPL with no downward mobilization. This can be achieved using neutral buoyancy surfactant flooding or SEAR-NB [23, 24]. SEAR-NB uses high concentrations of alcohol and viscosity controls to offset the DNAPL density and recovers the contaminant by solubilization alone. Thus, more surfactant may be required in comparison to a mobilization flood. Clearly, information about the capillary barrier is a necessary data element for the geosystem model used to design SEAR.

The costs of surfactant flooding are somewhat constrained by the cost of injected chemicals during scale-up and application at large sites. Under these conditions, surfactant recycling has the potential to reduce costs. The economics of surfactant recycling may also be favorable if adequate on-site facilities are not available to process the surfactant-rich wastewater generated during SEAR. Surfactant recycling first requires efficient contaminant separation from the contaminant and surfactant-rich effluent from the wells. Once the contaminants are removed, the surfactant may be re-concentrated using micellar enhanced ultra-filtration (MEUF) or nano-filtration techniques. MEUF techniques employ membranes to separate larger surfactant micelles, while nano-filtration has the ability to separate smaller surfactant micelles as well as concentrate surfactant monomers, i.e., unaggregated surfactant molecules. Recovery of dissolved contaminants from the effluent may be carried out by air stripping, steam stripping, or as in the case of the MCB Camp Lejeune, NC demonstration, pervaporation [22]. Air stripping and steam stripping are classical techniques and currently have the most potential for effective contaminant removal if an anti-foam agent is used to suppress surfactant foaming.

The recovery and re-concentration of surfactant by MEUF and nano-filtration will also enhance the concentration of the dissolved contaminants in the surfactant concentrate. For example, if the influent to MEUF has a dissolved contaminant concentration of 0.5 mg/L TCE and the MEUF concentrates the surfactant five-fold, the concentrated surfactant mix will have a TCE concentration of close to 2.5 mg/L. Therefore, the DNAPL removal step must be designed to accommodate this concentration factor so that injection can meet regulatory standards. At very large sites, surfactant recycling has the potential to greatly reduce chemical costs and therefore the cost of surfactant flooding. If the issue of treating the contaminant-rich effluent to an acceptably low concentration is resolved, recycling will be a practical option if a panel-by-panel approach for surfactant flooding is followed (i.e., recycle the surfactant from one panel and use it in the adjacent panel).

3.2 SURFACTANT SELECTION

Choosing a surfactant can greatly influence the performance of a surfactant flood. It can make the difference between a moderately successful flood that removes only a fraction of the DNAPL present and an extremely successful flood that removes >95% of the DNAPL in the swept pore volume. Thus it is useful to know what surfactant properties are desirable for optimum subsurface performance and the laboratory procedures by which appropriate surfactants are selected. First, a brief introduction to surfactants is provided to familiarize the reader with some of the basic terminology used to characterize surfactant properties. Then, surfactant selection criteria is discussed, followed by a description of phase behavior experiments to optimize surfactant selection for the contaminant alone, and discussion of soil column experiments to ensure surfactant compatibility with the contaminated soils.

At their critical micelle concentration or CMC, surfactant molecules will aggregate into structures called micelles [25]. In water, surfactant molecules will form micelles such that the hydrophilic end of the molecule faces the water, and the hydrophobic end of the molecule faces away from the water. Above the CMC, a significant enhancement in the solubility of the NAPL constituents and a significant reduction in the IFT are observed. Thus, surfactants used for DNAPL removal are injected at concentrations far above their CMC. Because surfactant properties will vary with the types of chemical constituents in their surroundings, a careful understanding of surfactant behavior in the presence of NAPL, groundwater constituents, and other organic and inorganic chemical agents is critical for the design of effective SEAR operations.

Three types of phase behavior are expected when NAPL and the surfactant solutions are mixed. These are Winsor Type behaviors I, II, and III. Winsor Type I behavior is observed when the surfactant solubilizes the DNAPL, and is generally associated with solubilization alone, although mobilization can be a possibility. The two phases that exist in a Winsor Type I system are microemulsion and NAPL. Winsor Type II behavior is observed when the surfactant is solubilized by the DNAPL. The two phases that exist in a Winsor Type II system are water and microemulsion. Winsor Type III behavior is observed when the aqueous, microemulsion and DNAPL phases coexist. Given the dynamic impact on phase behavior due to electrolyte concentration, it is very important that this aspect of a surfactant flood be understood and controlled. Sodium chloride and calcium chloride ($CaCl_2$) are examples of commonly used electrolytes. Co-solvents in the surfactant mixture prevent the formation of gels in the subsurface. Isopropyl alcohol (IPA) is an example of a commonly used co-solvent.

Surfactant selection criteria used to screen and select surfactants for applicability to a particular contaminant and site. Surfactant selection criteria for SEAR are described in detail in the literature [17, 26-29]

Whenever a surfactant and NAPL are physically mixed, a macroemulsion is formed with small NAPL droplets dispersed in a continuous aqueous phase containing the surfactant micelles. The excess NAPL in the macroemulsion coalesces together to form larger NAPL droplets, which eventually separate into an excess NAPL phase and a microemulsion phase with dissolved NAPL constituents. This phenomenon is termed coalescence. The most important parameter for surfactant selection is rapid coalescence of surfactant-NAPL mixtures into thermodynamically stable microemulsions. Surfactants with coalescence times of less than 24 hours are desirable. Poor surfactants form stable macroemulsions, have high coalescence times (greater than 40 hours) and therefore have the potential to cause mass transfer limitations between NAPL and surfactant micelles in the subsurface. This greatly impedes NAPL removal efficiency and most likely accounts for observations of mass transfer limitations reported by investigators such as Pennell et al. [30] and Mason and Kueper [31].

Poor surfactants may also reduce the permeability of the porous media due to pore plugging from the formation of macroemulsions, gels and liquid crystals [27, 29, 32]. As an illustration, sodium dihexyl sulfosuccinate, which was used by DE&S at HAFB OU2 [20, 21] to remediate a TCE-rich DNAPL, coalesced in less than 20 minutes [33]. Rapid coalescence can be promoted by adding a suitable co-solvent such as IPA or by using a surfactant with a branched hydrophobic tail.

Surfactants tend to precipitate out of solution to form liquid crystals, gels, and viscous macroemulsions. All of these phases will hinder transport through the aquifer, cause pore plugging and a reduction in the aquifer permeability. The addition of a co-solvent or co-surfactant will minimize the formation of these undesirable phases. Their formation may be further limited by using a surfactant with a branched hydrophobic tail [25]. The Alfoterra 145-4-PO-sulfate™ has two branches on its hydrophobe tail.

The candidate surfactant should be capable of enhancing the solubility of the contaminant several fold above the normal solubility of the contaminant in water alone. This will allow rapid dissolution and removal of the NAPL. The equilibrium solubilization of Hill DNAPL by an optimized solution of sodium dihexyl sulfosuccinate was 625,000 mg/L, which is 600 times the solubility of Hill DNAPL in water [17]. This formulation was used at the surfactant flood demonstration at Hill AFB, Utah in 1996 [20]. Similarly the equilibrium solubilization of tetrachloroethylene (PCE) by an optimized solution of Alfoterra 145-4-PO-sulfate™ was 460,000 mg/L, which is 1,900 times the solubility of PCE in water [23]. The Alfoterra 145-4-PO-sulfate™ surfactant, with lower than optimum salinity, was used for the surfactant flood demonstration at MCB Camp Lejeune, NC.

Many of the environmental impact criteria used to assess the acceptability of surfactants for use in laundry and cleaning products can be used for SEAR applications. The subsurface is the first point impacted by SEAR operations. Surface waters to which the treated effluent will be discharged should also be considered. When evaluating the risk associated with surfactant usage, the primary consideration is whether the concentrations of surfactant at the point of consideration exceed the "no-effects" concentration (i.e., the lowest concentration that will not result in chronic or acute toxicity to the most sensitive species at that point). Biodegradability and aquatic toxicity information are used to make this assessment.

As an illustration, the MCB Camp Lejeune demonstration used the Alfoterra 145-4-PO-sulfate™ surfactant, which is classified as an alcohol ether sulfate (AES). Specifically, the alcohol is branched at the beta position, and the ether portion is composed of repeating propylene oxide (PO) units. The terminal PO is esterified to sulfate which gives it the anionic designation. AES surfactants are the most widely used laundry product surfactants. They are readily biodegradable in aerobic and anaerobic (e.g., subsurface) environments, and aquatic toxicity is not a problem at the concentrations observed in the various environmental receptors that receive these compounds.

Phase behavior is the stability and quality of the immiscible water, microemulsion and NAPL phases that form when a surfactant solution and NAPL are mixed together. Phase behavior experiments quantify the behavior of NAPL-surfactant mixtures and are essential for screening and selecting surfactants for SEAR. These experiments have been described in detail [27, 29, 34-39].

Phase behavior experiments identify surfactants with acceptably high contaminant solubilization, rapid coalescence times (less than 24 hours), and minimal tendency to form liquid crystals, gels and emulsions. Volume fraction diagram experiments are usually conducted, where equal volumes of NAPL and surfactant solution are mixed and allowed to equilibrate. The surfactant (and co-solvent, if used) concentrations are held constant and the electrolyte concentration is varied between various samples. The effect of electrolyte on the phase behavior is carefully noted, especially the electrolyte concentrations that cause a transition between Winsor Type I to Type III to Type II behavior. In addition, the time taken for complete coalescence is measured. The time for complete coalescence is defined as the time when the aqueous, microemulsion and NAPL phases are in equilibrium with each other. Surfactants that have coalescence times of greater than 24 hours are considered unsuitable for use in SEAR.

To design successful surfactant floods, it is important to perform a variety of carefully designed column experiments under conditions similar to those to be encountered in the field. These conditions include using site NAPL, soil, and water to perform the tests at the average site groundwater temperature. These experiments provide information on potential adverse interactions between the surfactant, NAPL, and the aquifer sediments, such as surfactant adsorption, clay mobilization, or the formation of liquid crystal, gel, or macroemulsion phases. Another piece of information critical for design of these field tests that can be obtained from the column tests is the mean residence times required for surfactants in the subsurface to avoid potential non-equilibrium effects. Soil column experiments also evaluate the ability of surfactants to remediate NAPL-contaminated aquifer material. These experiments help determine the final NAPL saturation that can be achieved as a result of surfactant flooding. The details of performing soil column experiments with Ottawa sand and field soil are given in Dwarakanath et al. [27].

3.3 MODELING USING UTCHEM

To design and predict the performance of the SEAR without performing numerous field experiments, it is necessary and desirable to perform numerical simulations. The numerical model must simulate the advection, dispersion, and transformation of the different species (contaminant, surfactant, water, electrolytes, co-solvent, and possibly air) in the aquifer under various pumping and injection strategies.

UTCHEM is a 3-D, non-isothermal chemical compositional simulator. Properties such as density, IFT, capillary pressure, residual saturation, relative permeability, adsorption, viscosity, the full dispersion tensor, and cation exchange with the clays are modeled. Surfactant floods performed at HAFB OU1, HAFB OU2 [40], and DOE Portsmouth were all modeled and designed with UTCHEM.

The UTCHEM model has the ability to incorporate the influences of vadose zone drainage, multiple DNAPLs, rate-limited mass transfer, hysteresis in relative permeability and capillary pressure, and wettability on the design and performance of SEAR. Of particular importance to the SEAR process is the possible mobilization of residual DNAPL in an aquifer due to the reduced IFT caused by the introduction of a surfactant solution. In UTCHEM, IFT is used to compute the capillary number, and the residual phase saturations are computed based on the capillary desaturation curve. Relative permeability and capillary pressure are then adjusted according to the calculated residual saturations based on the capillary number. Initially either MODFLOW or UTCHEM can be used to input the stratigraphy to calibrate the hydraulic properties. As mentioned earlier, the advantages of MODFLOW are its efficiency and simplicity in setting up and calibrating the aquifer data. Once the aquifer properties are calibrated, the results such as permeability, porosity, and capillary barriers should then be transferred to UTCHEM for three-dimensional modeling of SEAR design.

Considering the complexity of the subsurface as well as of the surfactant properties, reflected in the large number of model parameters and the difficulty in accurately assessing them, it is necessary to determine the sensitivity of the SEAR design performance to these parameters. Doing so will help identify variables that are critical to the accuracy of the model. Once the SEAR simulation model is set up with the appropriate input parameters, sensitivity simulations should be run to study the behavior of SEAR under different conditions to formulate the optimum design strategy. There are a number of variables that have a major influence on the SEAR design but their exact values are uncertain. Sensitivity studies are normally conducted by adjusting these variables within their range of uncertainties. Results of these sensitivity studies are then used to select the surfactant injection scheme and to determine the mass of chemical needed, as well as the injection and extraction rates. The simulations are used to predict the duration of the SEAR, the extraction well effluent chemical concentrations, and the amount of injectate compounds as well as contaminant recovered at the end of SEAR operation. The following is a brief listing of the sensitivity variables:

- Heterogeneity in aquifer permeability
- Initial DNAPL volume and DNAPL saturation distribution
- Residual DNAPL saturation
- Number of pore volumes of surfactant needed and the optimum composition of the surfactant/co-solvent mixture taking into account mixing and dilution effects
- Pre- and post flush electrolyte concentration and duration
- Vertical to horizontal permeability ratio
- Concentration and pore volumes of polymer (if used) during and after surfactant injection for mobility control
- Balance of injection and extraction needed for hydraulic control as well as the need for water injections outside the wellfield i.e. hydraulic control wells
- Variations in petrophysical properties such as the relative permeability and adsorption to take into account uncertainty in these data
- DNAPL mobilization using different DNAPL trapping curves
- Mobility control using polymer

The UTCHEM simulator provides very comprehensive output information and the computed data for many parameters can be written to the output files for every gridblock and at different times during the simulation. There are two Graphical User Interfaces (GUI) available to post process the UTCHEM output. The utility called UTHIST converts the well history output files to a column format to be processed and viewed in Microsoft Excel. The utility called UTSURF generates 2-D maps of pressure, saturation, and species concentrations along with the simulation grid and well locations using Golden Software Surfer. These utilities are written in Visual Basic and they can only be run on Windows (NT, 95 or higher).

The data routinely examined in the SEAR simulation design study are:

- Total effluent concentrations of contaminant, surfactant, co-solvent (if any), and calcium as a function of time for each extraction well
- Recoveries of contaminant, surfactant, and co-solvent (if used) for each extraction well
- Recovery of contaminant as a free product for each well
- Concentration of contaminant, surfactant, and co-solvent (if any) at the multilevel sampling locations or monitoring locations (if installed at the site)
- The pressure/water level at the injection and extraction wells
- The saturation and volume of the DNAPL in the selected boundary comprising the wellfield both at the beginning and end of the demonstration for performance assessment

- 2-D images of total concentrations of contaminant and surfactant, saturation of DNAPL and microemulsion phases, and pressure at selected times during the simulation

The graphical displays of pressure, concentration, and saturation contours at different depths give valuable insights into the simulation results of the SEAR design. The areal extent of the injectate concentration at selected times gives a qualitative measure of the sweep efficiency and hydraulic control with depth. The examination of the DNAPL saturations within the wellfield provides a measure of the SEAR design performance and an idea about the DNAPL remaining. The contour maps of the surfactant concentration at the end of the demonstration provide crucial information on its final spatial distribution and whether the time is adequate to flush the surfactant out to a satisfactory level. In addition these contour maps allow the user to review the hydraulic gradient and frontal movement to ensure that no undesirable trends are developing.

3.4 WASTEWATER TREATMENT

SEAR requires the subsurface injection of surfactants, co-solvents (if used) and ionic solutions and the subsequent capture of these fluids. Injectate recovery minimizes the amount of residue left in the subsurface. With a proper design, high recoveries (> 90%) of the injected fluids can be accomplished. These fluids, also rich in DNAPL, require treatment once extracted from the subsurface prior to discharge to a wastewater treatment facility.

SEAR chemicals and contaminants are recovered in the extracted groundwater. At the peak, the extracted groundwater can contain several weight percent surfactant, co-solvent (if used), and contaminant. The presence of the contaminant alone classifies the extracted groundwater as hazardous waste, particularly at the elevated levels at which the contaminant is removed due to surfactant. Proper disposal of a large quantity of hazardous waste would be a significant cost expenditure. Treating the extracted groundwater can reduce the amount of hazardous waste generated and hence disposal costs. Treatment also provides the potential to recover and recycle the injected chemicals, further reducing costs.

The objectives of treating extracted groundwater from a SEAR application are to:

- Protect human health and the environment from further contamination
- Minimize the generation of hazardous waste
- Produce a waste stream which can be readily discharged to an existing treatment operation with minimal impacts
- Provide an effective approach for recovering injected chemicals

In general, the design of a treatment system to pre-treat SEAR effluents is very site specific. Federal, state and local regulations and the receiving facility will govern the discharge standards of the treatment system. The process flow rate will be governed by the hydraulic conductivity of the treatment zone and SEAR design. The concentrations of SEAR chemicals and contaminants in the effluent can be estimated before SEAR operations through numerical modeling. Furthermore, the process selection will be driven by budget limitations and time allowable in the field. A good design will incorporate the items discussed within the context of other site-specific issues which may have been overlooked.

3.5 ACKNOWLEDGEMENT

The collaboration and contributions of Jackie Avvakoumides, Larry Britton, Fred Holzmer, Richard Jackson, John Londergan and Laura Yeh to this section are gratefully acknowledged.

We would also like to acknowledge funding by several federal agencies and programs including AATDF, ESTCP, AFCEE, DOE and NSF of various research and field projects over the past several years that formed the basis for this overview of the SEAR process.

4. Bioremediation of Munitions

All of the commonly used military and industrial explosives are nitrated organic compounds. The most widely used explosives are 2,4,6-trinitrotoluene (TNT) and hexahydro-1,3,5-trinitro-1,3,5-triazine (RDX) (Figure 6). The explosives in Figure 6 along with feedstocks such as the mono- and dinitrotoluenes can be found as soil and ground water contaminants at manufacturing sites, load and pack facilities, and test ranges. In the United States, the US Army Environmental Center has identified and investigated the contaminated sites. Many have been cleaned up or are in the process of cleanup. Much of the heaviest contamination is in the soil, relatively near the surface so the most common cleanup strategy has been excavation followed by incineration or composting. In contrast, little has been done about permanent solutions for contaminated ground water and large areas of lightly contaminated soil. At present, contaminated ground water is treated exclusively by pump and treat systems that rely on carbon sorption. Recent advances in understanding of the biodegradation of explosives provide the basis for bioremediation, a cost effective strategy for complete destruction of the contaminants. Bioremediation has been used successfully for treatment of excavated soil, but has only very recently emerged as an option for in-situ treatment. Strategies such as phytoremediation, natural attenuation, biostimulation, landfarming, and biotreatment combined with iron slurry walls are in various stages of development. A recent book [42] provides a good review of what is known about the environmental behavior, biodegradation, and bioremediation of explosives. The following discussion will be a brief overview of several chapters from the book.

Microbes can attack nitro organic compounds (or any organic compounds) under a variety of different redox conditions and with a variety of outcomes (Figure 7). In some instances the compound of interest serves as a source of carbon and energy and the process is self-sustaining. Accordingly, most simple nitroaromatic compounds such as nitrobenzene, nitrotoluenes, and nitrophenols are considered to be “biodegradable”. Some explosives, such as TNT and RDX can serve as sources of nitrogen or as electron acceptors for microbes and, therefore, support growth without providing a source of carbon and energy. They are “biotransformed” in processes that provide a selective advantage to the microbes, so the processes are self-sustaining if an external electron donor or carbon source is available. Finally, some compounds are biotransformed in processes that provide no selective advantage to the microbes and the process can even be detrimental to the microbes if the process requires metabolic energy or produces toxic metabolites. Such biotransformation is the basis for composting and anaerobic slurry treatments of TNT. The latter process can be very effective for compounds that are not biodegradable as long as the final products are not toxic.

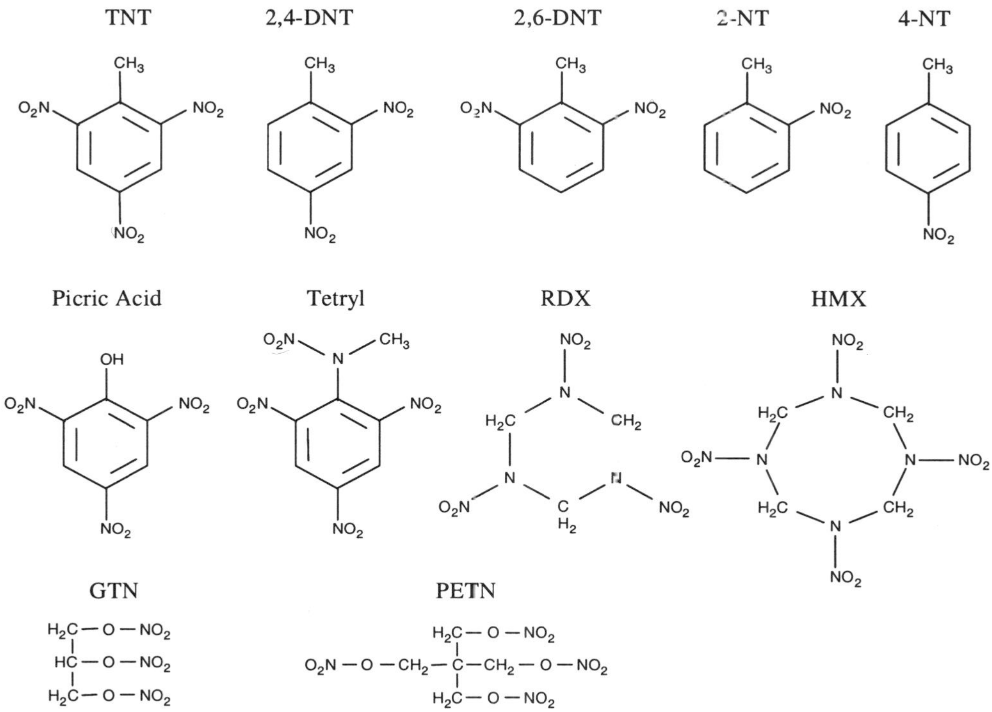

Figure 6 . Major military explosives and production intermediates [41].

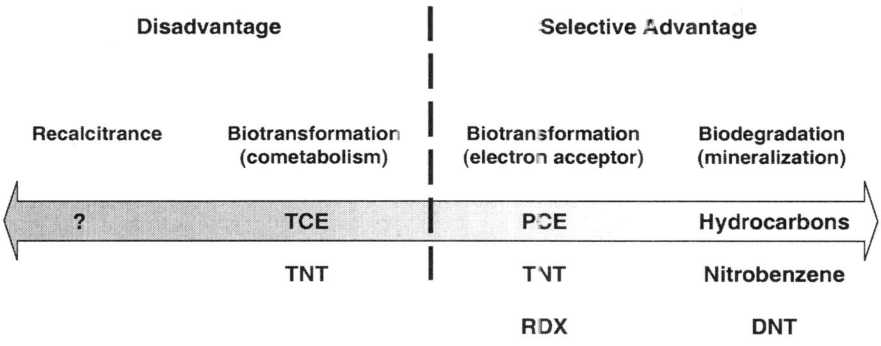

Figure 7. Contaminant-microbe interactions. modified from [43].

Microbes have evolved several strategies for removing nitro groups from nitroaromatic compounds (Figure 8) [44]. Complete reduction of the nitro group to an amino group can be catalyzed by virtually all biological systems. The amines are often not biodegradable, however, and the pathway can be a dead end. In contrast, reduction of the nitro group to the hydroxylamine can lead to productive pathways for biodegradation of a variety of compounds. If the parent nitroaromatic compound contains a hydroxyl group the nitro group can be eliminated by the action of flavoprotein monooxygenase enzymes such as those involved in the biodegradation of nitrophenols. Some polynitroaromatic compounds, such as TNT and picric acid can be attacked by enzymes that reduce the aromatic ring and form Meisenheimer complexes that can be further metabolized. Finally, many nonpolar nitroaromatic compounds are substrates for nitroarene dioxygenases that add two hydroxyl groups to the aromatic ring and eliminate the nitro group as nitrite.

All of the biodegradable nitroaromatic compounds listed in Figure 8 are excellent candidates for bioremediation. Former TNT manufacturing sites are often heavily contaminated with mono- and dinitrotoluenes, the feedstocks for TNT synthesis. Bacteria able to grow on mononitrotoluenes, 2,4-DNT and 2,6-DNT have been isolated from a variety of contaminated sites and the biodegradation mechanisms are well-established [44, 45]. The bacteria use the dinitrotoluenes as sources of carbon, nitrogen, and energy. Several field demonstrations have indicated that mono- and dinitrotoluenes can be removed from contaminated water and soil by bioremediation [46,47]. In-situ treatment of DNT contaminated water and soil is currently being explored at Badger Army Ammunition Plant in Wisconsin [48, 49]. More information about biodegradation of simple nitroaromatic compounds can be found in a recent review [44].

Cometabolic biotransformation has been used as the basis for bioremediation of explosives that do not serve as growth substrates for microorganisms. For example, TNT can be transformed under anaerobic conditions to intermediates that bind to soil and become unavailable (Figure 9). Under highly reducing conditions all of the nitro groups of TNT can be reduced. When electron donors are plentiful the reduction is rapid and can provide the basis for effective bioremediation. Several organizations have applied anaerobic slurry reactors for field scale treatment of TNT. Most of the reduction under anaerobic conditions is probably catalyzed by *Clostridium.*

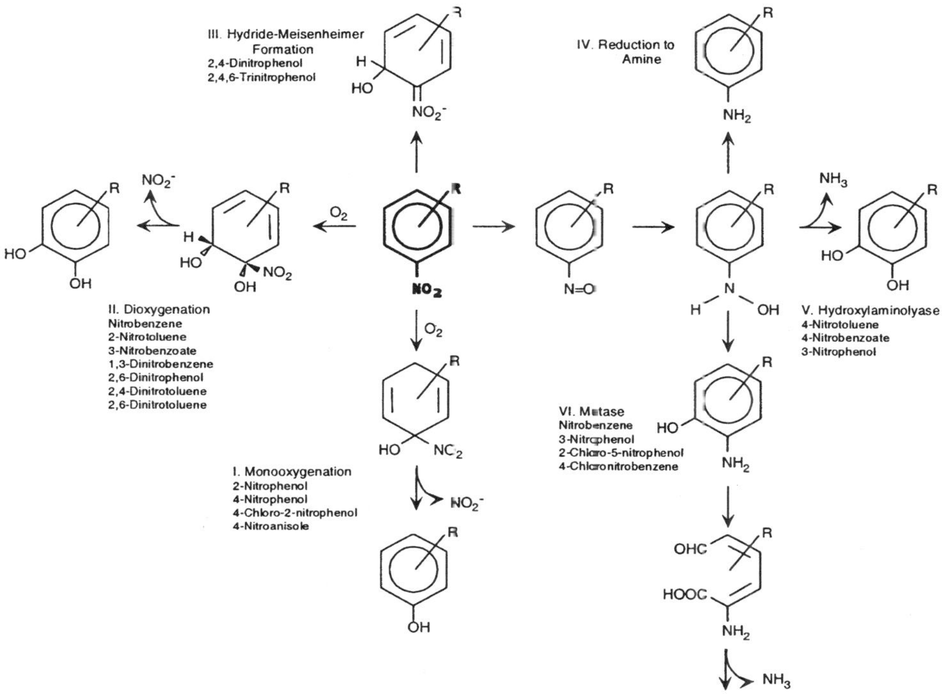

Figure 8. Microbial pathways for the degradation of nitroaromatic compounds, modified from [44].

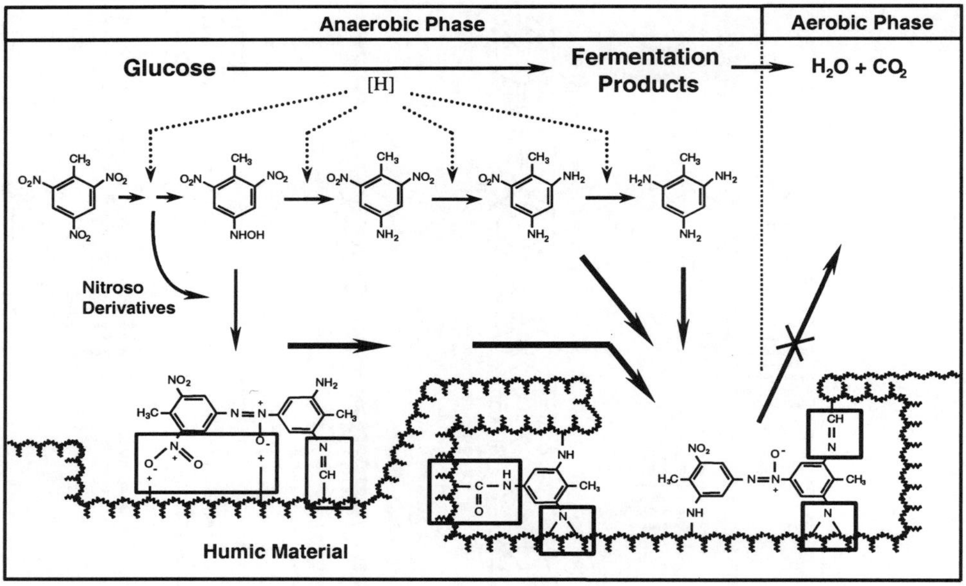

Figure 9. Cometabolic biotransformation of TNT in an anaerobic soil slurry, modified from [50].

During anaerobic treatment of TNT, very little of the molecule is reduced completely to triaminotoluene and even less is mineralized. Instead the intermediates interact with components of the soil or slurry and become bound. The nature of the soil and the conditions of treatment determine whether physical interactions, hydrogen bonding, ionic or covalent bonds are the most important. Clay minerals and soil organic matter are the major sorbents in soil. The interactions of TNT metabolites with the humic substances in soil seem to be more important than sorption to clay. Both the amino groups and hydroxylamino intermediates can participate in the binding. The reactions are complex and not fully understood. What is clear is that the initial binding is relatively weak and that TNT metabolites are still extractable from soil during the early stages of anaerobic treatment. Extended treatment leads to irreversible binding and the bound residues cannot be extracted by harsh reaction conditions including acid and alkali treatment.

TNT can also be transformed extensively under aerobic conditions and recent work indicates that it can serve as an electron acceptor and a source of nitrogen for some bacteria. The disadvantages of strategies that rely on cometabolic transformation include: the need for external electron donors, uncertainty about the identity of the final products, and requirement for relatively complex process controls.

Composting is a cometabolic transformation strategy that has been widely used for treatment of excavated explosives contaminated soil [51]. There have been a number of recent advances in the understanding of composting for destruction of TNT [52]. The process involves initial reduction of the molecule and eventual humification of the metabolites. It is highly effective and less expensive than most other bioremediation strategies and is widely accepted in the United States as a remediation strategy. The disadvantage of composting is that it increases the volume of the waste material. Several other technologies for cometabolic bioremediation of explosives are commercially available in the United States and have been tested at field scale. Their costs and performance have been compared and all are effective for removal of TNT [51].

Nitrate esters such as glycerol trinitrate, pentaerythritol tetranitrate, and cellulose nitrate can be denitrated by bacteria. The nitrite thus liberated can be assimilated and support the growth of the bacteria [53]. RDX and HMX can similarly be degraded and can serve as nitrogen sources for bacteria. In the case of RDX and HMX the parent molecules can be mineralized under both aerobic and anaerobic conditions but the details of the reactions are not known [54]. Although the nitrate esters and nitramines are now known to be susceptible to biodegradation there have been few if any attempts at practical applications involving bioremediation of such compounds.

There is a considerable amount of ongoing research in the area of explosives bioremediation. The fundamental microbial processes involved in biotransformation are still the subjects of many studies. The mechanisms and kinetics of binding of intermediates to soil and humic material are also just beginning to be understood. The factors that affect such binding are still not known. Several laboratories are working on the use of metabolic engineering to develop microbes that can mineralize TNT. Finally, there is a great deal of current work in the United States to develop strategies to predict and enhance in situ biodegradation of explosives.

4.1 ACKNOWLEDGEMENT

The authors of this section were supported by funds from the Strategic Environmental Research and Development Program.

5. Remediation of Cr(VI) And TCE in Groundwater Using an In Situ Redox Manipulation Barrier

A new method for creating permeable reactive groundwater treatment barriers has been developed at the Hanford Site in Washington State. Because these contaminant plumes at Hanford range from 15 to 90 meters below the ground surface, an alternative to "trench and fill" permeable reactive barriers was developed. This alternative, called In Situ Redox Manipulation (ISRM), can be installed through conventional groundwater wells.

The goal of the ISRM method is to create a permeable treatment zone in the subsurface to remediate redox-sensitive contaminants in the groundwater. A permeable treatment zone is created by reducing the ferric iron in the aquifer sediments to ferrous iron by injecting a reducing reagent and appropriate buffers (e.g. sodium dithionite and potassium carbonate). The ISRM approach extends the permeable treatment zone concept to sites where the groundwater contaminant plumes are too deep (tens of meters below the ground surface) to be treated by excavation or by trench-emplaced permeable barriers.

After successful bench-scale and smaller field-scale tests, a larger treatability test was installed at the Hanford 100-D Area beginning in fiscal year 1997. The test used five wells to create a permeable reactive barrier 46 meters long and 15 meters wide. Depth to groundwater in this area is about 26 meters. Recent monitoring data have shown that chromate concentrations in the reduced zone have decreased to below detection limits and chromate concentrations are decreasing in several down gradient monitoring wells.

Based on the success of the treatability test, DOE and the regulatory agencies decided to deploy the technology by expanding the barrier to approximately 700 meters in length. Construction began in fiscal year 2000 and will be completed in fiscal year 2002.

An ISRM proof-of-principle test was also conducted for dissolved TCE at Fort Lewis in Tacoma, Washington. The aquifer is in a complex glaciofluvial setting, with high hydraulic conductivity (700 ft/day), low aquifer temperatures (12°C), and baseline TCE concentrations of ~140 μg/L. Groundwater monitoring showed TCE concentrations of less than 95% of upgradient values within the reduced zone after 4 dithionite injections. Acetylene has also been detected in the wells with significantly lower TCE concentrations as further evidence of TCE destruction. However, the large amount of the dithionite reagent that was required to achieve the TCE destruction made the barrier uneconomic at this site.

5.1 INTRODUCTION

As a legacy from weapons production at the U. S. Department of Energy (DOE) Hanford Site in south central Washington State, several chromate groundwater plumes are currently impinging upon the Columbia River, potentially impacting aquatic receptors in the river. Pump and treat systems are operating at two of the plumes in the Hanford 100HR-3 Operable Unit. Because permeable reactive barriers appear to offer several advantages over pump and treat, DOE decided to test this concept on a third plume. Due to the depth of contaminant plumes in the Hanford 100 Areas (average 25 meters) below the ground surface, an alternative to "trench and fill" permeable reactive barriers was developed. This alternative, called In Situ Redox Manipulation (ISRM), can be installed through conventional groundwater wells. An independent study by Los Alamos National Laboratory has shown that deployment of the ISRM technology at the Hanford 100D Area will result in a 60% cost savings compare to pump and treat over a 10 year period [55].

5.2 TECHNOLOGY DESCRIPTION

The In Situ Redox Manipulation (ISRM) technology involves creating a permeable subsurface treatment zone to reduce mobile chromate in groundwater to an insoluble form. An unconfined aquifer is usually an oxidizing environment; therefore, most of the contaminants that are mobile in the aquifer are mobile under oxidizing conditions. If the redox potential of the aquifer can be made reducing, then a variety of contaminants could be treated (Figure 10). Redox-sensitive contaminants migrating through this treatment zone would be destroyed (organic solvents) or immobilized (metals).

The ISRM permeable treatment zone is created by reducing the ferric iron (Fe(III), present as surface oxides) to ferrous iron (Fe(II), on sediment surfaces as adsorbed Fe(II) and $FeCO_3$) within the aquifer sediments. This is accomplished by injecting sodium dithionite ($Na_2S_2O_4$) into the aquifer and withdrawing unreacted reagent and reaction products. The sodium dithionite serves as a reducing agent for iron, changing ferric iron to ferrous iron within the unconfined aquifer sediments. Sodium dithionite is a strong reducing agent that has a number of desirable characteristics for this type of application, including instability in the natural environment, with reaction and degradation products that ultimately oxidize to sulfate. This instability is beneficial because it means that the reaction period is rapid, and that after a period of several days, no dithionite will remain in the aquifer. Potassium carbonate/bicarbonate is added to the injection solution as a pH buffer to enhance the stability of dithionite during the reduction of available iron. Unreacted reagent and reaction products are pumped out of the aquifer through the same well used for injection, starting about two days after injection.

An ISRM permeable treatment zone is emplaced perpendicular to the groundwater flow to intercept the contaminant plume, as shown in Figure 10. This geometry is created by a series of overlapping injection/withdrawal wells. The width of the permeable treatment zone multiplied by the pore volumes of treatment capacity of the reduced zone determines the volume of contaminated groundwater that can be treated. The treatment capacity is a function of the amount of reducible iron in the sediment, the efficiency of the reduction by the field emplacement (dithionite concentrations and sediment reaction time), and the oxidizing potential of the groundwater (e.g., dissolved oxygen and chromate concentrations). Groundwater velocities and aquifer contamination concentrations at the site determine the longevity of the treatment zone. Other dimensions of the permeable treatment zone (i.e., length and depth) are determined by the extent of contamination requiring treatment.

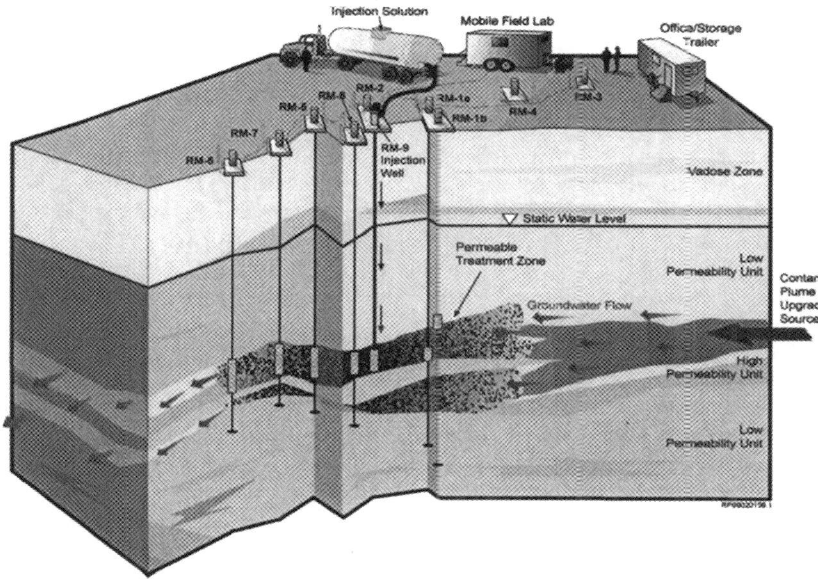

Figure 10. In situ redox manipulation process

5.3 RELEVANT CHEMISTRY OF CHROMIUM AND IRON IN THE SUBSURFACE

The target contaminant for this study is chromium in its hexavalent form (Cr(VI). In dilute, near neutral aqueous solutions, Cr(VI) is generally present as the highly soluble and mobile chromate anion: $HCrO_4^-$. Its concentration in groundwaters and surface waters in the United States is regulated under the Safe Drinking Water Act (Maximum Contaminant Level (MCL) = 0.100 mg/L as Cr) and the Clean Water Act (Ambient Water Quality Criteria (AWQC) = 0.0011 mg/L as Cr). The trivalent form of chromium is much less toxic and mobile in the environment. Upon reduction to Cr(III), Cr(III) in solution readily hydrolyzes and precipitates as $Cr(OH)_3(s)$ [56]. Under most groundwater conditions the solubility of $Cr(OH)_3(s)$ is well below the 0.100 mg/L MCL. When Cr(III) is precipitated in soils containing ferric iron, solid solutions with ferric iron can also form, (Cr, Fe) $(OH)_3(s)$. The aqueous Cr(III) concentrations in equilibrium with this phase are even lower than that of the pure $Cr(OH)_3(s)$ phase [56, 57].

Once the $Cr(OH)_3(s)$ or $(Cr,Fe)(OH)_3$ phase is formed, it is not readily re-oxidized back to chromate. Available information indicates that the rate of oxidation of Cr(III) by oxygen is very slow if it occurs at all [58, 59, 60]. Laboratory column experiments have shown that $Cr(OH)_3(s)$ remained immobilized in sediments even after 600 pore volumes of oxygen saturated water were injected [61]. Manganese oxides are the only other known naturally occurring oxidants capable of oxidizing Cr(III) [60]. Although Cr(III) oxidation readily occurs on Mn oxides, a number of factors have been found that greatly limit the progress of this reaction in the subsurface environment [60, 62, 63, 64, 65, 66, 67]. Therefore, a method which reduces hexavalent chromium to the trivalent form in groundwater should be effective as an aquifer restoration method.

The most abundant potential inorganic reductant in most groundwater aquifers is iron which occurs naturally in minerals and oxide coatings in aquifer sediments in both the divalent (ferrous) and trivalent (ferric) oxidation states. The reaction between chromate and reduced iron is illustrated by

$$HCrO_4^-(aq) + 3Fe^{2+}(s) + 8H_2O \rightarrow Fe_3Cr(OH)_{12}(s) + 5H^+$$

Although a number of common minerals contain ferrous iron in their internal structure, in oxidizing aquifers, the minerals and soil particles are often coated with ferric iron containing products of weathering such as amorphous and crystalline oxides and oxyhydroxides. It is this surface ferric iron, along with the ferric iron adsorbed in ditrigonal cavities in specific clay minerals, that is chemically reactive and available for use in forming a reductive treatment zone to react with dissolved chromate and other oxidized species.

5.4 TREATABILITY TEST SUMMARY

After successful bench-scale tests, a proof-of-principle field experiment was conducted in September 1995 at a chromate (Cr(VI)) contaminated groundwater site at 100H Area at the Hanford Site in Washington State. During this experiment chemicals were injected into a single-well to determine the feasibility of creating a reduced zone in an aquifer using the ISRM method, and to determine the longevity of the reduced zone in a natural environment. The test created a 15-meter diameter cylindrical treatment zone [68]. Monitoring data over the past three years later show that the treatment zone remains anoxic, chromium levels have decreased to below the detection limit of the analytical method (< 8μg/L) , and no significant permeability changes have been detected.

Beginning in fiscal year 1997, a larger treatability test facility (Figure 11) was installed at the Hanford 100-D Area(15). The test used five wells to create a permeable reactive barrier 46 meters long and 15 meters wide. Depth to groundwater in this area is about 26 meters.

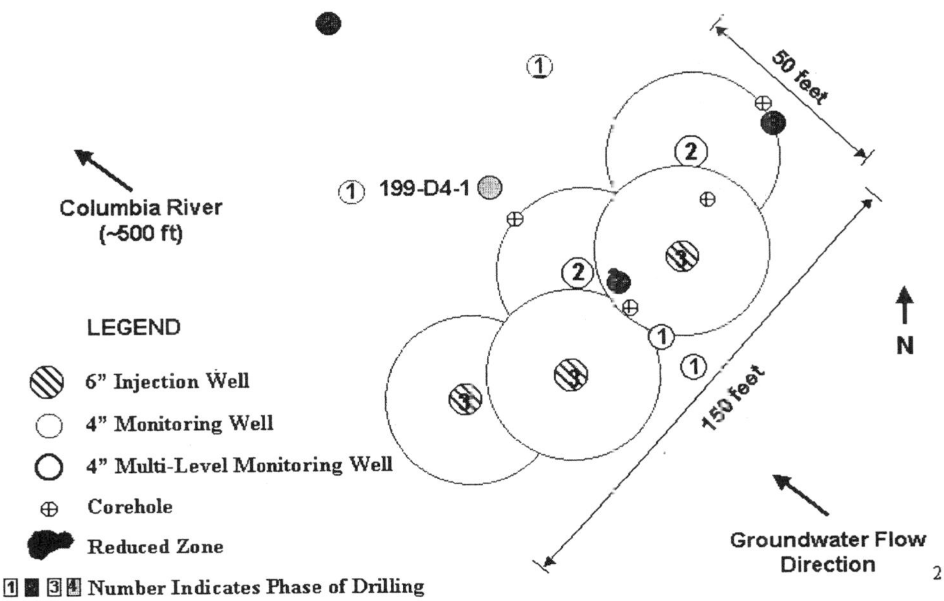

Figure 11. Well emplacement layout for the ISRM Treatability Test at the Hanford 100D Area.

Well installation for the 100-D Area ISRM Treatability Test started in the spring of 1997. Well installation and site characterization activities continued through the summer of and were completed in November 1997. Characterization activities included sampling the sediment, testing the aquifer, establishing baseline aqueous geochemistry, and conducting a bromide tracer test. The first dithionite injection/withdrawal took place during September and October of 1997; ground water monitoring was conducted from the fall of 1997 to the spring of 1998. The remaining four dithionite injection/withdrawal tests were conducted from May to July 1998.

As shown in Figure 11, these dithionite injections and withdrawals created a reduced zone in the aquifer approximately 48 meters long (perpendicular to groundwater flow), 15 meters wide, and ~5 meters thick, extending over the thickness of the unconfined aquifer. Laboratory analysis of sediment collected from the aquifer during the initial drilling measured an average treatment capacity of the sediment of 171 ± 46 pore volumes of groundwater from the site. Using 1 ft/day groundwater velocity (as measured at the site) and a 50 ft width, this treatment capacity results in a predicted longevity of the 100-D Area ISRM permeable treatment zone of 23 ± 6 years. If further treatment is required, the barrier can be re-established by re-injection of the dithionite reagent.

Groundwater monitoring at the site continued on a quarterly basis through FY99. Sediment samples were also collected from coreholes within the reduced zone to measure the reduction capacity of the sediment that was achieved by the treatability test emplacement.

5.5 PERFORMANCE RESULTS - GROUNDWATER QUALITY

Groundwater monitoring was conducted at the site on a monthly to bimonthly basis during FY98 and roughly quarterly during FY99. Analytes consisted of field parameters (EC, pH, and dissolved oxygen), chromate, anions, and trace metals. Results from the most recent complete sampling event (April 6, 1999) are shown in Tables II and III. Average values and ranges of selected parameters for this sampling event are given for wells within the treatment zone in Table II and for the downgradient wells in Table III. Hexavalent chromium (Cr(VI)) results are shown in plan view in Figures 12 and 13. These analyses are discussed below and compared with baseline data collected prior to any dithionite injection/withdrawal tests. Full results are shown in Williams et al. [69].

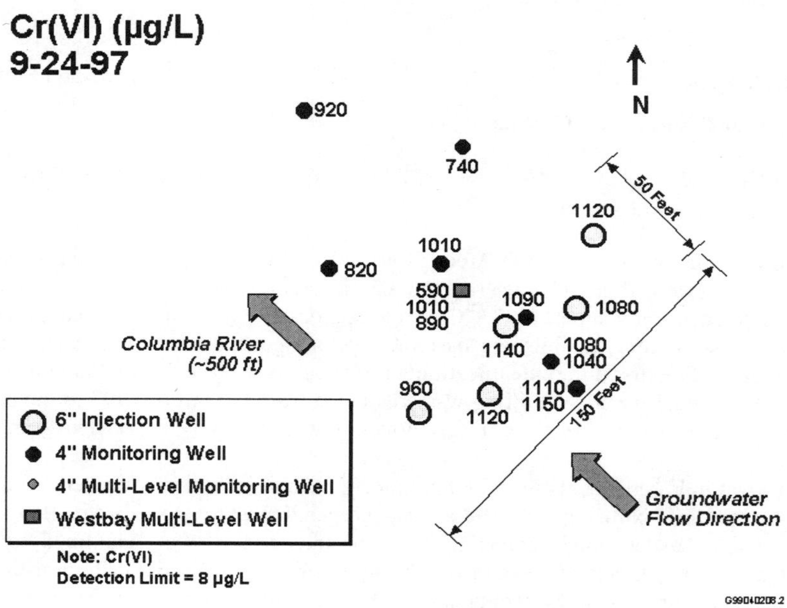

Figure 12. Cr(VI) concentrations (µg/L) at the Hanford 100D Treatability Test Site before dithionite injection (9/24/97).

Concentrations of Cr(VI) measured in the wells significantly influenced by dithionite during the emplacement process were all below the detection limits of the field analysis equipment (Hach DR-2000) of 7 µg/L (compare Figures 12 and 13) Average baseline Cr(VI) concentrations measured for all the wells in baseline sampling (September 1997) were 1000 µg/L.

Although concentrations of chromate and dissolved oxygen are significantly different from baseline values in the downgradient wells, the full impact downgradient from the treatment zone currently cannot be used to assess the final performance of the ISRM technology. While the Cr(VI) concentrations in the downgradient wells have showed a continuous decline from the emplacement through April, 1999, sufficient time has not elapsed since the last emplacement (mid-July, 1998) for the groundwater migrating through the treatment zone to fully influence the aquifer downgradient. Differences between baseline groundwater parameters within the treatment zone, in addition to the Cr(VI) concentrations discussed above, include dissolved oxygen, pH, electrical conductivity and sulfate. Dissolved oxygen reacts with the Fe(II) within the treatment zone. The oxidizing capacity of dissolved oxygen is greater than the Cr(VI) concentrations at the site and mainly determines the longevity of the treatment zone. Similarly to Cr(VI), water within the treatment zone is anoxic and the downgradient concentrations of DO have been generally decreasing since the barrier emplacement.

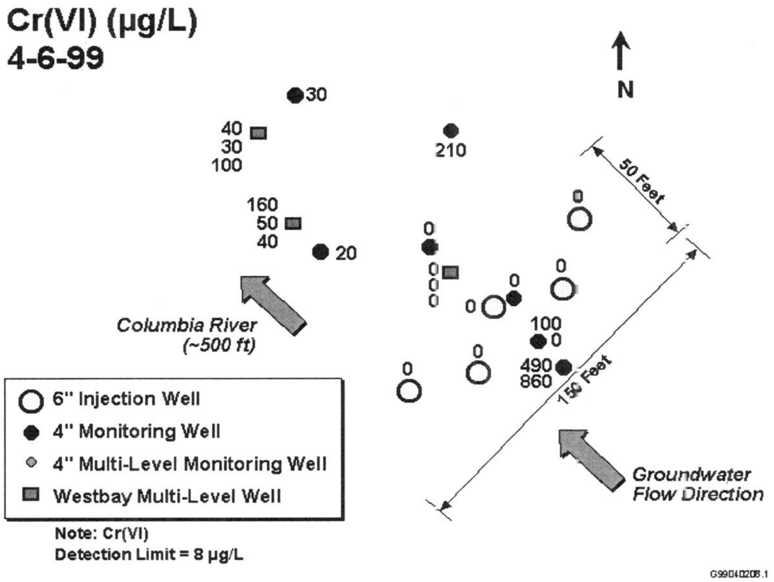

Figure 13. (VI) concentrations (µg/L) at the Hanford 100D Treatability Test Site after dithionite injection (4/6/99).

5.6 DEPLOYMENT AT HANFORD

Based on the success of the treatability test, and successful resolution of the reviewers' concerns, DOE and the regulatory agencies decided to deploy the ISRM technology for the treatment of Cr(VI) in the groundwater at the Hanford 100D Area. In order to accomplish the treatment, the barrier will be expanded from its present 46m length to approximately 700m in length. As required under the National Contingency Plan (40 CFR 300-399), a Proposed Plan describing the proposed remedy was prepared and submitted to the public for review in July of 1999. After the public review period was successfully completed, an Amended Record of Decision was prepared and signed in October 1999 by representatives of the U. S. Department of Energy, the U. S. Environmental Protection Agency, and the Washington State Department of Ecology. Design and planning was completed in FY 1999 and early fiscal year 2000. Construction began in fiscal year 2000, and is planned for completion in fiscal year 2002.

5.7 TCE TREATMENT

Bench-scale tests have shown that dithionite treated soils should also be effective for treatment of dissolved trichloroethylene (TCE) contamination. Based on the bench-scale results, a single well field test was conducted for TCE at Fort Lewis, Washington in November 1998. The aquifer is in a complex glaciofluvial setting, with high hydraulic conductivity (700 ft/day), low aquifer temperatures (12°C), and baseline TCE concentrations of ~140 μg/L.

5.7.1 *TCE Degradation*

At the Ft. Lewis site, the abiotic degradation of TCE and other organic contaminants is being tested using the In Situ Redox Manipulation treatment technology. In this case, the organic contaminants are electron acceptors. The degradation pathway of TCE by dithionite-reduced sediment has been investigated in other studies as well as in Ft. Lewis sediments. Degradation pathways for most organic compounds including TCE are complex, involving multiple and potentially parallel reaction steps. Of four possible abiotic degradation pathways for TCE, the two considered most common are reductive elimination and hydrogenolysis. Reductive elimination has been shown to be the major pathway in other studies using zero-valent and ferrous iron [70, 71, 72]

$$\text{TCE} + 2e^- \Leftrightarrow \text{chloroacetylene} + 2Cl^-; \qquad Eh = 0.60\ \text{v}$$

which describes the dechlorination of TCE to easily degraded (abiotically or biotically) chlorinated acetylene products. Abiotic degradation of these products by hydrogenolysis:

$$\text{chloroacetylene} + H^+ + 2e^- \Leftrightarrow \text{acetylene} + Cl^-; \qquad Eh = 0.50\ \text{v}$$

$$\text{acetylene} + 2H^+ + 2e^- \Leftrightarrow \text{ethylene}; \qquad Eh = 0.39\ \text{v}$$

apparently proceeds rapidly as chlorinated acetylenes are unstable. The degradation of TCE to ethylene by reductive elimination (or hydrogenolysis discussed below) involves the production of 6 moles of electrons, or 22 mg L^{-1} TCE needed to oxidize the equivalent mass of Fe^{II} as water saturated with dissolved oxygen (1.05 mmol L^{-1} Fe^{II}). Therefore, water containing partial oxygen saturation and ~1.0 mg L^{-1} TCE (as likely present in the Ft. Lewis aquifer with 0.3 mg/L in solution and 2x that mass adsorbed) means that TCE has an insignificant impact on Fe^{II} oxidation and remediation barrier lifetime. In the event that the reduced iron barrier is exhausted, previous laboratory studies with the Hanford 100D and 100H sediment have shown that sediment can be re-reduced with only a small (5% to 10%) loss in capacity.

Another pathway for the dechlorination of TCE involves sequential hydrolysis. This is the pathway by which microorganisms typically dechlorinate solvents. Hydrogenolysis reactions include:

$$\text{TCE} + H^+ + 2e^- \Leftrightarrow \text{1,1-DCE (more likely) or cis-DCE} + Cl^-; \qquad Eh = 0.51\ \text{v}$$

$$\text{1,1-DCE or cis-DCE} + H^+ + 2e^- \Leftrightarrow \text{vinyl chloride} + Cl^-; \qquad Eh = 0.41\ \text{v}$$

These reduction potentials are lower than reductive elimination, indicating they are less likely to occur abiotically. Activation energies and the specific electron transfer mechanism, which does involve the Fe^{III} oxide surface, may also influence which reactions actually do occur. Studies of TCE degradation pathways using zero-valent iron and various Fe^{II} minerals [72, 73, 74] indicate that reductive elimination is the major pathway, with minor amounts of DCE isomers and vinyl chloride produced from the hydrogenolysis pathway. One study also indicates that the DCE isomers and vinyl chloride slowly degraded to ethylene.

5.7.2 *Bench-Scale Tests for TCE Destruction*

The feasibility of chemically treating sediments from the Ft. Lewis, Washington, Logistics Center to develop a permeable barrier for dechlorination of trichloroethylene (TCE) was investigated in a series of laboratory experiments. The effects of temperature, partial iron reduction, and flow on these redox reactions were also studied to ascertain how to achieve viable TCE dechlorination rates at the field scale.

Dithionite-reduced Ft. Lewis sediments were shown to degrade TCE in Ft. Lewis groundwater at sufficiently fast rates (1.2 h to 19 h) during static and transport experiments to create a permeable barrier at the field scale. The TCE degradation rate can be calculated for all sediments from the product of the intrinsic degradation rate (0.0034/h mol) and the mass of reduced iron (range of 12 mol/g to 126 mol/g; averaged = 63 mol/g). Products of TCE dechlorination clearly show that 99.5% to 100% is occurring via reductive elimination, producing acetylene, ethylene, and chloroacetylene. The TCE degradation rate decreased up to 3 orders of magnitude in partially reduced sediment. This departure on fraction of reduced iron has significant implications, because uniform full sediment reduction is not possible at the field scale. Although minimally reduced sediment had nearly no TCE reactivity, >40% reduced sediment resulted in TCE reduction rates that were viable at the field scale (<65 h). The second-order dependence of the TCE dechlorination rate on the fraction of reduced iron demonstrates the significant role of the iron oxide surface (as a catalyst or for surface coordination) in addition to Fe^{II} as the electron donor for TCE dechlorination to proceed. Reduced sediment barrier longevity was demonstrated in a column in which TCE was degraded for over 230 pore volumes. The design of a field-scale reduced iron barrier should to be wide enough to allow the TCE to be degraded to below the maximum concentration level (MCL) during the groundwater transport time through the barrier (10 half-lives). Because few sites are homogeneous, barriers are typically designed wider than needed to account for the spatial variability in the iron content and the velocity variability resulting from hydraulic conductivity variability and temporal changes. Full results are available in Szecsody, et al. [75].

5.7.3 *Proof of Principle Field Test*

The objective of the ISRM proof-of-principle test was to determine the field-scale feasibility of the ISRM technology for the treatment of dissolved TCE contamination in the groundwater at the Fort Lewis Logistics Center. While the ISRM technology for TCE destruction was demonstrated in laboratory experiments using sediment from the Fort Lewis site, a field-scale test was required to determine the feasibility of applying the technology at a large scale in the complex hydrogeologic and geochemical conditions of the subsurface. Emplacement of the ISRM treatment zone was accomplished through a series of four separate dithionite injection tests conducted between November 10, 1998 and March 29, 2000. An extensive program of chemical monitoring was also performed before, during, and after each injection to evaluate ISRM's performance. Prior to emplacement of the ISRM treatment zone, the geologic, hydrologic, and geochemical properties of the site were characterized. Sediment core samples collected in connection with the characterization studies were analyzed in bench-scale column tests at PNNL to determine reducible iron content. These site-specific hydrogeologic and geochemical data were used to develop the emplacement design of the pilot-scale (i.e., single injection well) ISRM treatment zone.

Performance data obtained from the proof-of-principle test indicate that field-scale reductive dechlorination of TCE is feasible using the ISRM technology. A treatment zone was created in the subsurface that reduced TCE concentrations as much as 92% on the downgradient side of the reduced zone, from a background concentration of approximately 140 ppb to approximately 11 ppb. The appearance of the principal degradation product, acetylene, also demonstrated that TCE destruction was occurring via reductive dechlorination. Preliminary analysis of sediment samples collected from post-test boreholes showed a high degree of iron reduction, which helped confirm the effectiveness of the treatment zone.

Laboratory analysis of sediment from the Fort Lewis site showed that it contains a relatively large amount of reducible iron (0.25 wt%) and that a significant percentage (>30 to 40%) of iron reduction was required to achieve the necessary TCE degradation rates at the site given the low aquifer temperature (11°C) and high groundwater velocities (time-weighted average value of ~4 ft/d). The technical objectives of this test and reagent requirements to achieve this reduction led to a series of four dithionite injection/withdrawal tests with monitoring periods between them to evaluate the interim TCE degradation performance (Figure 5). After the first injection, which had very low iron reduction efficiency, a number of design modifications were incorporated into the injections. These modifications included diluting of the reagent to minimize density sinking effects and preheating the treatment zone by injecting warm, anoxic water to increase the Fe(III) to Fe(II) reduction reaction rate. The dithionite/sediment contact time during the injection and residence phases was also increased. Including these design modifications as a standard procedure for all subsequent injections substantially improved iron reduction efficiency during the remaining injection tests (dithionite Injections 2a, 2b, 3, and 4). Improved TCE degradation performance within the ISRM treatment zone was also observed after each dithionite injection/withdrawal test as the percentage of iron reduction at the site increased.

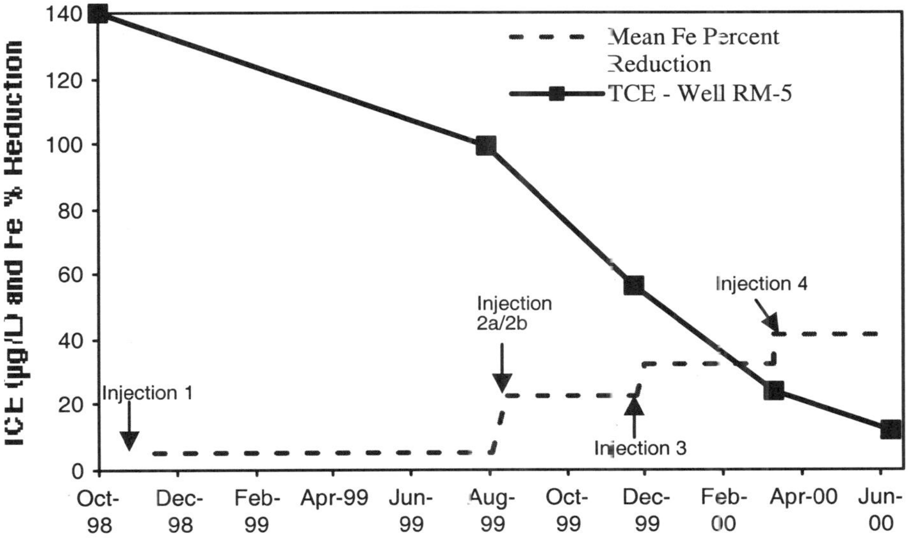

Figure 14. TCE destruction and iron reduction as a result of four sodium dithionite injections at Fort Lewis, Washington.

5.7.4 *TCE Treatment Results*

The primary objective of the ISRM proof-of-principle test was met. A single-well treatment zone was created and it was demonstrated that TCE could be reductively dechlorinated at the field scale. However, to quantify the performance and economic viability of a full-scale barrier deployment, a treatability test-scale demonstration is needed in which multiple injection wells are used to form an adequately sized linear barrier (e.g., 30 to 60 meters long). It is not practical to attempt to obtain this information from a single-well proof-of-principle test. Uncertainties in determining groundwater flow direction at the accuracy required for a small-scale single-well test and associated local-scale hydrogeologic heterogeneities make detailed interpretation of the downgradient performance difficult. Additionally, costs associated with a detailed small-scale proof-of-principle test, which requires an extensive sampling and analysis program, cannot be extrapolated accurately to evaluate the costs of a full-scale deployment (i.e., reduction in sampling, analysis, and interpretation during full-scale deployment results in economy of scale). Given the site-specific nature of the ISRM technology, the cost associated with full-scale deployment is most strongly affected by aquifer and contaminant plume characteristics at the selected barrier location (barrier length required, aquifer thickness and depth, reducible Fe(III), groundwater velocity). Full results are available in Vermeul et al. [76].

5.8 SUMMARY

An innovative technology called In Situ Redox Manipulation (ISRM) has been developed and deployed for the treatment of Cr(VI) in groundwater at Hanford, Washington. The new technology is both effective and economical for Cr(VI) treatment. As a result, ISRM is currently being used to treat Cr(VI) in groundwater at the 100D site at Hanford. ISRM has also been developed and tested for TCE contamination in groundwater at Fort Lewis, Washington. Bench-scale tests showed ISRM to be effective for TCE treatment. However, due to a variety of both basic and site-specific factors, deployment in the field for TCE at Fort Lewis has been more difficult.

6. References

1. National Research Council. (1993) *In Situ Bioremediation: When Does It Work?*, National Academy Press, Washington, D.C.
2. National Research Council. (1994) *Alternatives for Groundwater Cleanup*, National Academy Press, Washington, D.C.
3. National Research Council. (1999) Groundwater & Soil Cleanup: Improving Management of Persistent Contaminants, National Academy Press, Washington, D.C.
4. Ward, C.H., Cherry, J.A., and Scalf, M.R. (1997) *Subsurface Restoration*, Ann Arbor Press, Inc., Chelsea, MI.
5. Carr, C.S. and Hughes. J.B. (1998) Enrichment of high-rate PCE dechlorination and comparative study of lactate, methanol, and hydrogen as electron donors to sustain activity, *Environ. Sci. Technol.* **32**, 1817-1824
6. DiStefano, T.D., Gossett, J.M., and Zinder, S.H. (1991) Reductive dechlorination of high concentrations of tetrachloroethene to ethene by an anaerobic enrichment culture in the absence of methanogenesis, *Appl. Environ. Microbiol.* **57**, 2287-2292
7. DiStefano, T.D., Gossett, J.M., and Zinder, S.H. (1992) Hydrogen as an electron donor for dechlorination of tetrachloroethene by an anaerobic mixed culture, *Appl. Environ. Microbiol.* **58**, 3622-3629
8. deBruin, W.P., Kotterman, M.J.J., Posthumus, M.A., Schraa, G., and Zehnder, A.J.B. (1992) Complete biological reductive transformation of tetrachloroethene to ethane, *Appl. Environ. Microbiol.* **58**, 1996-2000
9. Fennell, D.E., Gossett, J.A., and Zinder, S.H. (1997) Comparison of butyric acid, ethanol, lactic acid, and propionic acid as hydrogen donors for the reductive dechlorination of tetrachloroethene, *Environ. Sci. Technol.* **31**, 918-926
10. Freedman, D.L. and Gossett, J.M. (1989) Biological reductive dechlorination of tetrachloroethylene and trichloroethylene to ethylene under methanogenic conditions, *Appl. Environ. Microbiol.* **55**, 2144-2151
11. Holliger, C. and Schumacher, W. (1994) Reductive dehalogenation as a respiratory process, *Antonie van Leeuwenhoek* **66**, 239-246
12. Lee, M.D., Quinton, G.E., Beeman, R.E., Biehle, A.A., Liddle, R.L., Ellis, D.E., and Buchanan, R.J. (1997) Scale-up issues for in situ anaerobic tetrachloroethene bioremediation, *Journal of Industrial Microbiology and Biotechnology* **18**, 106-115
13. Seagren, E.A., Rittmann, B.E., and Valocchi, A.J. (1993) Quantitative evaluation of flushing and biodegradation for enhancing in situ dissolution of nonaqueous-phase liquids, *Journal of Contaminant Hydrology* **12**, 103-132
14. Seagren, E.A., Rittmann, B.E., and Valocchi, A.J. (1994) Quantitative Evaluation of the Enhancement of NAPL-Pool Dissolution by Flushing and Biodegradation, *Environ. Sci. Technol* **28**, 833-839
15. Carr, C.S. and Hughes. J.B. (1999) Response to 'Comment on "Enrichment of High-Rate PCE Dechlorination and Comparative Study of Lactate, Methanol, and Hydrogen as Electron Donors To Sustain Activity"', *Environ. Sci. Technol* **33**, 2683-684
16. Cope, N. and Hughes. J.B. (2001) Biologically-Enhanced Removal of PCE from NAPL Source Zones, *Environ. Sci. Technol* **35**, 2014-2021.

17. Pope, G.A. and Wade, W.H. (1995) Lessons from Enhanced Oil Recovery Research for Surfactant Enhanced Aquifer Remediation, ACS Symposium, Sabatini, D.A., Harwell, J., and Knox, R.C. (eds.).
18. INTERA. (1997) Demonstration of Surfactant Enhanced Aquifer Remediation of Chlorinated Solvent DNAPL at Operable Unit 2, Hill AFB, Utah, prepared for the Air Force Center of Environmental Excellence. Technology Transfer Division, Brooks AFB, Texas.
19. INTERA. (1997-November) AATDF Surfactant/Foam Process for Aquifer Remediation, prepared for the Advanced Applied Technology Demonstration Facility, Rice University, Houston, TX.
20. Duke Engineering & Services (DE&S). (1998-January) Demonstration of Surfactant-Enhanced Aquifer Remediation of Chlorinated Solvent DNAPL at Operable Unit 2, Hill AFB, Utah, prepared for the Air Force Center for Environmental Excellence Technology Transfer Division, Brooks Air Force Base, Texas, and OO-ALC/EMR, Hill Air Force Base, Ogden Utah.
21. Duke Engineering & Services (DE&S), Rice University, University of Texas at Austin, and Radian. (1998) Surfactant/Foam Process for Aquifer Remediation – Draft Final, prepared for the Advanced Applied Technology Demonstration Facility, Rice University, Houston, Texas.
22. Duke Engineering & Services (DE&S). (2000) Surfactant Enhanced Aquifer Remediation Demonstration at Site 88, Marines Corps Base, Camp Lejeune North Carolina, prepared for Naval Facilities Engineering Service Center, Port Hueneme, California.
23. Shook, G.M., Pope, G.A., and Kostarelos, K.. (1998) Prediction and minimization of vertical migration of dnapls using surfactant enhanced aquifer remediation at neutral buoyancy, *J. Contaminant Hydrology,* **34** 363-382.
24. Kostarelos, K., Pope, G.A., Rouse, B.A., and Shook, G.M. (1998) A new concept: The use of neutrally-buoyant microemulsions for DNAPL remediation *J. Contaminant Hydrology* **34**: 383-397.
25. Bourrel, M. and Schechter, R.S. (1988) *Microemulsions and Related Systems*, Marcel Dekker, Inc., New York.
26. Shiau, B., Rouse, J.D., Sabatini, D.A., and Harwell, J.H. (1995) surfactant selection for optimizing surfactant enhanced subsurface remediation, in *Surfactant Enhanced Subsurface Remediation Emerging Technologies*, Sabatini, D.A., Knox, R.C., Harwell J.H., (eds), ACS symposium series 594, American Chemical Society, Washington, DC, pp. 65-81.
27. Dwarakanath, V., Kostarelos, K., Pope, G.A., Shotts, D., and Wade, W.H. (1999) Anionic surfactant remediation of soil columns contaminated by nonaqueous phase liquids, *J. Contaminant Hydrology* **38**, 465-488.
28. Dwarakanath, V. and Pope, G.A. (1999) Surfactant Enhanced Aquifer Remediation, in *Surfactants: Fundamentals and Applications in the Petroleum Industry*, Schramm, L.L. (ed.), Cambridge University Press, Cambridge, UK.
29. Dwarakanath, V. and Pope, G.A. (2000) Surfactant phase behavior with field degreasing solvent, *Environ. Sci. Technol* **34**, 4842.
30. Pennell, K.D., Abriola, L.M., Weber Jr, W.J. (1993) Surfactant enhanced solubilization of residual dodecane in soil columns 1. experimental investigation, *Environ. Sci. Technol.* **27**, 2332-2340.
31. Mason, A., and Kueper, B.H. (1996) Numerical simulation of surfactant-enhanced solubilization of pooled DNAPL, *Environ. Sci. Technol.* **30**, 3205-3215.
32. Renshaw, C.E, Zynda, G.D., and Fountain. J.C. (1997) Permeability reductions induced by sorption of surfactant, *Water Resour. Res.* **33**, 371-378.

33. Mayer, A.S., Zhong, L., and Pope, G.A. (1999) Measurement of mass-transfer rates for surfactant-enhanced solubilization of nonaqueous phase liquids, *Environ. Sci. Technol.* **33**, 2965-2972.
34. Baran, J.R. Jr., Pope, G.A., Wade, W.H., and Weerasooriya, V. (1994) Phase behavior of water/perchloroethylene/anionic systems, *Langmuir* **10**, 1146.
35. Baran, J.R. Jr., Pope, G.A., Wade, W.H., Weerasooriya, V., and Yapa, A. (1994) Microemulsion formation with chlorinated hydrocarbon liquids, *J. Colloid Interface Sci.* **168**, 67.
36. Baran, J.R. Jr., Pope, G.A., Wade, W.H., Weerasooriya, V., and Yapa, A. (1994) Microemulsion formation with chlorinated hydrocarbons of different polarity liquids, *Environ. Sci. Technol.* **28**, 1361-1366.
37. Baran, J.R. Jr., Pope, G.A., Wade, W.H., and Weerasooriya, V. (1996) Surfactant systems for soil and aquifer remediation of JP4 jet fuel, *J. Dispersion Sci. Technol.* **17**, 131-138.
38. Baran, J.R. Jr., Pope, G.A., Schultz, C., Wade, W.H., Weerasooriya, V., and Yapa, A. (1996) Toxic Spill Remediation of Chlorinated Hydrocarbons Via Microemulsion Formation, in *Surfactants in Solution*, Chattopadhyay A.K., and Mittal, K.L., (eds), Marcel Dekker Inc, Amsterdam, pp. 393-411.9
39. Baran, J.R. Jr., Pope, G.A., Wade, W.H., and Weerasooriya, V. (1996) Water/Chlorocarbon Winsor I-III-II microemulsion phase behavior with alkyl glucamide surfactants, *Environ. Sci. Technol.* **30** 2143-2147
40. Brown, C.L., Delshad, M., Dwarakanath, V., Jackson, R.E., Londergan, J.T., Meinardus, H.W., McKinney, D.C., Oolman, T., Pope, G.A., and Wade, W.H. (1999) Demonstration of Surfactant Flooding of an Alluvial Aquifer Contaminated with Dense Nonaqueous Phase Liquid, in *ACS Symposium Series 725, Innovative Subsurface Remediation, Field Testing of Physical, Chemical, and Characterization Technologies.*
41. Bruns-Nagel, D., Steinbach, K., Gemsa, D., and von Löw, E. (2000) Composting (humification) of nitroaromatic compounds, in Spain, J. C., Hughes, J. B., and Knackmuss, H.-J. (eds.), Biodegradation of nitroaromatic compounds and explosives, Lewis Publishers, Boca Raton, pp. 357-393.
42. Cuffin, S.M., Lafferty, P.M., Taylor, P.N., Spain, J.C., Nishino, S.F., and Williams, K.A. (2001) Bioremediation of dinitrotoluene isomers in the unsaturated/saturated zone, abstract in Poster Session B1, Sixth International In Situ and On-Site Bioremediation Symposium.
43. Hawari, J. (2000) Biodegradation of RDX and HMX: from basic research to field application, in Spain, J. C., Hughes, J. B., and Knackmuss, H.-J. (eds.), Biodegradation of nitroaromatic compounds and explosives, Lewis Publishers, Boca Raton, pp. 277-310.
44. Jerger, D.E., and Woodhull, P. (2000) Applications and costs for biological treatment of explosive-contaminated soils in the United States, in Spain, J. C., Hughes, J. B., and Knackmuss, H.-J. (eds.), Biodegradation of nitroaromatic compounds and explosives, Lewis Publishers, Boca Raton, pp. 395-423.
45. Lenke, H., Achtnich, C., and Knackmuss, H.-J. (2000) Perspectives of bioelimination of polynitroaromatic compounds, Spain, J. C., Hughes, J. B., and Knackmuss, H.-J. (eds.), Biodegradation of nitroaromatic compounds and explosives, Lewis Publishers, Boca Raton, pp. 91-126.
46. Nishino, S.F., Paoli, G., and Spain, J. C. (2000) Aerobic degradation of dinitrotoluenes and pathway for bacterial degradation of 2,6-dinitrotoluene. Appl. Environ. Microbiol. 66, 2139-2147.
47. Nishino, S.F., and Spain, J. C. (2001) Technology status review: bioremediation of dinitrotoluene (DNT). Internet source. http://www.estcp.org/documents/techdocs/DNT_Report.pdf

48. Nishino, S.F., Spain, J. C., and He, Z. (2000) Strategies for aerobic degradation of nitroaromatic compounds by bacteria: process discovery to field application, in Spain, J. C., Hughes, J. B., and Knackmuss, H.-J. (eds.), Biodegradation of nitroaromatic compounds and explosives, Lewis Publishers, Boca Raton, pp. 7-61.
49. Spain, J.C. (1997) Synthetic chemicals with potential for natural attenuation. Bioremed. J. 1, 1-9.
50. Spain, J.C. (2000) Introduction, in J. C. Spain, J. B. Hughes, and H.-J. Knackmuss (ed.), Biodegradation of nitroaromatic compounds and explosives. Lewis Publishers, Boca Raton, pp. 1-5.
51. Spain, J.C., Hughes, J.B., and Knackmuss, H.-J. (eds.). (2000) Biodegradation of nitroaromatic compounds and explosives, Lewis Publishers, Boca Raton.
52. Spain, J.C., Nishino, S.F., Green, M.R., Forbert, J.E., Nogalski, N.A., Unterman, R., Riznychok, W.M., Thompson, S.E., Sleeper, P.M., and Boxwell, M.A. (1999) Field demonstration of FBR for treatment of nitrotoluenes in groundwater, in Alleman, B. C. and Leeson, A. (eds.), Bioremediation of Nitroaromatic and Haloaromatic Compounds, Battelle Press, Columbus, OH, pp. 7-14.
53. Williams, R.E., and Bruce, N.C. (2000) The role of nitrate ester reductase enzymes in the biodegradation of explosives, in Spain, J. C., Hughes, J. B., and Knackmuss, H.-J. (eds.), Biodegradation of nitroaromatic compounds and explosives, Lewis Publishers, Boca Raton, pp. 161-184.
54. Zhang, C., Nishino, S.F., Spain, J.C., and Hughes, J.B. (2000) Slurry-phase biological treatment of 2,4- and 2,6-dinitrotoluene: role of bioaugmentation and effects of high dinitrotoluene concentrations. Environ. Sci. Technol. 34, 2810-2816.
55. Cummings, M.A. and Booth, S.R. (1997) Cost Effectiveness of In Situ Redox Manipulation for Remediation of Chromium-Contaminated Groundwater LA-UR-97-165 Los Alamos National Laboratory, Los Alamos, NM.
56. Rai, D., Eary, L.E., and Zachara, J.M. (1989) Environmental chemistry of chromium. *Sci. Total Environ.* **86**, 15-23.
57. Sass, B.M. and Rai, D. (1987) Solubility of amorphous chromium(III) iron(III) hydroxide solid solutions, *Inorg. Chem.* **26**, 2228-2232.
58. Schroeder, D.C. and Lee, G.F. (1974) Potential transformation of chromium in natural waters, *Water Air Soil Pollut.* **4**, 355-365.
59. Van Der Weijden ,C.H. and Dreith, M. (1982) Chromium(III)-Chromium(VI) interconversions in seawater, *Mar. Chem.* **11**, 565-572.
60. Eary ,L.E. and Rai, D. (1987) Kinetics of chromium (III) oxidation to chromium(VI) by Reaction with Manganese Dioxide, *Environ. Sci. Technol.* **21**, 1187-1193.
61. Szecsody, J.E., Krupka, K.M., Williams, M.D., Cantrell, K.J., Resch, C.T.,and Fruchter, J.S. (1998) Uranium Mobility During In Situ Redox Manipulation of the 100 Areas of the Hanford Site, PNNL-12048.
62. Fendorf, S.E., Zasoski, R.J., and Burau, R.G. (1993) Competing metal ion influences on chromium(III) oxidation by birnessite, *Soil. Sci. Soc. Am. J.* **57**, 1508-1515.
63. Fendorf, S.E., Fendorf, M., Sparks, D.L., and Gonsky, R. (1992) Inhibitory mechanisms of Cr(III) oxidation by δ-MnO_2, *J. Colloid and Interface Sci.* **153**, 37-54.
64. Fendorf, S.E. and Zasoski, R.J. (1992) Chromium(III) oxidation by δ-MnO_2, *Environ. Sci. Technol.* **26**, 79-85.
65. Manceau ,A.A. and Charlet, L. (1992) X-ray absorption spectroscopic study of the sorption of Cr(III) at the oxide-water interface: II. Adsorption, coprecipitation, and surface precipitation on hydrous ferric oxide, *J. Colloid Interface Sci.* **148**, 443-458.
66. Johnson, C.A. and Xyla, A.G. (1991) The oxidation of chromium(III) to chromium(IV) on the surface of manganate (γMnOOH), *Geochim.Cosmochim. Acta* **55**, 2861-2866.

67. Amacher, M.C. and Baker, D.A. (1982) Redox reaction involving chromium, plutonium, and manganese in soils, Final Rep. DE-ASO8-77DPO4515, Inst. for Res. on Land and Water Resour., Penn. State Univ.
68. Fruchter, J.S., Amonette, J.E., Cole, C.R., Gorby, Y.A., Humphrey, M.D., Istok, J.D., Spane, F.A., Szecsody, J.E., Teel, S.S., Vermeul, V.R., Williams, M.D., and Yabusaki, S.B. (1996) In Situ Redox Manipulation Field Injection Test Report - Hanford 100H Area, PNNL-11372, Pacific Northwest National Laboratory, Richland, Washington.
69. Williams, M.D., Vermeul, V.R., Szecsody, J.E., Fruchter, J.S., and Cole, C.R. (1999) 100-D Area In Situ Redox Treatability Test for Chromate-Contaminated Groundwater: FY-1998 Year-End Report, PNNL-12153, Pacific Northwest National Laboratory, Richland, Washington.
70. Sivavec, T., and Horney, D. (1995) Reductive Dechlorination of Chlorinated Ethenes by Iron Metal and Iron Sulfide Minerals, in *Emerging Technologies in Hazardous Waste Management VII*, American Chemical Society, Atlanta, Georgia, p. 42-45.
71. Orth, W., and Gillham, R. (1996) Dechlorination of trichloroethene in aqueous solution using Fe^{o}, *Environ. Sci. Technol.* **30**, 66-71.
72. Sivavec, T., Mackenzie, P., Horney, D., and Baghel, S. (1996) Redox-Active Media for Permeable Reactive Barriers, Research Report, General Electric Research and Development Center, Schenectady, New York, p. 753-759.
73. Roberts, A., Totten, L., Arnold, W., Burris, D., and Campbell, T. (1996) Reductive elimination of chlorinated ethylenes by zero-valent metals, *Environ. Sci. Technol.* **30**, 2654-2659.
74. Thornton, E., Szecsody, J., Cantrell, K., Thompson, C., Evans, J., Fruchter, J., and Mitroshkov, A. (1998) Reductive Dechlorination of TCE by dithionite, in *Physical, Chemical, and Thermal Technologies for Remediation of Chlorinated and Recalcitrant Compounds*, Wickromanayake, G. and Hinchee, R., (ed.), p. 335-340.
75. Szecsody, J.E., Fruchter, J.S., Sklarew, D.S., and Evans, J.C. (2000) In Situ Redox Manipulation of Fort Lewis, Washington: Iron Reduction and TCE Dechlorination Mechanisms, PNNL-13178, Pacific Northwest National Laboratory, Richland, Washington.
76. Vermeul, V.R., Szecsody, J.E., Williams, M.D., Bjornstad, B.N., Evans, J.C. and Liikala, T.L. (2000) In Situ Redox Manipulation Proof-of-Principle Test at the Fort Lewis Logistics Center, Final Report, PNNL-13357, Pacific Northwest National Laboratory, Richland, Washington.

PCB – APPROACHES TO REMOVAL FROM THE ENVIRONMENT

Current Status of Bioremediation in the Czech Republic

K. DEMNEROVÁ[1], M. MACKOVÁ[1], J. PAZLAROVÁ[1], M. VOSAHLÍKOVÁ[1], H. NOVÁKOVÁ[1], E. JINDROVÁ[1], E. RYSLAVÁ[1], T. MACEK[2], N. VRCHOTOVA[3], V. BRENNER[4], L. PAVLŮ[4], S. TOTEVOVÁ[4], T. KRISTOFFER[4], D. D. FOCHT[5], F. FAVA[6], D. DI GIOIA[6], L. MARCHETTI[6], J. S. FLETCHER[7], M. B. LEIGH[7], P. KUCEROVÁ[1], H. STIBOROVÁ[1], V. MATEJU[3], M. SOBOTKA[4], F. KASTANEK[9], P. KASTANEK[1], L. KASÁK[10]

[1]*Department of Biochemistry and Microbiology, Institute of Chemical Technology, Prague, Czech Republic*

[2]*Institute of Biochemistry and Organic Chemistry of the Academy of Sciences of the Czech Republic, Prague, Czech Republic*

[3]*Institute of Landscape Ecology of the Czech Academy of Sciences, Ceske Budejovice, Czech Republic*

[4]*Institute of Microbiology Czech Academy of Sciences, Prague, Czech Republic*

[5]*University of California, Dept. of Plant Pathology, Riverside, USA*

[6]*DICASM, Faculty of Engineering, University of Bologna. V.le Risorgimento, 2. I-40136 Bologna, Italy*

[7]*Department of Botany and Microbiology, University of Oklahoma, Norman, OK, USA*

[8]*ENVISAN-GEM a.s., Biotechnological Division, Budova VUPP, Radiova 7, CZ-102 31 Prague 10, Czech Republic*

[9]*Institute of Chemical Process Fundamentals, Czech Academy of Sciences, 165 02 Prague 6- Suchdol, Czech Republic*

[10]*IDOS Prague LTD, Center of Innovative Technologies, Czech Republic*

1. Abstract

Polychlorinated biphenyls (PCBs) belong to the group of the most recalcitrant compounds with proved negative effect on the human health. They may be removed from the environment by chemical, physical and biological methods, while the latter ones through microorganisms or plants represent more economical way of removal. This study describes use of bacteria isolated in the Czech Republic from PCB contaminated sites. Several physiological characteristics of biodegradation and different ways of process improvement are documented. The utilization of microbial consortia of chlorobiphenyl and chlorobenzoate degrading bacteria is proposed. The beneficial effect of natural compounds of plant origin for the induction of PCB upper degradation pathway is also discussed.

Depending on the chloride substitution, bacteria use different routes to complete their mineralisation. In this study, we have also investigated a possibility for the complementation of both *ortho*- and *meta*-cleavage pathway for chlorocatechols - intermediates of lower PCB metabolic pathway in one strain and its potential impact for degradation of chlorobenzoates, which are known to accumulate during degradation of polychlorinated biphenyls (PCBs). Two lab-scale studies addressed to enhance the aerobic bioremediation of an aged PCB-contaminated soil (containing about 350 mg/kg of a variety of PCBs) through its amendment with defined PCB-bioavailability enhancing agents are reviewed.

D. Reible and K. Demnerova (eds.), Innovative Approaches to the On-Site Assessment and Remediation of Contaminated Sites, 217–263.

The aerobic bioremediation of the aged-contaminated soil employed was found to be significantly and differently influenced by the presence of Triton X-100 (TX-100), Quillaya Saponin (QS), Hydroxypropyl-β-Cyclodextrin (HP-β-CD) and γ-Cyclodextrin (γ-CD) in the soil slurry-phase and fixed-phase aerobic reactors used in the experiment. Among the tested PCB-solubilizing agents, the biogenic products QS, γ-CD and in particular HP-β-CD were found to be very promising bioremediation stimulating agents.

2. Introduction

Removal of hazardous compounds and hazardous wastes from the environment is still the crucial problem of many countries around the world. At least three different approaches could be applied based on chemical, physical and biological methods. The latter ones represent through living organisms, especially microorganisms, white-rot fungi, plants, more economical way of the removal of environmental contaminants. Many of them have been found to possess the ability to mineralize numerous xenobiotics. Oil-derived substances, namely n-alkanes belong to the easiest degradable compounds, whereas polychlorinated biphenyls (PCB) are considered to be the most persistent pollutants.

Polychlorinated biphenyls are a family of compounds with a wide range of industrial applications in the heat transfer, dielectric and hydraulic fluids, flame-retardants, plasticizers, and solvents. In the United States and in the United Kingdom complex PCB mixtures were manufactured under the trade name Aroclor. In the Czech Republic a commercial name for the similar mixture was Delor. All these mixtures consist from a number of congeners, which differ, in the number and in the distribution of chlorines attached to the biphenyl ring. According to the IUPAC nomenclature more than 200 possible congeners were described while about 150 congeners have been reported to be present in the environment. In the Czech Republic two different industrial mixtures were produced, Delor 103 and Delor 106, respectively. After the environmental health risk of PCBs had been discovered, their production was stopped in the USA in late seventies and in the former Czechoslovakia in 1984. Meanwhile PCBs have entered into soil and sediment environments as a result of the wrong disposal of industrial PCB wastes and also by the leakage of PCBs from electrical transformers. Their thermal and chemical stability, resistance to chemical corrosion, and general inertness have contributed to their persistence in the environment [1]. Chemically stable lipophilic PCBs are easily transported through the animal food chain. The concentration sometimes reaches thousands of ng of PCB/l of water, which is often contaminated also by petroleum products or by chlorinated hydrocarbons. Recently, a study of PCB presence [2] in butter, used as a sampling matrix, reflecting regional and global scale of PCBs in different countries, proved high concentrations of these compounds, even 14 ng/g of lipid, in butter distributed in the Czech Republic. This study clearly showed that the highest concentrations of PCBs were detected in butter samples from areas with previous frequent PCB production and utilization. In the Czech Republic and in the other countries (Italy, USA) PCBs contaminants presented in the soil, groundwater, and sediment are still common.

Among non-metabolic method of PCBs decomposition hydrolysis, photolysis and thermal desorption could be included. In the following part the experience obtained during the work on the development of an efficient PCBs decontamination technological unit is described.

2.1. PRINCIPLES OF SYNERGIC SYSTEM OF DECONTAMINATION OF WASTES CONTAINING PCBS

The whole decontamination project has been focused to the development of a complex of mutually corresponding technological units - thermal desorption unit for separation of PCBs from contaminated soils and sediments and other contaminated solid wastes, - extraction of PCB from soils and solid particles by ecological friendly aqueous solution of selected detergents. Output streams from mentioned technologies, condensates with desorbed PCB from thermal desorption and extraction wastes waters from extraction unit are consequently treated by chemical methods: - destruction of PCBs with activated hydroxyl radicals enhanced by UV radiation (so called Photo-Fenton reaction in the case of aqueous emulsions of PCBs)-dehalogenation of PCBs by the reaction of elemental sodium with atoms of Cl from the molecules of PCBs in the case of waterless condensates containing PCBs and /or waste contaminated oils.

The aim of technological function of a whole system is to obtain a clean decontaminated soil which could be re-cycled to the environment and to obtain decontaminated oils which could be used as a fuel (for thermal desorption unit) and/or recycled to the following technological use (transformer dielectric liquids).

Mentioned technologies are operating in IDOS Facility and are located at company IDOS LTD in the town Pribram (approx. 60 km from capital Prague) and currently treat contaminated soils and liquids on pilot scale. This technology operates under the trade names Termidos, Chemidos and Bifidos.

2.1.1. Termidos

It is a batch low temperature thermal desorption process for the desorption of organics and PCB from soil Chemidos – is a reductive dechlorination operation for destruction of PCB in liquids Bifidos – is a washing process for sandy soils that includes a water treatment and recycle operation. The processes operate in a closed loop configuration with effluents circulating among the unit processes.

The Termidos process uses a batch chamber reactor where soils are placed in trays and heated indirectly to 480°C - 500°C for approx. 4 days. PCB is swept from the reactor using a nitrogen purge and condensed. During the first several hours of heating, the moisture in the soil is converted to a stream and strips out about 5 percent of the PCB in the soil. This moisture is condensed and routed to activated carbon filtration and the Bifidos water treatment process. The contaminated activated carbon filter media can be recycled for treatment in the Termidos unit. After the first several hours of heating, there is no more moisture in the soils and a relatively pure PCB and organic phase evaporates from the soil. The organic phase, which represents about 95 % of the total PCBs removed from the soil, is condensed into an emulsion that is suspended in a mineral oil for treatment in the Chemidos process. The separation of the early aqueous PCB phase from the later organic PCB phase is necessary because water would react violently with the elemental sodium used in the Chemidos process. The treated soils from Termidos process contain a residual concentration bellow of about 1 mg/kg PCB, independently on the original concentration of PCBs in treated contaminated soil. Generally, current average concentrations of PCB in contaminated soil, which are available in Czech contaminated localities, are about 300 mg/kg PCB. However, we have demonstrated high efficiency of decontamination (over 99,9%) even in the soil with original contamination about 30,000 mg/kg. At present, the regulatory level for unrestricted disposal of treated soil is 0.05 mg/kg PCB. It is unlikely that lower residual PCB levels can be achieved in the Termidos unit because of the fixed bed configuration. It may be possible to lower the residual levels, however, by incorporating reductive dechlorination reaction within the Termidos process. The existing Termidos unit occupies an area of about 15 m^2 and has a batch capacity about 28 metric tonnes of the soil. The heating and cooling cycle takes about 5 days. On this basis, the annual soil treatment capacity for one unit is about 1,600 tonnes per year operating 7 days per week, 40 weeks per year. The downtime would be used for system maintenance and equipment repairs. 3 unit are in operation at present. Above-mentioned Termidos technology is very suitable for regeneration and reactivation of contaminated activated carbon because of the simple possibility of accurate control of a temperature history and composition of the atmosphere in the chamber. Treatment capacity for activated carbon regeneration is about half of the soil treatment rate because of the longer cycle times (due to the longer time of the cooling of the hot activated carbon). The throughput capacity of the Termidos process is low compared to other thermal treatment technologies because of the poor heat transfer conditions inherent in the fixed bed design. On the other hand, the process has some advantages in that additional treatment capacity can be added on a modularized basis. The inert nitrogen atmosphere in the chamber substantially suppresses the generation of highly toxic intermediates like dibenzodioxins and dibenzofurans in the course of thermal operation. Preferentially, existing Termidos units could be used for onsite carbon regeneration, decontamination of metallic packing of electrical condensers and its paper-aluminum rolls, or shipped offsite for treating small quantities of PCB contaminated soil at the source of the spill.

2.1.2. Chemidos

The Chemidos process uses reductive dechlorination to destroy PCBs by exchanging the chlorine on the biphenyl molecule with hydrogen, forming biphenyl and NaCl. Chemidos uses activated elemental sodium as the electron donor required for the reaction. As an activation matter is used sodium dihydrido-bis (2-methoxyethoxy)aluminate. Chemical reduction of PCBs using elemental sodium requires anoxic and anhydrous conditions and is primarily applicable to PCB contaminated oils, such as transformer oils. The Chemidos process currently is used by IDOS to treat both PCB contaminated oils shipped to the facility and dielectric fluids from the electric condensers and PCBs that are generated by the Termidos process as an organic PCB emulsion. It is a batch process consisting of two reactor vessels. The PCB contaminated oil is placed in the first reactor vessel, which is then purged of air. Particulate sodium that meets a maximum size specification of 10 microns is placed in the reactor at the ratio of five moles of sodium for every mole of chlorine on the PCBs, to complete the dechlorination reaction. The rate of reaction is relatively low. The current reaction sequence in the existing Chemidos equipment (for inlet concentration of PCBs in the oil of about 1000 mg/l) is: Heat to 80°C for 12 hours, stir for approximately 4 hours at temperature, heat to 120°C for 6 hrs, transfer the contents to the second vessel where ethanol is added to quench the reaction. In the final step the ethanol reacts with the unreacted sodium, forming sodium ethanolate. Once the reaction is quenched, the treated oil is disposed of or used to suspend the organic PCB- emulsion from Termidos. The process generally results in the effluent PCB concentration of approximately 10 mg/l. At this PCB level, or less, the oil might be blended for energy recovery in the burners of the Termidos unit. The existing Chemidos equipment has a capacity of 600 litres. The cycle time per batch is about 22 hours giving the total treatment capacity of about 200,000 litres per year of contaminated oil. The maximum inlet concentration of PCBs in treated oils in the existing equipment was 4,000 mg/l. Clearly, the Chemidos plant will have to be significantly expanded or replaced. The existing system reportedly requires some modification to reduce heating times and metallic sodium addition rates that will increase throughput rates and reduce cost.

2.1.3. Bifidos

Bifidos is a soil washing process where PCB contamination washed from sandy or gravel soils is oxidized and destroyed so that the product water can be re-used in the washing process. As it is currently configured, the Bifidos process requires a large working area for the washing part of the process. Currently the soils are spread out to the depth of 1 meter and washed over a period of 4 weeks. At the beginning of washing process, the soil is sprayed with the aqueous solution of $(NH_4)_2SO_4$ and salts of phosphorus with the aim to enhance the liberation of sorbed organic from pores of particles of soil to the water phase. In the second period of washing, the soil is sprayed with aqueous solution of environmental-friendly detergents. The liberated PCB is destroyed in a UV/OX process with activated carbon polishing and the treated water is recycled for further washing. In comparison with thermal desorption process, the efficiency of separation of PCBs from sandy- clay soils is much lower, of about 87%. The current batch capacity of Bifidos unit is about 300 tonnes of sandy-clay soil per 2 month. The UV/OX and filtration system has a net throughput capacity of between 2 and 10 litres/min. The efficiency of destruction of PCB is relatively high, above 99,9% (from the inlet PCB) content in the treated water of approx. 10,000 ng/l. The UV/OX system is a viable method of destroying contamination in wastewater. Modularised units can be installed as needed to meet the effluent treatment requirements. The Bifidos process and processing area should probably be retained for washing screened oversize material and debris separated from the incoming feed to the IDOS Facility.

2.1.4. Product Disposal or Recycle

The products generated by the treatment operations at the facility include soils and possibly oil if an indirect-fired thermal treatment process is used. The soil would be blended with municipal solid waste residues (MSW) for use in non-agricultural application, specifically land reclamation. The blended material truly has a value as a raw material product and is accepted the disposal method as being legitimate. The oil is treated to a level where it can be used onsite as a fuel supplement in the thermal treatment unit. Soil disposal in a landfill facility currently falls under the following PAH, PCB, and TPH landfill standards. The Chemidos process, or an equivalent process, is comfortably able to meet the 10 mg/l limit in product oil.

2.2. BIOLOGICAL TRANSFORMATION OF PCB

Chemical and physical processes of PCB removal, described above, are efficient but price of the treatment is high. For this reason in the suitable environmental conditions the biotransformation of PCBs are preferred. Living organisms could decompose PCB under both aerobic and anaerobic conditions. The dechlorination of highly chlorinated congeners to less chlorinated compounds by anaerobic bacteria has been documented [4], [5], [6], [7]. The less chlorinated PCBs could than be degraded by aerobic bacteria [8].

Furukawa et al [9] and Bedard et al [10] were among the first who reported aerobic microbial biodegradation of these compounds. Later on many aspects of bacterial PCB degradation were described. The catabolic pathway for the total degradation of PCBs is encoded by two different sets of genes that are not usually found in the same organisms. First as called "upper pathway" is responsible for chlorobiphenyl transformation to chlorobenzoate and chlorinated 5-C aliphatic acid. The "lower pathway" is responsible for the transformation of chlorobenzoates through chlorocatechols to complete mineralization. It was shown that principal route of the chlorobiphenyl pathway in the most soil bacteria appears to involve 2,3-dioxygenase [11, 29] while in several cases also 3,4-dioxygenase system was detected. 2,3-dioxygenase pathway was further very well studied. The initial step in the aerobic microbial biodegradation of aromatic ring is usually the introduction of two hydroxyl groups into the aromatic nucleus, forming *cis*-dihydrodiols or *cis*-diol carboxylic acids. For this reaction, the enzyme system coded by the biphenyl operon (*bph*ABCD) [12,13, 30], usually located on chromosome, was found to be responsible. The first enzyme of this complex, biphenyl-2,3-dioxygenase, is responsible for the substrate specificity [12], [13]. The regiospecifity of dioxygenation by the initial enzyme is of particular importance, because the site (or sites) of dioxygenation of a PCB determines the sites of attack of the subsequent enzymes of the catabolic pathway [14]. So far, only several strains were found to be capable to use mono chlorinated biphenyl as a sole carbon and energy source [15], [11], [13]. The more chlorines are substituted at biphenyl rings, the lower is the degradation. The majority of PCB degrading strains demand another organic compound for obtaining carbon and energy, usually biphenyl.

Chlorobenzoates (CBs) (the lower pathway) are usually transformed to chlorocatechols (CCs), which may be further metabolized by either an *ortho*- or *meta*-cleavage pathway. Depending on the chloride substitution, bacteria use different routes to complete their mineralization. In our studies, the possibility for the complementation of both *ortho*- and *meta*-cleavage pathway for chlorocatechols in one strain and its potential impact for degradation of chlorobenzoates was investigated.

In the beginning of ninetieths the effort to construct recombinant PCB degraders by pooling of available pathways from different bacteria has been accelerated, however these constructions were oriented exclusively on the complementation of biphenyl and chlorobenzoate pathways [41, 42, 45]. To facilitate such efforts, the need for better characterization of existing pathways and subsequent molecular mechanisms, have been recognized [51, 30, 39].

PCB biodegradation may be accomplished either by the consortia of bacteria or by the mixture of bacteria and plants or by construction of recombinant PCB degraders by pooling of available pathways from different bacteria. Nowadays a practical use of mentioned types of consortia seems to be a more efficient way for PCB transformation.

Consortia of indigenous microorganisms, isolated from the contaminated sites in the Czech Republic, and capable of the effective PCB biodegradation were described in 1994 by Bokvajová et al. [16]. Kaštánek et al. [17] published the data based on the application of combined microbial treatment and further sorption of PCB polluted groundwater. In our laboratory several individual strains were isolated from five types of co-cultures. Their further identification involved the NEFERM Test identification kit (Lachema, Czech Republic). The isolates were mostly classified as members of genus *Pseudomonas.* Tested strains were capable to use biphenyl as the sole source of carbon and energy. Among these strains, isolate Ps. sp. P2 was found to be the most efficient PCB degrader [18].

3. Methods

3.1. CULTIVATION OF BACTERIAL STRAINS

50 ml of minimal medium [10] containing biphenyl (50 mg) in 250 ml Erlenmeyer flasks were inoculated with bacterial cells of *Pseudomonas* sp. P2 and than incubated at 28°C for 12 days.

Strain *Burkholderia cepacia* P166 was isolated by enrichment on biphenyl (Bp) from an industrial sewage effluent in Panama City, Republic of Panama [34]. *Pseudomonas fluorescens* S12 was isolated by enrichment on biphenyl from PCB contaminated soil in Žamberk, Czech Republic. *Pseudomonas fluorescens* B3 was isolated by enrichment on 3-CB from PCB contaminated sediment in Milevsko, Czech Republic [47]. Recombinant strains *Burkholderia cepacia* P166/C and *Pseudomonas fluorescens* S12/C were constructed in this study.

Burkholderia cepacia P166 and *Pseudomonas fluorescens* S12 were cultivated in liquid mineral (MM) medium [32] with biphenyl (500 mg/l) as the sole carbon source at 28°C, while recombinant strains *Burkholderia cepacia* P166/C and *Pseudomonas fluorescens* S12/C were grown with biphenyl (500 mg/l) and kanamycin (100 μg/ml). In the case of chloride release measurements chlorobenzoates (3-CB, 4-CB) were added into the liquid MM medium.

3.2. TAXONOMICAL IDENTIFICATION OF BACTERIA

Bacteria from the soil contaminated with PCBs were isolated by the extraction with phosphate buffer pH 7.0 after 2 hours of shaking. Extracts were plated on Plate Count Agar (Oxoid) (for the estimation of total counts of microbial cells) and on minimal medium [2] with biphenyl. Bacteria growing on solid minimal medium with biphenyl were characterized for their biochemical activities by Gram staining and also by Nefermtest (Lachema).

3.3. DNA MANIPULATIONS

DNA ligation, transformation, digestion with restriction endonucleases, plasmid isolation, and separation of DNA by gel electrophoresis was done as described previously [52].

3.4. BIPARENTAL MATING

20 μl of minimal medium containing acceptor cells (*B. cepacia* P166, *Pseudomonas fluorescens* S12) cultivated with Bp and grown to the exponential phase were dropped on LB agar plates. After drying off the cell suspension, the spot was covered with 10 μl of donor cells (*E. coli* S17-1) and cultivated at 30°C. Conjugated cells were resuspended in 200 μl of MM, diluted in 900 μl, and a 100-μl aliquot was used as inoculum on plates supplied with selective medium (MM with Km and Bp).

3.5. ENZYME ASSAYS

Cells were grown aerobically in 100 ml MM medium containing either 500 mg/l of biphenyl or 500 mg/l of biphenyl and 500 mg/l of chlorobenzoates, respectively, at 28°C on a reciprocal shaker, harvested by centrifugation, and washed twice with phosphate buffer (0.1 M, pH 7.5). Cells were resuspended in the same buffer, and then disrupted by sonication (6 x 15 s on ice). Cell debris was removed by centrifugation (20,000 rpm, 20 min, 4°C). Enzyme activity was assayed spectrophotometrically (at 260 nm and 375 nm) in the reaction mixture composed of 20 μl of crude extract, 970 μl of phosphate buffer, and 10 μl of substrate (0.5 mM stock solution of catechol, 3-chlorocatechol, and 4-chlorocatechol). Protein concentration was determined by the method of Bradford with BSA as a standard.

3.6. CHLORIDE RELEASE MEASUREMENTS

Chloride release was measured electrochemically by ion-selective electrode (ISE) (model 96-17B, Orion, Boston, MA, USA). 5 ml sample of cell culture was diluted with 50ml distilled water with the addition of 0.5 ml low-level ISA (ionic strength adjustor, 1M $NaNO_3$). The concentration of released chlorides was determined from a calibration curve.

3.7. ANALYSIS OF BACTERIAL DEGRADING ABILITIES BY GC

After cultivation bacterial cells were heat-killed (90°C) and after subsequent sonication, the contents of the flasks were extracted by 10 ml of hexane at 20°C on a rotary shaker for 2 h. Following phase separation, the upper, hexane layer was subjected to GC analysis. Samples were analysed using a Hewlett-Packard 5890 gas chromatograph with an electron capture detector and a fused silica capillary column (30 m, 0.20 mm inner diameter) coated with 0.25 µm immobilised phase SE-54 with nitrogen as the carrier gas (flow rate 1 ml/min). Using GC analysis with EC detector, 22 of the Delor 103 congeners were assigned to peaks with areas larger than 0.5 % of the total area of all 59 individual chromatographic peaks.

For the calculation of the residual amount of PCBs the above mentioned 22 chromatographic peaks were used. These 22 congeners represent 80-90% of the total sample amount. Controls containing heat-killed cells were included to establish that observed changes in the content of congeners were dependent exclusively on the activity of living cells. Using standard conditions for the preparation of samples for GC analysis, the accuracy of the results obtained was within 15%. Results were calculated from the residual amounts of each congener peak of the sample, comparing to the respective peaks of the controls.

TABLE 1: List of followed peaks in Delor 103

Peak No.	IUPAC	Substitution		
1	5, 8	2,3	2,4´	
2	15, 18	4,4´	2,5,5´	
3	17	2,4,2´		
4	16, 32	2,3,2´	2,6,4´	
5	26	2,5,3´		
6	31	2,5,4´		
7	28	2,4,4´		
8	20, 33, 53	2,3,3´	3,4,2´	2,5,6,2´
9	45	2,3,6,2´		
10	52, 69	2,5,2´,5´	2,4,6,3´	
11	49	2,4,2´,5´		
12	47, 75	2,4,2´,4´	2,4,6,4´	
13	48	2,4,5,2´		
14	44	2,3,2´,5´		
15	37, 42, 59	3,4,4´	2,3,2´,4´	2,3,6,3´
16	41, 64	2,3,4,2´	2,3,6,4´	
17	96	2,3,6,2´,6´		
18	74	2,4,5,4´		
19	70	2,5,3´,4´		
20	66, 88, 95	2,4,3´,4´	2,3,4,6,2´	2,3,6,2´,5´
21	101	2,4,5,2´,5´		
22	77, 11	3,4,3´,4´	2,3,6,3´,4´	

3.8. CULTIVATION OF BACTERIAL MIXED CULTURES

A mixture of two strains *Ps. aeruginosa* JB2 [31] (chlorobenzoate degrader) and *Ps.*sp. P2 (PCB degrader) was cultivated in liquid mineral medium supplemented with biphenyl (5000mg/l) and Delor 103 (0.04mg/l). The incubation medium was inoculated with 4 ml of strain *Ps.*sp. P2 cells in the exponential growth and after 4 days the inoculum of *Ps. aeruginosa* JB2 was added. Mixed culture was incubated at 28°C with shaking for 14 days.

3.9. CULTIVATION OF PLANT CELLS AND EXTRACTION OF PHENOLICS

As a model plant species the hairy root culture SNC90 was chosen for the purposes of the project due to its good growing ability and also efficient PCB transformation ability [19, 20]. Plant cells were cultivated for 14 days in standard MS (Murashige and Skoog) medium without hormones. The incubation was interrupted when medium became brown (due to exudation of phenolics), and biomass and medium were divided and separately extracted with methanol. Part of both extracts was hydrolysed with HCl to cleave the glycoconjugates and obtain free phenolic compounds. The concentration of phenolic compounds was measured according to Amorim et al. [25].

4. Results and Discussion

4.1. DIFFERENT ASPECTS OF BACTERIAL PCB DEGRADATION AT LABORATORY CONDITIONS

Strain *Pseudomonas* sp. P2 isolated from the soil contaminated with PCB was at laboratory conditions able to grow on simple mineral media with biphenyl as the sole source of carbon and energy and degrade mixture of congeners present in Delor 103 rather efficiently 60-70% [18]. Various conditions influencing the growth and degradation efficiency of bacterial cells of *Ps*. sp. P2 were tested for better understanding of the strain properties and their comparison with other PCB degraders. Figure 1 documents degradation abilities of the bacterial strain to degrade PCBs at different pH levels. The lowest content of PCB added (the best degradation) was analysed after cultivation of the cells at pH 7,5.

Final metabolites of upper pathway of PCB metabolism were qualitatively and quantitatively analysed and identified [28] as chlorobenzoic acids (CBA) with one to three chlorine atoms. According to several reports certain CBAs [21], [11], [22] [23] have been considered as the repressors of the enzymes participating on PCB degradation and their further mineralisation in the environment was shown to be essential for efficient chlorobiphenyl degradation. We have chosen the most abundant metabolite obtained from Delor 103 – 2,5-dichlorobenzoic acid and we tested its effect both on the growth and on the PCB degradation of *Ps*. sp. P2. Figure 1 shows the effect of various amounts of 2,5-dichlorobenzoic acid on the growth of *Pseudomonas* sp P2.

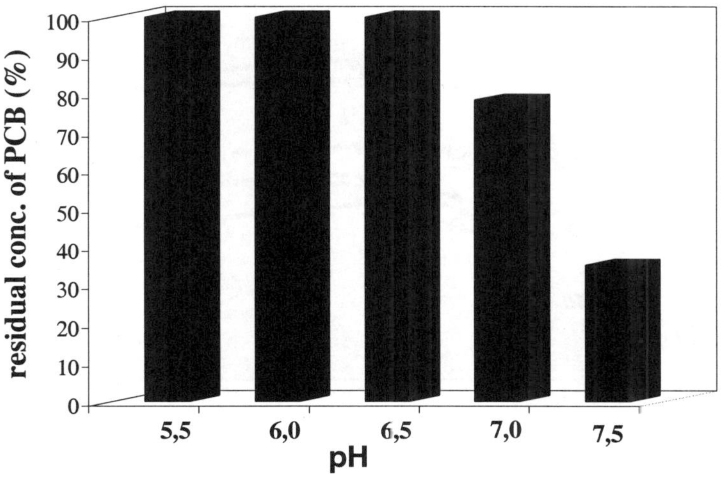

Figure 1. Residual content of PCB after degradation of Delor 103 by *Ps.* sp. P2 at different pH conditions

The first immediate effect on the prolongation of the time of lag was followed. The addition of 50 ppm and higher concentrations (to 2000 ppm) of 2,5-dichlorobenzoic acid decreased maximal growth rate and thus the maximal reached culture density. Even the highest tested concentration did not cause the entire growth inhibition. The effect of the same concentrations of 2,5 dichlorobenzoic acid on the degradation of Delor 103 by the same strain is documented in Table 2. In this case the negative effect of 2,5-dichlorobenzoic acid on the degradative efficiency was not proved. Even if we consider that the results from GC analysis may vary up to 15%, the effect of CBA presence is rather lower. At natural conditions there are always numerous consortia of microorganisms that gradually use PCB and its metabolites, so this situation is rather improbable. It can be concluded that enzymes participating on PCB degradation (upper pathway) are resistant to the repression by their products.

TABLE 2. The effect of 2,5-diCBA addition on Delor 103 biodegradation*

CBA (ppm)	0	50	500	1000	2000
Residual PCBs after biodegradation (%)	38,6	41,4	31,7	36,4	42,2

* biphenyl added in concentration of 5 g/l in all flasks

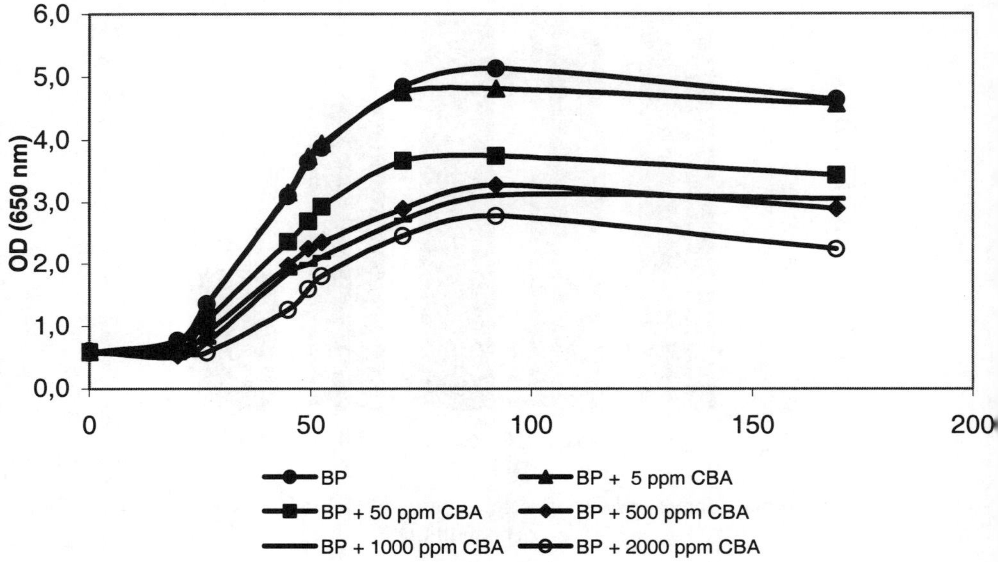

Figure 2. Growth of *Pseudomonas* sp. P2 on biphenyl in the presence of various concentrations of 2,5-di-chlorobenzoic acid.

To follow the energy balance of bacterial cells the additions of various carbon sources to incubation medium were tested. Table 3 summarizes the efficiency of biodegradation in presence of tested carbon sources.

Metabolites containing 3 atoms of carbon (glycerol and pyruvate) negatively influenced the process of PCB biodegradation (Figure 4). The sucrose surprisingly had no negative effect on the biodegradation, while a main objective was probably the absence of enzymatic system for the sucrose decay. Two natural polymers, cellulose and agar, added to the medium, are completely metabolically inert (Figure 3) and their beneficial effect could be ascribed to the fact, that they served as a carriers for the cells and support physiological parameters.

TABLE 3. Residual PCB content after degradation of Delor 103 by *Pseudomonas* sp. P2 growing on biphenyl in the presence of various other carbon sources

Carbon source	Residual PCB content After degradation (%)
Biphenyl	20
Biphenyl+glycerol	70
Biphenyl+pyruvate	60
Biphenyl+saccharose	15
Biphenyl+cellulose	30
Biphenyl+agar	10

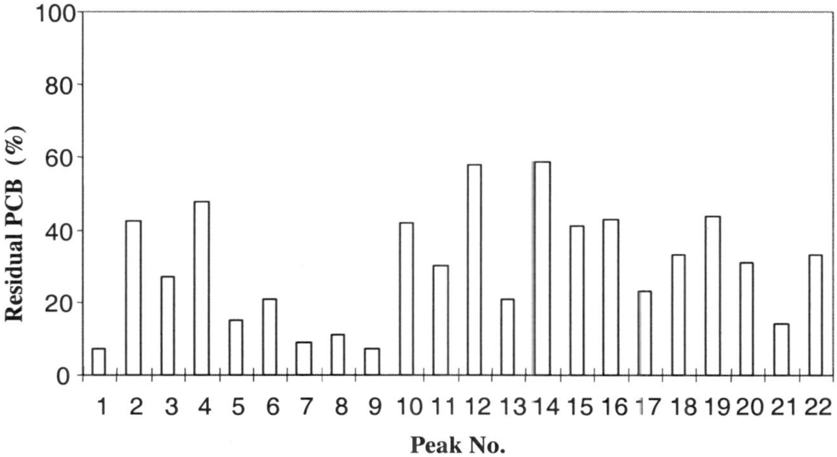

Figure 3. Degradation of Delor 103 by *Pseudomonas* sp. P2 in mineral medium with biphenyl and 1% cellulose

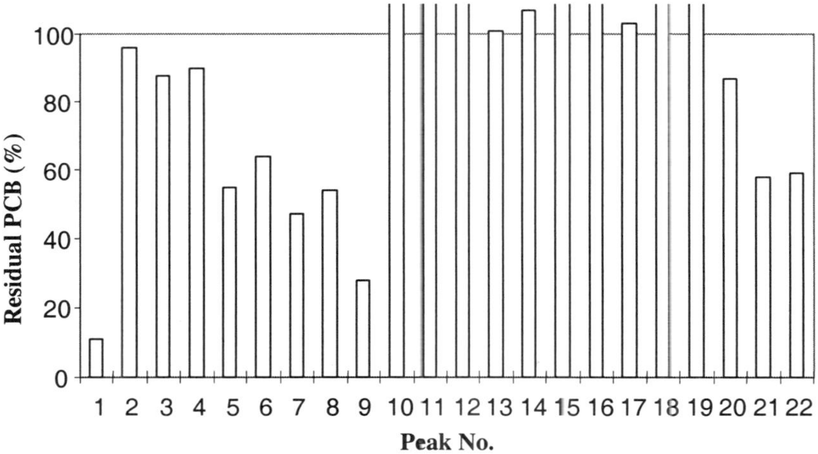

Figure 4. Degradation of Delor 103 by *Pseudomonas* sp. P2 in mineral medium with biphenyl and 1% pyruvate

Previously it was shown that some phenolics, flavonoids or terpenes of plant origin are able to induce PCB degradative pathway instead of biphenyl [26], [27]. Original hypothesis was based on the fact that these compounds produced by plants and released into the soil by plant roots could serve as natural carbon sources and thereby stimulate PCB degradation as biphenyl. Other benefit of the presence of these compounds is their less or non-toxic origin. The efficiency to degrade PCBs by strain Ps. sp. P2 was tested when various phenolic compounds and flavonoids were used as a sole carbon and energy source (Table 4)

The data summarised in Table 4 proved that several of naturally occurring plant compounds supported the growth of bacterial strain *Ps.* sp. P2 and the organism retained its ability to efficiently metabolise PCBs. This phenomenon is quite important because these compounds can serve as natural bioaugmenting agents and as natural substrates for the induction of *bph* genes instead of biphenyl.

TABLE 4. Degradation of PCB present in Delor 103 by *Ps.* sp. P2 in presence of various compounds of plant origin

Carbon source (inducer)	Residual PCB (%)
Biphenyl	47
Ferulic acid	23
vanillic acid	34
Cinnamic acid	43
Quercetine	28
Morin	46
S (+)-limonene	68
R (+)-limonene	83
Cumarin	81

The cells of the strain *Pseudomonas* sp. P2 were cultivated in medium containing 25 ppm of Delor 103 and phenolic extract from plant cells obtained from the cells of hairy root culture of *Solanum nigrum* as a sole carbon and energy source. We used plant extracts (non-hydrolysed extract contained mostly conjugated phenolics and other monomer compounds, hydrolysed extract after acid hydrolysis contained monomers of phenolics). Degradation of PCBs after 14 days was analysed (see Figure 5). Data demonstrate the possibility of natural plant cell extracts to support growth and degradation of PCBs by *Ps.* sp. P2. HPLC analysis of the extract (data not shown) proved that mostly vanillic and hydroxybenzoic acids are present, also traces of ferulic and cinnamic acids were detected. Data support the previous hypothesis that naturally occurring plant compounds, for example those excreted to the soil by roots, can induce *bph* operon and thereby degradation of PCBs even in lower concentrations (than were used above) which occur in nature. On the other hand only a few plant species may have desired properties and awareness of such species would be extremely valuable. From the results published earlier it was demonstrated that plant cell are able to transform PCB [19], [20]. This study confirms that plant-derived compounds can also enhance PCB degradation by bacterial cells and thus plant-microbe systems have the potential of providing inexpensive bioremediation possibilities in real polluted areas.

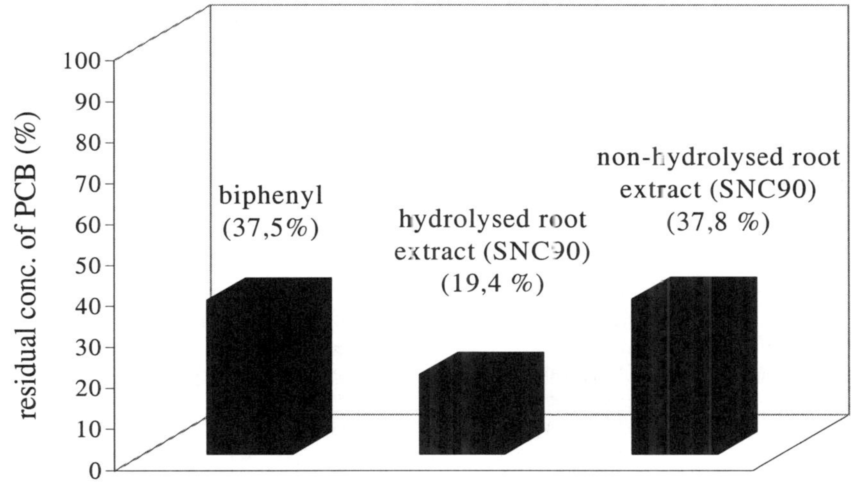

Figure 5. Bacterial degradation of PCBs in presence of biphenyl, and plant extracts obtained from hairy root culture of *Solanum nigrum* SNC 90. Concentrations of phenolics in non-hydrolysed plant extract was 0,2 mg/ml.

For the enhancement of PCB degradation, the mixtures of two strains - *Pseudomonas sp.* P2 (biphenyl degrader) and *P. aeruginosa* JB2 (chlorobenzoate degrader), were used. The minimal medium containing biphenyl (5000 mg/l) and Delor 103 (0.04 mg/l) was inoculated, first with biphenyl strain and after 3 days of cultivation, inoculum of the chlorobenzoate strain was added. In the second experiment the glass beads (1mm, 5g/50ml) were added to obtain a better contact between the cells and poorly soluble PCB. The degree of degradation with both strains was compared with the degradation using only biphenyl degrader. The results of both experiments summarized in the Table 5 are the average of three parallel experiments.

TABLE 5. Comparison of PCB degradation by individual chlorobiphenyl degrading strain and consortia of chlorobiphenyl and chlorobenzoate consortium

Microorganism	Residual concentration of PCB [%]
A_1- both strains	37.2
A_2- biphenyl degrader	46.5
B_1- both strains	21.0
B_2- biphenyl degrader	54.4

A - without glass beads
B- with glass beads

Obtained results show that the mixture of two strains carrying the enzymes complementing both biphenyl and chlorobenzoate metabolic pathway enhanced the PCB degradation. The addition of glass beads improved the process only in the case of the mixture of both strains.

Finding of the optimal conditions for PCB biodegradation is under further investigation but we suggest that the combination of strains carrying complemented metabolic pathway for PCB mineralisation with potential inducing effect from plant exudates could serve as a model for the development of the efficient bioremediation system.

The present study follows different characteristics and properties of PCB degrading strain *Pseudomonas* sp.P2 isolated from PCB contaminated site in the Czech Republic. Beside basic knowledge about this strain certain approaches, based on knowledge of mutual relationships among different organisms, for enhancement of bioremediation process are shown. It can be seen that use of microbe-microbe or plant-microbe consortia have the potential to provide inexpensive, naturally derived, bioremediation systems.

4.2. EFFICIENCY OF CHLOROCATECHOL METABOLISM IN NATURAL AND CONSTRUCTED CHLOROBENZOATE AND CHLOROBIPHENYL DEGRADERS

Chlorocatechols are central intermediates in metabolism of chlorinated aromatic compounds [48]. For some time, they were considered to be metabolized only through a modified *ortho*-cleavage pathway (intradiol cleavage) because *meta* fission (extradiol cleavage) of 3-chlorocatechol creates suicidal acyl halides [50, 36]. However, *meta* fission of 4-chlorocatechol does not produce an acyl halide. Thus, *Comamonas acidovorans* M3GY, a constructed hybrid strain, was able to utilize 3,4'-dichlorobiphenyl, by conversion to 4-chlorobenzoate, and subsequent *meta* fission of 4-chlorocatechol [45]. *Burkholderia cepacia* P166, a biphenyl degrader, also metabolized 4-chlorocatechol by *meta* fission [34, 35]. In addition, Kaschabek et al. [43] have reported a new type of extradiol-cleaving enzyme, isolated from *Pseudomonas putida* GJ31, in which 3-chlorocatechol was actually the preferred substrate for chlorocatechol 2,3-dioxygenase.

Among the most significant pollutants producing chlorocatechols as their intermediates are polychlorinated biphenyls (PCBs). Although PCBs as a group are generally regarded as recalcitrant to biodegradation, bacterial isolates metabolizing specific congeners have been described [15, 33, 37, 39, 42, 51] The chlorocatechol *meta*-cleavage products were demonstrated to interfere with PCB transformation by inhibiting of 2,3-dihydroxybiphenyl dioxygenase, one of the enzymes from the upper part of PCB pathway [21, 32]. In this chapter we have focused on the complementation of alternative *ortho*- and *meta*- pathways for chlorocatechol degradation into one strain to follow effects on the degradation of chlorobenzoates.

Burkholderia cepacia P166 transforms chlorobenzoate isomers resulting from the degradation of the 3 existing monochlorobiphenyls to the corresponding chlorocatechols. 3-Chlorocatechol is produced from 2- and 3-chlorobiphenyls. The formation of acyl halide inactivates proteins and therefore chlorobiphenyl metabolism. In our experiment, plasmid pDSK519-*clc* containing genes responsible for *ortho*-cleavage of chlorocatechol, was introduced into two biphenyl degraders, *Burkholderia cepacia* P166 and *Pseudomonas fluorescens* S12, by biparental mating. *Ortho*- and *meta*-cleavage enzyme activities of both wild type and recombinant strains were compared with exclusively *ortho*-cleavage degrading strain *Pseudomonas fluorescens* B3. The ability of the constructed and parental strains to degrade 3- and 4-chlorobenzoate, respectively, was compared by monitoring chloride release in liquid

The *clc* operon, carrying genes for *ortho*-cleavage of chlorocatechols from *Pseudomonas putida* P51 [53], was subcloned on a 4 kb *Pst*I fragment into the polylinker of the broad host range vector pDSK 519 [44]. The resulting plasmid pDSK-*clc* was transferred separately into two biphenyl-utilizers, *Burkholderia cepacia* P166 [34] and *Pseudomonas fluorescens* S12, neither of which possessed significant *ortho* fission activity with 4-chlorocatechol. Biparental mating and electroporation methods were used in both cases. Because growth of hybrid strains was better from biparental mating, all data presented herein are from these strains. Plasmid pDSK-*clc* with inserted *clc* operon was confirmed by the isolation and further digestion with *Pst*I restriction endonuclease (data not shown).

Catechol and chlorocatechol dioxygenase activities of all wild types and hybrid strains were considerably variable when grown on biphenyl in the presence of a chlorobenzoate. In general, *ortho* fission activity was more pronounced with catechol as substrate. In the case of 3-chlorocatechol, only the hybrid strain *P. fluorescens* S12/C possessed any significant activity, and this was exclusive only to *ortho* fission activity (Fig. 5). Strains with either *ortho* (*P. fluorescens* B3, *P. fluorescens* S12/C) or *meta* (*P. cepacia* P166), or both *ortho*- and *meta*- (*P. cepacia* P166/C) cleavage abilities with regard to chlorocatechols, were compared (Fig. 7, 8, 10, 11). Biphenyl degrading strain *P. fluorescens* S12 was surprisingly found to possess moderate *meta*-cleavage ability regarding 4-chlorocatechol (Fig. 9).

Burkholderia cepacia P166, which possesses *meta*-cleavage activity with 4-CCs, degrades 4-chlorobiphenyl and 4-chlorobenzoate without product inhibition, unlike its degradation of 3-chlorobiphenyl [35] (Fig. 7). *Pseudomonas fluorescens* B3, which possesses only *ortho*-cleavage activity on CCs, degrades 3-CB without product inhibition, but has lower activity on 4-CB (Fig. 11).

Recombinant strain P166/C has gained no apparent advantage from the mating of P166 and B3, as it is unable to dehalogenate 3-CB, and possesses similar activities towards dehalogenation of 4-CB (Fig. 12) as its parental strain P166. Moreover, it has no significant advantage in 3-CB and 4-CB degradation (Fig. 8). However, transfer of *clc* operon from strain B3 into strain S12 resulted in increased degradation of 3-CC in hybrid strain S12/C (Fig. 9, 10).

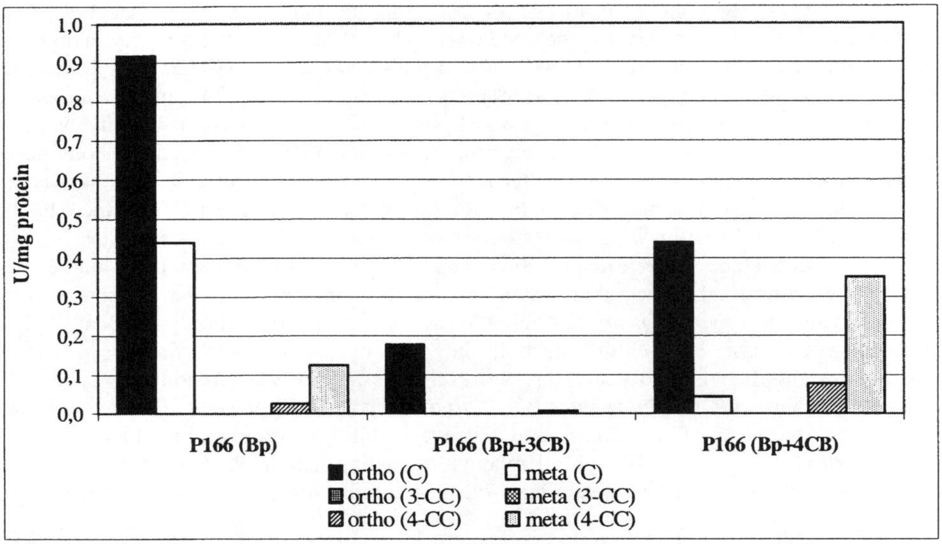

Figure 7. Specific enzyme activities of catechol *ortho-* and *meta-*cleavage enzymes from the strain *Burkholderia cepacia* P166.

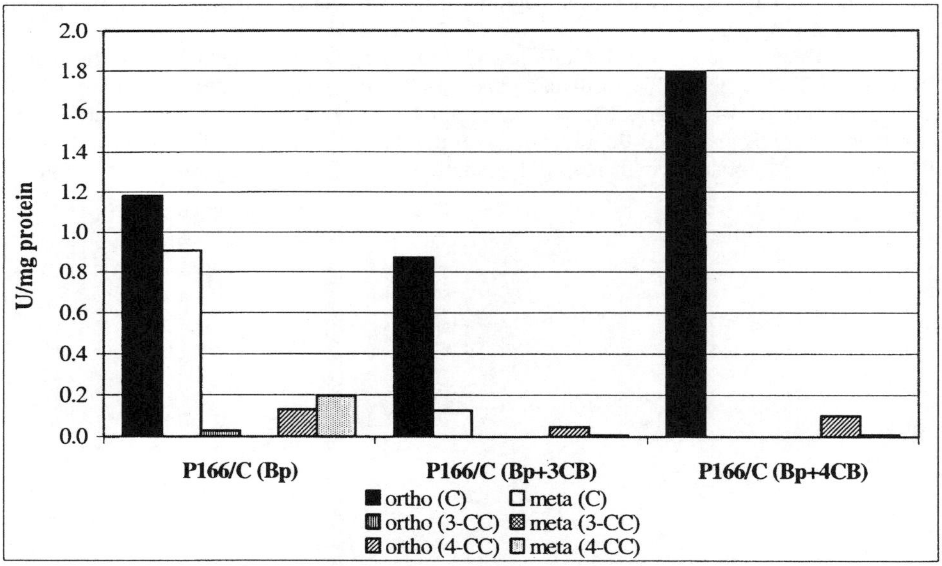

Figure 8. Specific enzyme activities of catechol *ortho-* and *meta-*cleavage enzymes from the strain *Burkholderia cepacia* P166/C.

The great majority of reports characterizing growth of bacteria on chlorinated biphenyls have described chlorobenzoates as dead-end products. In the case of biphenyl degrader *P. cepacia* P166, however, chlorobenzoates created by growth on each of the three monochlorobiphenyls, are slowly metabolized to toxic metabolites that can restrict growth.

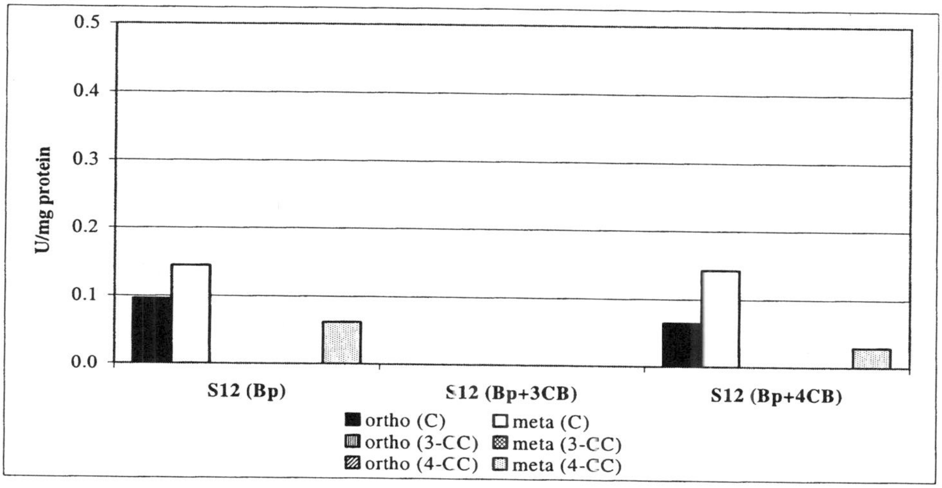

Figure 9. Specific enzyme activities of catechol *ortho*- and *meta*-cleavage enzymes from the strain *Pseudomonas fluorescens* S12.

These toxic interactions were all associated with the *meta*-cleavage of 3-chlorocatechol [34, 35]. 3-Chlorobenzoate is usually transformed by benzoate dioxygenase followed by a dehydrogenase into 3-chlorocatechol, which, upon *meta*-cleavage, give rise to a reactive acylchloride inactivating the ring cleavage enzyme [36]. The only escape from that scheme was recently reported by Kaschabek et al. 1998 [43]. On the other hand 4-chlorocatechol degradation through the 3-oxoadipate pathway was shown to produce toxic intermediate called protoanemonin [38]. The authors proved that PCB cometabolizing microorganisms have been negatively affected by the formation of this antibiotic. Also indigenous soil microflora in that study was shown to convert 4-chlorobenzoate into a toxic product, most probably protoanemonin.

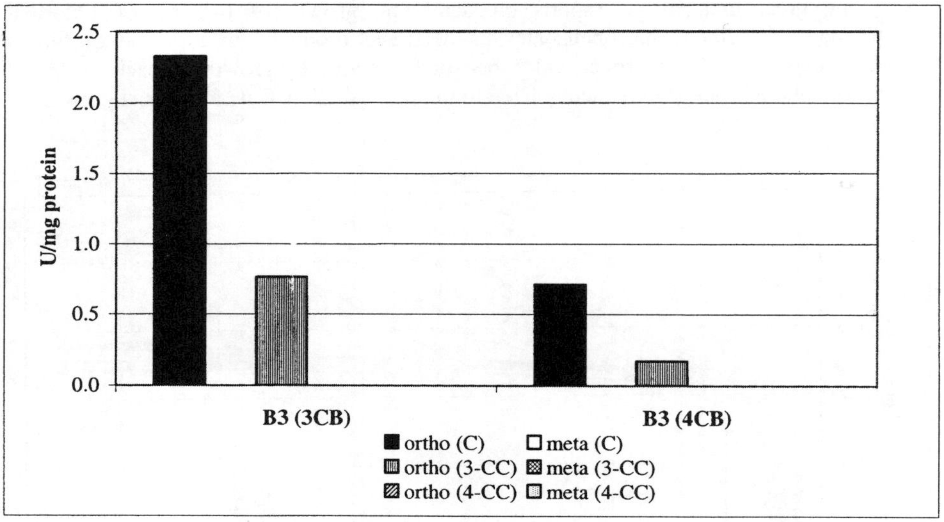

Figure 10. Specific enzyme activities of catechol *ortho-* and *meta-*cleavage enzymes from the strain *Pseudomonas fluorescens* S12/C.

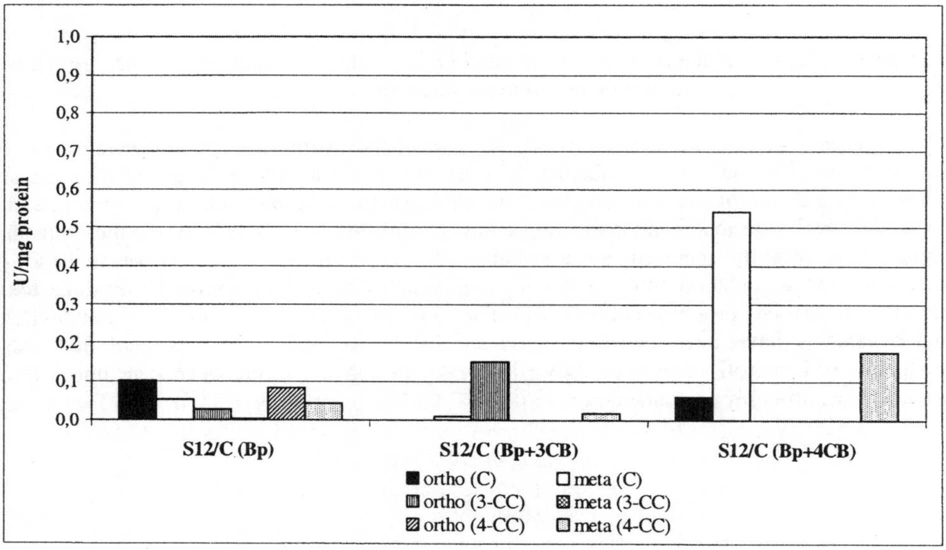

Figure 11. Specific enzyme activities of catechol *ortho-* and *meta-*cleavage enzymes from the strain B3.

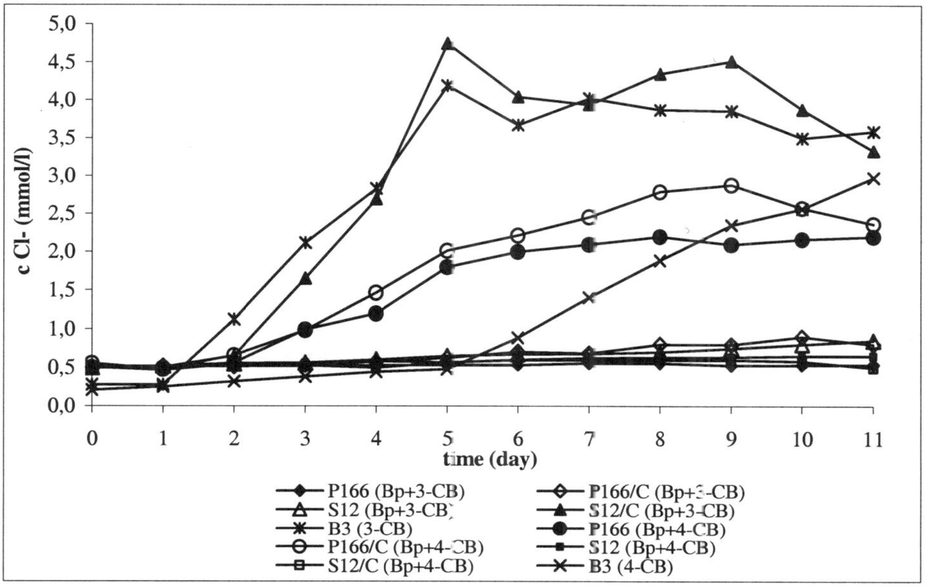

Figure 12. Chloride release during cultivation on chlorobenzoates.

In developing recombinants utilizing chlorobiphenyls many strain constructions have focused on pooling *in vivo* the biphenyl pathway of PCB-cometabolizers with the chlorobenzoate pathway of chlorobenzoate degraders [32,41,42,45,46]. Such recombinants, however, have been quite unstable and not very effective in degrading mixtures of PCB congeners, as found in Aroclor 1242. Inhibition from 3-chlorocatechol, the metabolite of 3-chlorobenzoate, is one case in point [32]. Aside from product inhibition, chlorobenzoate degradation requires two operons: one is involved in the conversion of chlorobenzoates to chlorocatechols; the other (*clc* operon) encodes a modified *ortho*-cleavage pathway. Strain P166 converts 2- and 3-chlorobiphenyls directly to 3-chlorocatechols [34, 35]. Introduction of the *clc* operon should facilitate routing of 3-chlorocatechol through the modified *ortho* pathway in obviating the formation of acylhalides, and thus enhancing degradative potential (Fig. 13).

While introduction of the 3-CC pathway into the strain S12 was beneficial, there was, however, no such clear observation regarding degradation of chlorocatechols into the strain P166 which would fit our original idea for re-routing non functional *meta* pathway. The degradation of catechol in that strain, however, increased with the introduction of *clc* operon and hence supporting hypothesis about the evolution of *clc* operon from the catechol one [40]. Pooling of both classical *meta* and *ortho* pathways into one organism will not bring an explicit solution to the problem of suicidal inactivation because the former pathway is essential to the conversion of PCBs to chlorobenzoates. Therefore we will focus in further experiments on introducing another type of *meta*-cleavage pathway, namely that one discovered by Kaschabek et al. 1998 [43], into the strain which already carries *ortho*-cleavage pathway.

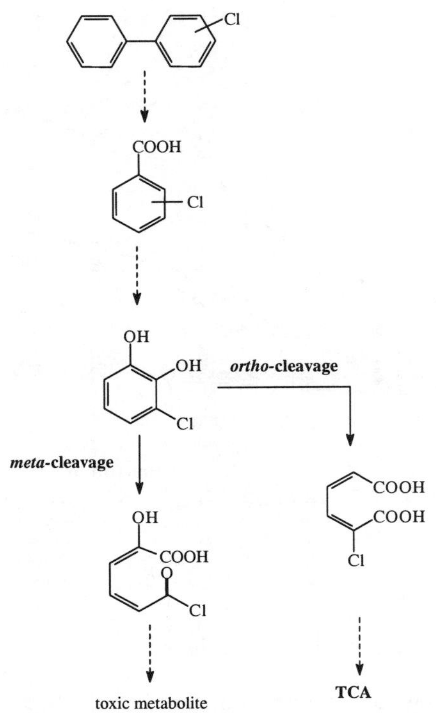

Figure 13. Recruitment of *ortho*-cleavage chlorocatechol pathway in strain *B. cepacia* P166.

4.3. SOIL MODIFICATION IN THE AEROBIC BIOREMEDIATION OF POLYCLORINATED-CONTAMINATED SOILS

4.3.1 Bioremediation of PCB-contaminated soils and sediments: lab successes and difficulties in the technology transfer.

Several laboratory studies have shown that lightly-chlorinated PCBs occurring in aged-contaminated soils can be aerobically biodegraded through the concerted action of indigenous aerobic PCB-co-metabolising bacteria and chlorobenzoic acid (CBA)-mineralising bacteria [54]. The aerobic biodegradation of PCBs was also found to occur in Hudson River sediments, where important depletions of "indigenous" low-chlorinated PCBs, the presence of CBAs and of PCB- and CBA-degrading microbes have been observed [55,56,57]. The occurrence of PCB-reductive dehalogenation has been demonstrated in many anaerobic environments, including soils as well as freshwater, estuarine and marine sediments [57, 58]. This process can remove chlorines from higher chlorinated PCBs, producing mono- through tetra-chlorinated biphenyls, which subsequently can be aerobically dehalogenated and mineralised. Therefore, for efficient bioremediation processes, it is suggested that an anaerobic reductive dehalogenation process should be followed by an aerobic treatment phase, which may mediate the final removal of chlorinated products resulting from the previous step [57, 58].

A number of lab-scale technological experiments have demonstrated that bioremediation is a feasible option for remediating PCB-contaminated soils, sediments and aquifer materials [59], and several bioremediation methods consisting of a single anaerobic treatment [58], aerobic treatment [57, 60, 61, 62, 63] or of a sequential anaerobic-aerobic treatment [64, 65, 66] have already been proposed for this purpose. However, in spite of these promising laboratory-scale findings, to the best of our knowledge, no field large-scale bioremediation technology for PCB-contaminated soils and sediments has been developed up to now. This may be ascribed to the fact that a technological process for the complete biological removal of PCBs from such matrixes may be very complicated, as it has to consist of the integration of two distinct treatment stages, each characterized by specific hydraulic retention time and co-factor requirements, i.e., *i)* an anaerobic treatment, to reductively dechlorinate high-chlorinated PCBs into the corresponding low-chlorinated congeners and *ii)* an aerobic treatment, to oxidatively biodegrade the resulting low-chlorinated PCBs into CBAs and other chlorinated compounds (through the action of aerobic PCB-co-metabolizing bacteria) and then to biodegrade the latter compounds into Krebs's cycle intermediates and chlorine ions. The former is generally a process characterized by a high-hydraulic retention time [67] and specific nutrient requirements [58], whereas the latter process typically exhibits low residential times [67], the requirement of defined availability of exogenous biphenyl, O_2 [68], and (sometimes) specialized bacteria [69], as well as of a high degree of mixing [70]. This long list of technological requirements has probably hindered the formulation and the scale-up of industrial plants for the bioremediation of PCB-contaminated environments. Furthermore, the evidence that PCBs are biodegraded slowly, with low yields and in general with the production of a large array of known and un-known chlorinated (and therefore potentially toxic and stable) water-soluble intermediates [54, 58], indicates that any types of *in-situ* techniques should be avoided in the biological restoration of PCB-contaminated areas, for which, therefore, a treatment of the contaminated area-deriving soils or sediments in dedicated large-scale bioreactors is greatly recommended [59, 67].

4.3.2. Biodegradation of PCBs in aged-contaminated matrixes: the bioavailability problem and strategies to mitigate ti. Introductive remarks

Some lab-scale studies have shown that the bioremediation of aged-PCB-contaminated soils and sediments is often affected by the low bioavailability of the pollutants, which, due to their very low solubility and dissolution rate in water, tend to adsorb onto the soil organic matter, thus becoming poorly available in the soil-water phase where the pollutant-metabolising microorganisms are in general mainly located and active [68, 71, 72]. One possible approach to improve the bioavailability and, in turn, the biodegradation of PCBs in such contaminated matrixes may consist in amending them with PCB-mobilizing/solubilizing agents [68, 72]. This strategy has been applied in the field of oil recovery and then in the aerobic biodegradation of hydrocarbon-contaminated soils [20]. However, only little has been done up to now in order to test its effectiveness in the biodegradation of PCBs in contaminated soils and sediments [58, 72, 73, 74, 75]. The few solubilizing agents tested in this field are synthetic non-ionic surfactants; when applied at concentration higher than their CMC, they have generally exhibited interesting PCB-solubilizing properties, but also a certain recalcitrance and important toxic effects *vs.* the specialized bacteria occurring in the contaminated matrixes [63, 72, 74, 76]. Bacterial surfactants appeared to be promising PCB-biodegrading enhancing agents, not only for their PCB-solubilizing properties, but also for their biodegradability, non-toxicity and biological origin [8,24]. Unfortunately, little is known about their chemical properties, and they are not still available on the large scale and at low prices, and this currently precludes their application in the bioremediation industry [61, 77]. Other biogenic agents, such as cyclodextrins and phytogenic surfactants, which are extensively used as hydrophobic compound-solubilizing agents in the food and the pharmaceutical industry, also resulted promising PCB-biodegradation enhancing agents as they, similarly to the bacterial surfactants, could conjugate important PCB-solubilizing activities to other properties, such as non-toxicity, biodegradability and biological origin, which are generally not exhibited by the synthetic surfactants and that can make their application in the soil bioremediation socially more acceptable [62, 63, 72]. In this chapter, the main data collected by treating an aged-PCB-contaminated soil in slurry- and fixed-phase conditions in the presence of the biogenic agents hydroxypropyl-β-cyclodextrin (HP-β-CD), γ-cyclodextrin (γ-CD) or Quillaya saponin (QS) along with the synthetic non-ionic surfactant Triton X-100 (TX-100)[9,10] are reviewed.

4.3.3. Effects of biogenic solubilizers VS. TX-100 on the aerobic bioremediation of an aged PCB-contaminated soil.

The capability of cyclodextrins (CDs) of enhancing the water-solubility of a large array of hydrophobic organic compounds is well-documented in the literature [79]. CDs are cyclic D-glucose oligosaccharides produced biotechnologically from starch; in general, they complex hydrophobic molecules by carrying them into the water-phase where they can be spontaneously released and made available [80]. Fava and Grassi [81] demonstrated that the bioavailability and the biodegradation rate and extent of Aroclor1221 PCBs in liquid cultures of *Pseudomonas* sp. strain CPE1 can be enhanced by the addition of HP-β-CD. In the same period, the positive effects of two biogenic and biodegradable surfactants, i.e., QS and Soya Lecithin, on the aerobic biodegradation of phenanthrene and fluoranthene in liquid cultures of specialised bacteria was reported in the literature [82]. Thus, it appeared interesting to test some of these biogenic solubilizers in the aerobic biodegradation of PCBs in soils. As TX-100, one of the most investigated synthetic surfactants used in the bioremediation of hydrocarbon contaminated soils [73], had been found to enhance the aerobic biodegradation of PCBs in liquid media [83] and in soil/water systems [84], it was also interesting to investigate its effects on the same process. An aged-contaminated soil obtained from an Italian dump site was used in the study.

The soil, which contained 350 mg/kg of PCBs and about $1x10^5$ CFU/g of indigenous PCB- and CBA-degrading aerobic bacteria, was amended with biphenyl (4g/kg, to induce and support the activity of the indigenous aerobic PCB-biodegrading bacteria), inorganic nutrients (to have a final C:N:P ratio of 200:10:1) and treated in aerobic slurry-phase reactors (70 g of soil suspended in 700 ml of water in 1l baffled Erlenmeyer flasks, which were closed with steel cups and then incubated in a dark chamber at 150 rpm and 28°C) and fixed-phase reactors (column loop reactors packed with 1.1 kg of soil saturated with 515 ml of water; aerobic conditions were provided through daily injections of H_2O_2) [62, 63]. TX-100, the phytogenic surfactant QS, and the two analytical-grade CDs, HP-β-CD and γ-CD, were added to the bioreactors at the concentration of 10 g/l (referred to the water-phase actually available in the two reactor systems) at the 39th or 42nd day of treatment. Both surfactants, TX-100 and QS, were thus applied at a concentration higher than their CMCs.

QS, HP-β-CD, γ-CD were found to be extensively metabolised by the soil aerobic microorganisms which were found to grow on them when applied at concentrations from 0 to 20 g/l as the sole carbon and energy sources in shake cultures of mineral medium. On the contrary, TX-100 not only did not sustain the growth of the soil microorganisms, but it also inhibited their growth in a rich liquid medium, such as Tryptic Soy medium, when applied at concentrations higher than 2.5 g/l [62, 63]. Significant depletions of the original soil PCBs were attained both in the presence and in the absence of the solubilizers, both in slurry- and in fixed-phase conditions (Tables 1 and 2). In spite of several attempts to prepare sterile reactors through heat or chemical sterilization, no abiotic controls could be obtained in this study, and this did not allow us to quantify the actual impact of abiotic losses on the observed depletion percentages. However, some experimental observations, such as the persistence of

PCB-co-metabolising and CBA-degrading bacteria as well as the production of CBAs and of chloride ions (Figure 14) in the soil reactors throughout the treatment strongly suggested that the soil PCBs were mainly removed through aerobic biodegradation processes. In the slurry-phase reactors supplemented with QS, HP-β-CD and γ-CD, important enhancements of the yields of biodegradation (of 13, 26 and 22%, respectively) and dechlorination (Figure 14) of the soil PCBs were recorded. On the contrary, TX-100 was found to inhibit the same processes in the same treatment conditions (Table 6, Figure 14).

TABLE 6. Characterization and average depletion percentages of the main GC peaks (ascribed to PCBs) revealed in the soil treated in slurry-phase reactors. The depletion data shown (±SD) were calculated using the depletion percentages determined on the samples collected from each reactor at the 119th, 131st and 141st day of treatment (n.i.=non-identified peak)(after [62,63]).

GC peak no.	Peak identification (substitution pattern)	Average peak depletion % in the slurry-reactors after 141 days				
		None	*TX-100*	*QS*	*HP-β-CD*	*γ-CD*
1	n. i.	44±18	37±11	75±12	82±5	80±4
2	n. i.	57±25	60±13	89±3	77±8	77±3
3	2,4'	100±0	75±8	78±8	82±3	88±8
4	n. i.	-4±18	0.2±16	59±9	62±15	54±10
5	4,4'/2,2',5/2,2',4	100±0	100±0	100±0	100±0	100±0
6	n. i.	100±0	58±12	90±5	100±0	100±0
7	2,2',3/2,4',6	3±27	52±9	66±3	80±8	78±4
8	n. i.	69±4	65±7	82±7	87±7	85±3
9	2,2',5,5'	59±7	25±9	69±2	84±7	79±3
10	2,2',4,5'	60±5	53±2	78±11	86±7	82±3
11	2,2',3,4/2,3,4',6	73±5	60±5	80±2	88±11	84±2
12	2,3',4',5	63±9	49±12	76±5	86±4	83±1
13	2,3',4,4'/2,2',3,5',6	87±4	73±8	85±7	96±7	93±5
14	n. i.	51±7	53±7	76±4	85±7	87±6
15	2,2',3,4',5/2,2',4,5,5'	100±0	62±3	79±13	88±10	83±2
16	2,2',4,4',5	74±5	48±6	79±3	91±11	89±7
17	2,2',3',4,5	78±1	28±4	70±4	93±10	90±8
18	2,2',3,4,5'	55±3	32±12	70±6	67±6	76±7
19	2,3,3',4',6	62±2	32±6	71±9	87±6	86±3
20	n. i.	70±3	33±2	73±5	85±11	83±1
21	n. i.	69±5	41±4	75±8	80±4	86±2
22	2,3',4,4',5/2,2',3,4',5',6	95±4	31±6	78±7	95±6	91±2
23	n. i.	69±3	50±2	74±10	88±5	83±3
24	2,3,3',4,4'/2,2',3,3',4,6'/2,2',4,4',5,5'	85±8	55±3	80±2	90±11	79±4
25	2,2',3,4,4',5'/2,3,3'4,4',6	69±4	56±5	78±3	86±7	84±4
26	2,2',3,3',4,5/2,2',3,3',5,5',6	82±8	53±7	81±7	92±10	88±3
27	n. i.	68±23	51±3	72±4	83±16	69±7
	Average % of depletion:	**68**	**46**	**77**	**86**	**83**

QS, HP-β-CD, γ-CD significantly enhanced the availability of aerobic cultivable heterotrophic bacteria in the reactors without influencing appreciably that of the PCB- and CBA-degrading bacteria revealed in the same reactors. On the contrary, TX-100 caused a premature disappearance of the heterotrophic and the CBA-degrading bacterial biomass. With the exception of γ-CD, which was found to sequester the soil PCBs, QS and HP-β-CD both enhanced significantly (from 0 to 54 and from 15 to 96 %, respectively) the PCBs availability in the reactor water-phase; comparable bioavailability enhancing effects (from 20 to 78 %) were also exhibited by TX-100. These findings suggested that the biogenic agents enhanced the PCB-biodegradation and dechlorination by increasing the availability of both PCBs and indigenous bacteria in the soil reactors and that the adverse effects exhibited by TX-100 may be ascribed to the negative impact of this agent on the availability of the heterotrophic and CBA-degrading bacteria in the amended soil. TX-100 effects are comparable with those reported by Viney and Bewley [84], who employed this surfactant in studies of PCB bioavailability and biodegradation in a spiked soil, and by other authors who tested it in the biodegradation of hydrocarbons in soils [72, 73]. γ-CD and in particular HP-β-CD also exerted important positive effects (enhancements of the PCB-biodegradation yield of 35 and 84%, respectively) on the soil bioremediation in the fixed-phase reactors, where biodegradation percentages even greater that those recorded in slurry-phase were observed (Tables 6 and 7)

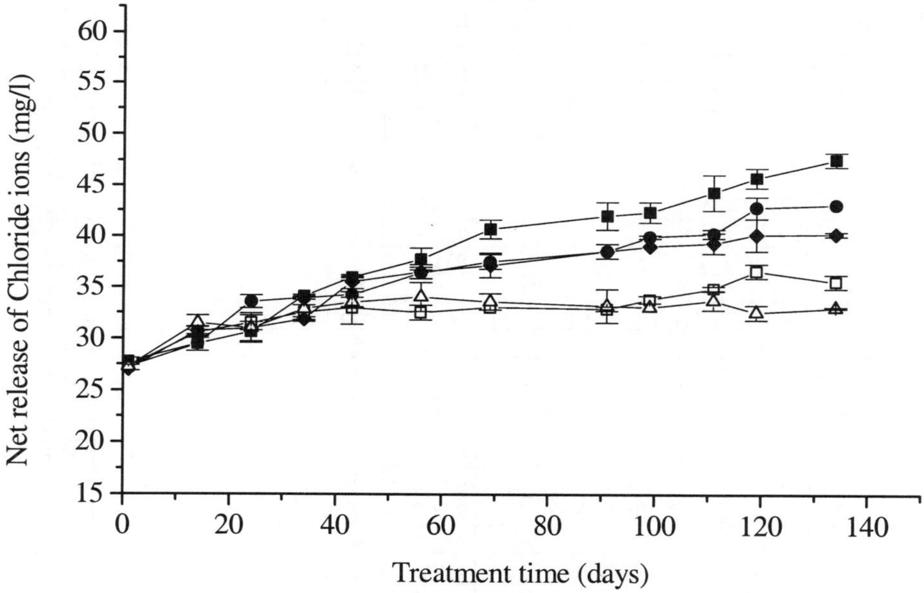

Figure 14. Net releases of chloride ions (chloride ion concentration detected in the reactors minus that due to the exogenous chloride ions carried by the added QS and HP-□-CD) in the slurry reactors in the absence of agents (□), in the presence of TX (△),QS (♦), HP-□-CD (□) or □-CD(□) (after [62, 63]).

TABLE 7. Characterization and average depletion percentages of the main GC peaks (ascribed to PCBs) revealed in soil treated in the aerobic fixed-phase bioreactors. The depletion results shown (±SD) were calculated using the data obtained from 3 soil sample collected from each reactor at the end of the experiment (n i.=non-identified peak) (after [62, 63]).

GC peak no.	Peak identification (PCB substitution pattern)	Average peak depletion % in the fixed-phase reactors after 141 days				
		None	TX-100	QS	HP-β-CD	γ-CD
1	n. i.	34±4	88±5	33±7	90±9	52±5
2	n. i.	59±3	89±8	100±0	100±0	65±6
3	2,4'	28±14	87±11	42±5	73±7	-1±5
4	n. i.	50±3	87±3	95±2	90±4	59±14
5	4,4'/2,2',5/2,2',4	100±0	100±0	100±0	100±0	100±0
6	n. i.	49±5	88±2	40±12	100±0	64±9
7	2,2',3/2,4',6	-72±9	84±4	15±5	85±7	-35±4
8	n. i.	37±4	87±2	25±7	96±9	56±7
9	2,2',5,5'	47±7	78±9	40±2	93±7	62±5
10	2,2',4,5'	57±4	84±7	59±1	97±5	69±6
11	2,2',3,4/2,3,4',6	43±8	82±5	40±7	95±9	69±7
12	2,3',4',5	49±3	84±9	37±2	92±7	53±3
13	2,3',4,4'/2,2',3,5',6	100±0	85±4	61±1	100±0	100±0
14	n. i.	44±3	82±2	39±5	90±2	50±5
15	2,2',3,4',5/2,2',4,5,5'	100±0	81±7	45±4	100±0	100±0
16	2,2',4,4',5	51±3	82±1	42±3	96±9	78±9
17	2,2',3',4,5	62±8	82±3	60±9	96±3	77±5
18	2,2',3,4,5'	37±4	78±3	38±2	91±8	38±4
19	2,3,3',4',6	57±4	81±2	56±3	94±7	74±7
20	n. i.	45±8	79±9	42±5	92±9	65±6
21	n. i.	41±4	73±8	39±3	92±7	61±8
22	2,3',4,4',5/2,2',3,4',5',6	89±5	76±11	39±4	99±1	100±0
23	n. i.	48±4	76±2	49±3	94±4	73±5
24	2,3,3',4,4'/2,2',3,3',4,6'/2,2',4,4',5,5'	78±5	74±6	45±7	99±1	100±0
25	2,2',3,4,4',5'/2,3,3'4,4',6	37±5	74±9	43±2	98±2	87±7
26	2,2',3,3',45/2,2',3,3',5,5'6	44±10	71±3	48±1	99±2	100±0
27	n. i.	62±7	61±8	44±3	96±4	90±9
	Average % of depletion:	**51**	**80**	**49**	**94**	**69**

QS was not effective in these treatment conditions, where, surprisingly, TX-100 resulted very effective on the PCB-biodegradation (which was enhanced of about 57%) (Table 7) and not inhibitory *vs.* the soil cultivable bacteria population. The inactivity of QS as well as the unexpected effectiveness of TX-100 may be ascribed to their lower availability in the soil treated in fixed-phase conditions, where they were applied at an initial concentration about 20 times lower (< 0.5% w/w, referred to the soil weight) than that used in slurry-phase conditions and at lower degree of mixing and soil homogeneity. In other words, it seems to be possible that QS, which displayed low PCB-solubilizing properties and high biodegradability, was ineffective as it was insufficiently available in the soil, and that TX-100, which resulted an excellent PCB solubilizing agent and not biodegradable into the soil, was effective as here it was available at concentrations, with respect to the slurry-phase conditions, which were not toxic on the soil biomass but still able to improve the PCB bioavailability in the soil.

Conclusions and research perspectives. HP-β-CD and γ-CD enhanced significantly the aerobic bioremediation of the aged PCB-contaminated soil used in the study both in slurry- and in fixed-phase conditions. On the contrary, QS influenced positively the soil bioremediation in the sole slurry-phase conditions, whereas TX-100 was effective in the sole fixed-phase reactors. HP-β-CD was the most versatile and effective agent among those tested in the study. TX-100 was very effective in fixed-phase conditions, i.e. in treatment conditions having great practical interest [85]; however, on the basis of the results presented here and those available in the literature [72, 73, 84], its use in the soil bioremediation appears to be precluded by its recalcitrance and toxicity *vs.* the soil microorganisms. On the contrary, QS, γ-CD and in particular HP-β-CD, appear promising PCB-bioavailability enhancing agents, not only for their positive effects on the soil bioremediation, but also for their biodegradability and non-toxicity. In addition, their biogenic nature makes their incorporation into soil socially more acceptable.

Further studies are necessary to *i)* confirm the results presented here by testing the proposed biogenic agents on different PCB-contaminated soils; *ii)* assay technical-grade products of the same agents and to optimise their use in order to make their application cost-effective in the soil bioremediation. Studies addressed to test and to optimise the use of a cheap industrial mix of randomly methylated-β-cyclodextrins (RAMEB) in the aerobic bioremediation of different PCB-spiked- and aged-contaminated soils are in progress within the NATO-*Science for Peace* research project No.973720; the data currently available seem to confirm that RAMEB can have a great potential as bioremediation promoting additive in the remediation of PCBs and in particular of PAH-contaminated soils. In addition, in a recent study [78] it has been shown that a Soya Lecithin industrial product may be another effective biogenic PCB-biodegradation enhancing agent which merits to be further investigated in the bioremediation field.

4.4. RHIZOSPHERE REMEDIATION OF POLYCYCLIC AROMATIC SOIL CONTAMINANTS

Underpinning rhizosphere remediation is the realization that although plants in natural ecosystems produce polycyclic aromatic compounds (flavonoids, coumarins, etc.) in their leaves, stems and roots, these compounds have not accumulated in the soil to concentrations reflective of their annual production over thousands of years. There is only a limited understanding of the production and recycling of carbon associated with naturally occurring polyaromatic compounds in terrestrial ecosystems (i.e. tannins in oak forests). However, it is apparent that since they have not accumulated to astronomic amounts over the last millennium, mechanisms do exist within nature to degrade and recycle carbon present in thousands of naturally occurring polyaromatic compounds many of whose structures resemble those of recalcitrant pollutants such as PCBs and PAHs (Figure 1). The obvious question is, "Will the biological mechanisms in nature that degrade natural polyaromatic compounds also degrade recalcitrant organic pollutants (i.e. PCBs and PAHs)?" If so, what soil ecosystems associated with what plant species are most active against pollutants? How should we introduce and manage these natural, multi-organism systems to optimise their degradative properties towards recalcitrant pollutants? All of these questions deserve attention and should be resolved in order to develop dependable, sustained rhizosphere remediation technology. First, however, there is a need for a better understanding of the plant rhizosphere with regard to plant-driven mechanisms that stimulate xenobiotic degradation. The rationale for placing emphasis on mechanistic studies is that the results provide a level of understanding that is necessary to improve the performance of rhizosphere remediation and to develop monitoring tools to facilitate field implementation.

There are three important plant driven mechanisms that may operate within the rhizosphere to degrade recalcitrant polyaromatic compounds: (1) direct metabolism by either endogenous or exogenous plant enzymes, (2) indirect stimulation of the rhizosphere microflora to degrade contaminants, and (3) release of plant compounds that solubilize contaminants whereby they are more available to both plant and microbial degradative enzymes. Evaluation of these plant-driven processes and their application in phytoremediation requires three important types of data: (A) proof that the mechanisms exist, (B) identification of which plant species drive which mechanisms and (C) field validation that the mechanisms operate *in situ* at contaminated sites. Laboratory studies can be used to address the issues of proof and plant identification. However, field studies are necessary to validate the operation of these mechanisms and in some respects, as described later, may be a preferred starting point for research on PCBs through forensic examination of revegetated PCB sites. Research addressing plant-driven mechanisms 1 and 2 listed above will be discussed in this paper, and plant surfactants will be discussed by other authors participating in this conference.

TABLE 8. Hydroxylated derivatives of monochlorobiphenyls formed from individual monochlorobiphenyls during 14 days incubation with plant cells of tobacco, black-nightshade, horseradish and alfalfa

Silylated sample	Number of products	Retention time	Standards
Tobacco (PCB 1)	4	11:13 11:18 11:32 11:37	NI NI NI NI
Tobacco (PCB 2)	3	10:24 12:50 13:38	NI NI NI
Tobacco (PCB 3)	3	12:04 14:03 14:13	NI 4-Cl-4′-OH NI
Black-nightshade (PCB 1)	5	8.18 10.06 10.28	NI NI 2-CL-5-OH

		10.46 12.59	NI NI
Black-nightshade (PCB 2)	4	9:12 9:29 11:54 12:43	3-Cl-6-OH 3-Cl-4-OH NI NI
Black-nightshade (PCB 3)	2	13:03 13:15	NI 4-Cl-4´-OH
Horseradish (PCB1)	5	11:00 11:17 11:25 11:37 11:44	3-Cl-2-OH NI 2-Cl-5-OH NI NI
Horseradish (PCB 2)	4	12:43 12:50 13:32 13:38	3-Cl-4-OH NI NI NI
Horseradish (PCB3)	4	12:37 12:45 13:45 13:54	NI NI 4-Cl-4´-OH NI
Alfalfa (PCB 1)	0	no hydrox. products	-
Alfalfa (PCB 2)	0	no hydrox. products	-
Alfalfa (PCB 3)	1	13:56	4-Cl-4´-OH

NI – THE STRUCTURE OF HYDROXYLATED DERIVATIVE WAS NOT IDENTIFIED, PCB1 – 2-CHLOROBIPHENYL, PCB 2 – 3-CHLOROBIPHENYL, PCB 3 – 4-CHLOROBIPHENYL

Direct metabolism of polyaromatic contaminants (PCBs) by plant enzymes was initially reported by Fletcher et al. [87, 88, 91] and has since been more fully described by several other laboratories as reported by other investigators at this meeting. A major concern in the continued work with plant systems has been characterization of the metabolic products. At the Institute of Chemical Technology in Prague it has been shown with axenic plant cultures of four different plant species that plant enzymes are capable of hydroxylating PCBs (Table 8).

Hydroxylation of PCBs by enzymes of plant roots growing in contaminated soil may be of great importance in the combined action of rhizosphere mechanisms (consult paragraph 1) since hydroxylated PCBs produced by plants may make the contaminants more soluble and/or more vulnerable to microbial attack.

However it was also observed by Kucerova et al. [94] that hydroxylated chlorobiphenyls produced by plant cultures were more toxic to photoluminescent bacteria (*Photobacterium phosphoreum*) than corresponding non-hydroxylated monochlorobiphenyl congeners (Figure 15, 16). These findings and common knowledge that several microbial breakdown products of PCBs (i.e. chlorobenzoic acids) inhibit early enzymatic steps of PCB degradation [99, 23] prompt reservations concerning the potential for biological systems to mineralize PCBs. It is important to note that the inhibition of PCB-breakdown products has been demonstrated with monocultures, whereas in the rhizosphere of a single or multiple plant species the combined uptake and metabolic features of the consortium of organisms present may keep the concentration of inhibitory breakdown products beneath their threshold of effect, at least during some seasons of the year. Testing such possibilities in the laboratory would be difficult and the most expedient way to evaluate these possibilities would be to start with forensic studies at revegetated contaminated sites.

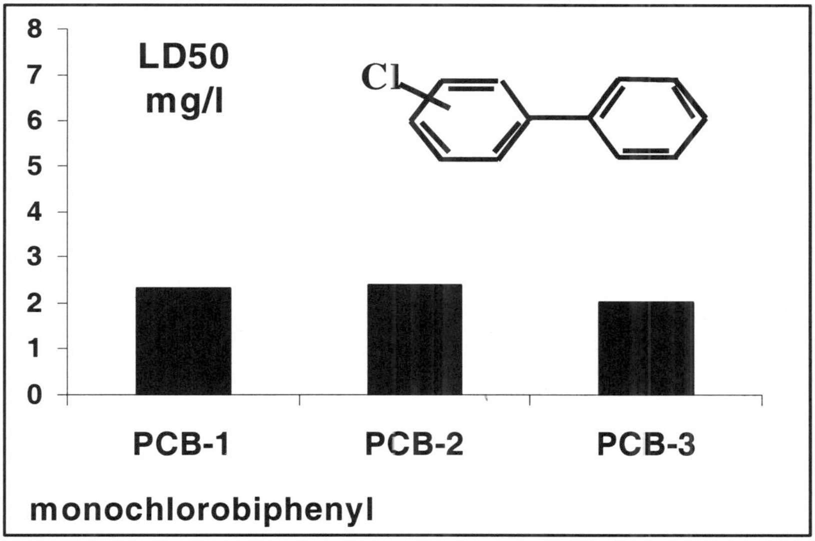

Figure 15. The toxicity of monochlorobiphenyls 2-chloro, 3-chloro- and 4-chlorobiphenyls to luminescent bacteria *Photobacterium phosphoreum*, LD 50 is expressed as the amount of toxicant which allows survival of 50% of bacterial cells.

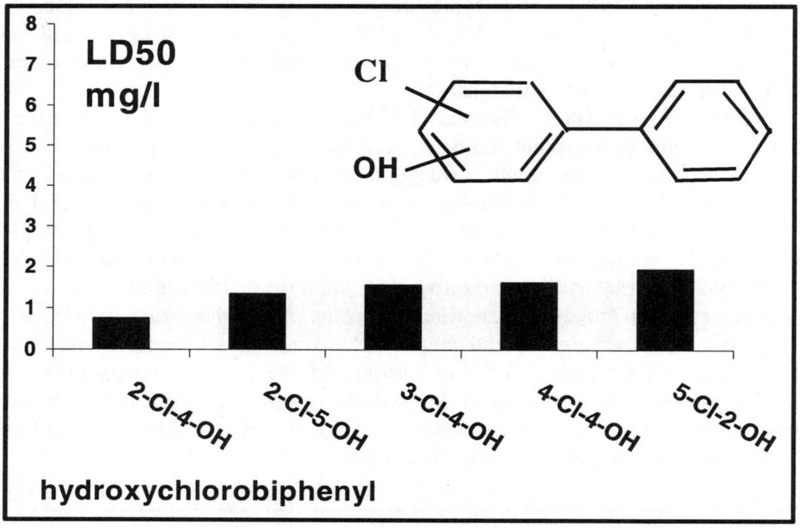

Figure 16. The toxicity of hydroxychlorobiphenyls to *Photobacterium phosphoreum*, LD 50 is expressed as the amount of toxicant which allows survival of 50% of bacterial cells

TABLE 9. Degradation of benzoate and chlorobenzoates by *Arthrobacte*r sp. UH82 and strains *Pandoraea* sp. UH133, UH222, UH1411, UH1613

Isolate	Decline of substrate (%)								
	BZ	2-CB	3-CB	4-CB	2,3-DCB	2,4-DCB	2,5-DCB	2,6-DCB	2,4,5-TCB
UH82	100[1]	0	0	100	nd	0	nd	nd	nd
UH133	100	100	100	0	100	0	100	0	0
UH222	100	100	100	0	100	0	100	0	0
UH1411	100	100	100	0	100	0	100	0	0
UH1613	100	100	100	0	100	0	100	0	0

Degradation of substrates in concentration of 1 mM was tested with non-growing cells concentrated to OD_{600}= 2,0. Substrate decline expressed in percents was analysed by HPLC after 24 hours cultivation at 28°C. BZ-benzoate, CB-chlorobenzoate, DCB-dichlorobenzoate, TCB-trichlorobenzoate, nd-not determined.

Chlorobenzoate-degrading bacteria were isolated from naturally revegetated soil contaminated with PCBs. From the microbial consortium collected at this site, five strains were able to survive after several passages on medium containing chlorobenzoic acids as sole carbon sources. Isolates showed different substrate specificities for an assortment of benzoic acid compounds (Table 9). These results suggest that the diverse microbial communities present in vegetated contaminated soil may limit the accumulation of inhibitory PCB-degradation products.

Indirect stimulation of rhizosphere microflora to degrade recalcitrant soil contaminants (PCBs and PAHs) has been examined by conducting experiments to test the hypothesis that: "Roots of some plant species enhance the degradation of recalcitrant, organic contaminants (i.e., PCBs and PAHs) by releasing cometabolites and facilitating soil aeration, both a result of fine root turnover".

Published results supporting the "cometabolite hypothesis" and providing a mechanistic understanding of rhizosphere degradation include: (1) purified natural plant compounds (i.e. flavonoids) stimulated the growth and activity of PCB degrading bacteria [26], and the stimulator compounds will vary depending on the microbe examined [98]. (2) plant roots released phenolic compounds that supported the growth of PCB degrading bacteria, but not all plant species were equally effective [89, 90, 93]; (3) flavonoid compounds accumulated in aging/dying fine roots of mulberry [95], and these compounds were later found to support the growth of PCB and PAH - degrading bacteria (Leigh, manuscript in preparation); (4) forensic studies have shown that the root of a 16-year old mulberry tree reached depths of 1 meter and fine roots penetrated PAH-contaminated sludge [100].

The combined interpretation of experimental results pertaining to the plant stimulation of rhizosphere microflora outlined in the previous paragraph is that the roots of some plant species are capable of growing to immobile soil contaminants (PCBs and high molecular weight PAHs) and delivering cometabolites (i.e. flavonoids) upon fine root death. These natural cometabolites foster the growth and activity of degradative microbes. The dead/decayed roots also create soil cavities that facilitate soil aeration [97]. Thus, in order for roots to foster the degradation of immobile soil contaminants (PCBs and PAHs) it is not necessary for the water insoluble contaminants to move to the root, because fine roots (<1.0 mm in diameter) grow to the contaminants, and upon root death serve as injectors of bacterial cometabolites and facilitators of soil aeration. Based on this mechanistic understanding, the performance of rhizosphere remediation can be improved by increasing both root synthesis of cometabolites and the rate of fine root turnover.

Although there is substantial evidence that plant driven mechanisms exist in the rhizosphere for the degradation of recalcitrant contaminants (PCBs and PAHs), field validation remains an issue because of the slow nature of the factors involved. It is conceivable that because of the low volume of soil occupied by roots (<1%) and the random growth pattern of root systems, statistically significant decreases at randomly sampled field sites may require 15-20 year treatment periods. Since field investigations of this length are not compatible with the desire to use low-cost rhizosphere remediation in the near future, an alternative is forensic examination of contaminated areas that have had volunteer vegetation growing for a period of years. Forensic studies have been performed in PAH contaminated areas [101, 102], but have not previously been performed in PCB contaminated sites. PCBs are ideally suited for forensic investigations since a comparison of the ratio of congeners present provides a signature indicative of biological metabolism and/or degradation as demonstrated by General Electric in river sediment studies [86, 92]. The presence of chlorine may also be a useful means of following partially degraded PCBs into the immobile humic material of terrestrial soil. Use of these PCB features are a means of identifying plants that effectively drive PCB degradation processes.

It is conceivable that all three plant-driven rhizosphere mechanisms (consult paragraph 2) may operate in the rhizosphere of a single plant species However, it is equally possible that only two, one, or none operate in the rhizosphere of some plant species. The interaction of these mechanisms and their potential seasonal fluctuations may be such that forensic data will provide evidence for PCB degradation that will not appear under laboratory conditions. These possibilities provide incentive to use forensic studies as a starting point to screen plant species for their potential to drive rhizosphere remediation processes with laboratory investigations held in reserve to gain an understanding of the organisms and mechanisms that have already been effectively operating in the field. Instead of developing laboratory technology that may not work in the field, start with the identification of what has already worked in the field (forensic data) and perhaps improve the system through understanding gained in the laboratory.

A collaborative forensic study is underway at two PCB-contaminated sites in the Czech Republic where various plant species have been growing for up to 40 years in PCB-contaminated soil. A total of 51 different plant species representing 23 families have been identified growing naturally in soil contaminated with up to 470 ppm (mg/kg) PCBs [96]. Figure 3 indicates the locations of the tree species and PCB concentrations of the soil at two contaminated sites. Preliminary microbial results show that plant species significantly affects the quantity of PCB-degrading bacteria in the root zone soil and on the rhizoplane. These results illustrate how field screening methods can be useful for identifying promising plant species for rhizosphere remediation of PCBs. Such field studies may be followed by laboratory studies to better understand the mechanisms involved and improve rhizoremediation performance.

4.5. CURRENT STATE OF BIOREMEDIATION IN THE CZECH REPUBLIC

The research into bioremediation as well as the application of bioremediation methods in full-scale operations have a long time history in the Czech Republic and more recently former Czechoslovakia. The research into bioremediation and biodegradation is undertaken at universities and research institutes (Table 10), applied research is firstly carried out in companies engaged in bioremediation business.

Bioremediation is defined in many ways [103, 104, 105], but all definitions say, that it is a use of living organisms to destroy hazardous compounds in the environment or in hazardous 2 wastes, in fact. This definition means that in the term „bioremediation“ all living organisms can be included. Despite the definition of bioremediation processes studied and used in the Czech Republic employed bacteria, white-rot fungi and plants only. Even if phytoremediation should be included in bioremediation methods according to definition we drop this technique in the following discussion because of its very limited significance in the Czech Republic up to now.

TABLE 10. Research Institutes and Universities engaged in bioremediation research in the Czech Republic

Institution	Type of Research
University of Chemical Technology, Prague	Biodegradation and bioremediation of PCBs, PAHs, oil hydrocarbons, halogenated hydrocarbons, phenols, MTBE Ecotoxicity of bioremediated materials
Institute of Microbiology, Czech Academy of Sciences, Prague	Biodegradation and bioremediation using bacteria and white-rot fungi Target pollutants: PCBs, PAHs, dyes, recalcitrant pollutants Ecotoxicity of bioremediated materials
University of Ostrava, Faculty of Natural history, Ostrava	Mutagenicity of bioremediated materials
Masaryk´s University, Faculty of Natural History, Brno	Biological dehalogenation, Biological catalysts for bioremediation
Charle´s University, Faculty of Natural History, Prague	Enzymes for Biodegradation
Institute of Organic Chemistry and Biochemistry, Czech Academy of Sciences, Prague	Phytoremediation of PCBs, PAHs, explosives Bioaccumulation of Heavy Metals
University of Pardubice, Dept. of Theory and Technology of Explosives, Pardubice	Phytoremediation of explosives

4.5.1. Brief Overview of Bioremediation Research in the Czech Republic

Institute of Microbiology. The research at this institute is aimed to possibility to use bacterial strains and white-rot fungi or their combination for bioremediation. The pollutants studied are polychlorinated biphenyls (PCBs), polyaromatic hydrocarbons (PAHs), azodyes and some other recalcitrant pesticides, herbicides etc.

Research of bacterial degradation of PAHs utilises strains isolated from contaminated sites and model compounds such as benzo(a)pyrene and tries to enhance biodegradation rate and efficiency of the process by optimisation of environmental conditions and by bioaugmentation using specialised bacteria mostly isolated from the contaminated site [106]. PCB degradation is studied using genetically modified bacteria in the Laboratory of Physiology of Microorganisms.

Laboratory of Experimental Mycology carries out very intensive research of bioremediation using composting, white-rot fungi, bacteria and two-step process employing white-rot fungi and bacteria. Pollutants of interest include PAHs, azodyes, PCBs and some other recalcitrant pollutants. For wastewater treatment containing azodyes a packed-bed bioreactor study are performed. The two-step process employing white-rot fungi in the first step as a pretreatment and bacterial treatment as a polishing second step gave very encouraging results.

Institute of Chemical Technology. At the Department of Biochemistry and Microbiology the research on bioremediation of PAHs in suspension system is proceeding using various bacteria (*Pseudomonas putida* 8368, *Sphingomonas* sp.) and white-rot fungi (*Irpex lacteus, Pleurotus ostreatus, Phanerochaete chrysosporium*). As model compounds are used various

PAHs and their mixture (phenathrene, fluoranthene, anthracene) [107]. Another research is aimed to bacterial transformation of PCBs and study of ecotoxicity and mutagenity of biotransfromation products.

At the Department of Fermentation and Bioengineering the research of bioremediation is aimed to study biodegradation pathways of acetone, acetonitril, and phenols using bacteria *Comamonas acidovorans, Rhodococcus equi* and fungi *Candida maltosa* and *Fusarium proliferatum* sorbed to solid particles. The emphasis are put to formation of active biofilm [108].

The Department of Water and Environmental Technology studies very intensely the biodegradability of various compounds and correlates the biodegradability with possible biological treatment [109].

Masaryks University. The intensive research into biodegradation of halogenated hydrocarbons and enzymatic systems involved in this process is carried out in the Laboratory of Structure and Dynamics of Biomolecules. Haloalkane dehydrogenases are the enzymes of interest because of their broad range of substrate specifity [110, 111, 112]. The main aim of the research is modification of enzymes to increase their reaction rates using mutations [113].

4.5.2. Bioremediation in the field

History. Pollution by organic compounds is huge in the Czech Republic. It is a consequence of intensive industrial activities without regard to the environment. The other source of pollution are military bases operated from 1968 to 1991 by Soviet Army.

Soviet troops used more than 80 military bases placed in the Czech Republic. Most of these bases were heavily polluted with oil hydrocarbons (like jet fuel, diesel fuel, gasoline and lubricants), various solvents (dichloro- and trichloroethene, perchloroethylene etc.) and other pollutants. The total amount of contaminated soil has been estimated in the range of 1.5 to 2.0 million tons [114]. The largest pollution was found in Hradcany and Milovice bases (approx. 500,000 to 600,000 tons of contaminated soil at each site). The concentration of oil hydrocarbons in contaminated soil ranges from 1,000 to 20,000 $mg.kg^{-1}$, somewhere it was found as high as 350,000 $mg.kg^{-1}$.

The contamination of groundwater is large as well. For example polluted area of groundwater around the air force base in Milovice is 58 ha. The thickness of layer of jet fuel free product above groundwater level was up to 2.6 m. In another site, Hradcany, polluted area of groundwater is smaller than in Milovice but the thickness of jet fuel free product layer reached 6.0 m [115].

There exist approx. 250 industrial polluted sites in the Czech Republic. Some of these polluted sites experienced pollution more than 100 years. The most abundant pollutants are oil hydrocarbons, PAHs, raw oil, halogenated hydrocarbons, phenols, coal tar, creosote and many more. Heavily polluted industrial areas are in some cases very large. For example in the Northern Moravia industrial agglomeration around Ostrava possesses hundred of hectares of polluted sites. Such huge pollution was neglected a long time by state authorities of communist regime. First cleanup projects started in the mid eighties when drinking water sources in villages in the neighbourhood of Soviet air force bases have been polluted with jet fuel and became useless leaving the whole villages without a drinking water source. In the late eighties the first bioremediation company has emerged on the remediation market in the former Czechoslovakia. After „velvet revolution" new government took responsibility for decontamination of the environment and started funding of remediation projects. The boom for remediation companies has started in the nineties. During two years (1990-1992) emerged six bioremediation companies in the Czech market. Than the number of bioremediation companies steadily increased reaching approx. 25 at the beginning of the year 2000.

In the late eighties and early nineties all bioremediation companies used only one technique. They employed land farming with bio augmentation and they treated excavated soil polluted with oil hydrocarbons only. These technologies were very simple, low efficient, needed large areas for treatment and possess another disadvantages. The general scheme was as follows: excavated soil was placed onto a decontamination plateau in a layer 0.2 to 0.3 m thick, suspension of bacterial preparation was used for inoculation, nutrients were added and during the treatment the soil was turned, ploughed or mixed. The control of the process nearly does not exist. For all that during the years 1990 to 1992 approx. 50,000 tons of soil contaminated with oil hydrocarbons were treated using bioremediation techniques.

Because of large extent of polluted sites and limited funds for remediation the priorities are related to risks arising from a contaminated site and the use of lower input technologies is stressed. Biodegradation techniques belong to lower input technologies and it is cheaper in comparison with other remediation techniques [116, 117]. Such a success of bioremediation, huge market for bioremediation in the heavily polluted Czech Republic, state funding of cleanup projects and nearly absent legislation to control the use and full-scale application of bioremediation technologies have encouraged many companies (having nothing to do with remediation business at all) to start bioremediation business in the Czech Republic during early nineties. The quality of services offered by some companies was very low if any.

In the course of the year 1993 to 1995 emerged some new bioremediation companies with a very good scientific and technological backup. These companies have brought to the market innovative technologies and they have started application of bioremediation *in-situ*. These companies used bioremediation for treatment of various organic pollutants different from oil hydrocarbons. These companies are the contemporary leaders in the bioremediation business in the Czech Republic During these years also a lot of bioremediation companies have disappeared from the market in the Czech Republic.

Current state. There is nearly no regulatory legislation for full-scale application of bioremediation techniques. The only mandatory condition to be fulfilled is the obtaining an approval of hygienic safety of a bioremediation technology issued by the Institute of Public Health. This approval is issued in accordance with the general act of public health protection. The other legislation which could be applied to bioremediation is a regulation for manipulation, treatment and disposal of hazardous waste.

There is no official obligation to test the efficacy of the bioremediation technology before its practical use in the Czech Republic. What is more interesting: the leading investors of remediation projects (The Fund of National Property - the organization which funds the remediation projects of contaminated industrial sites - and the Ministry of the Environment, which funds the remediation projects of former Soviet army military bases) are not interesting in the quality, efficiency and reliability of the technologies which are used for treatment of contaminated sites they fund. This is the reason for many failures in bioremediation application in the Czech Republic in the late eighties and early nineties In some cases there were used very peculiar technologies for bioremediation.

The first bioremediation technology on the market in the former Czechoslovakia was Russian technology which used the bacterial strain *Pseudomonas putida* as a biodegradation enhancing agent. This first bioremediation technology was on the market from the year 1988 till 1997 and it was used to treat excavated soil contaminated with oil hydrocarbons *ex-situ*. This bioremediation technology had a lot of positive results.

In the early nineties an American bioremediation technology was introduced into the Czech Republic. This technology used a „confidential“ biocatalyst which according to company claim, supplied oxygen for biological oxidation of organic pollutants in soil and groundwater, lowered bacterial biomass production, increased overall efficiency of the biodegradation process and enhanced biodegradation rate etc. Because of failure of this bioremediation technology in more applications the scientific evaluation of the process was funded by one unsatisfied client. It came out from the scientific evaluation made by three independent research institutes that the technology is absolutely inefficient and that the „confidential" biocatalyst was a freeze dried dung imported from the United States.

Another American company started its bioremediation activity in the former Czechoslovakia by supplying its bacterial preparation to a Czech company which had no specialist for bioremediation. This fact caused many problems in the very beginning. The lack of qualified personal at this company caused problems from time to time. Sometimes the company had to withdraw from the remediation project because of big problems.

More recently a French company tried to place on the market in the Czech Republic bioremediation technology using white-rot fungi for biodegradation of PAHs as claimed by the company. Because the company claimed every detail of the technology as confidential the pilot tests were ordered by the Institute of Public Health to verify the hygienic safety of the process. During the pilot tests it was found that white-rot fungi, even if inoculated into the treated soil, survived only a few days from seeding without any significant change of PAHs concentration. The results of pilot tests proved that the inoculation of contaminated soil with white-rot fungi according to protocol of French company was inefficient for PAH biodegradation in soil.

A subsidiary of a German bioremediation company introduced its bioremediation technology on the Czech market in the year 1991. This technology was based on a very careful homogenisation of soil to be treated, separation of stones and building debris and amendment of nutrients and selected bacterial strains produced in a bioreactor. Sometimes this technology employed lignocellulose materials to improve soil structure. During treatment the soil was turned and if necessary water was supplied. The results of this technology were very good. The company withdraw from the market in the year 1998.

Czech companies started their activities in bioremediation business in the year 1991, some companies with success, some with failures. It took three to four years to introduce innovative bioremediation technologies, to improve efficiency and to start new application of bioremediation as *in-situ* treatment, soil flushing with biological polishing, enhancement of avaiability using surface active compounds etc.

Current status of bioremediation in the Czech Republic. As mentioned above about 25 companies are engaged in bioremediation business in the Czech Republic these days. Only two companies are specialised in bioremediation which is their main activity. Only these two companies possess their research and development and apply innovative and efficient technologies.

The rest of companies started bioremediation as an inferior business to their main activities in hydrogeology. These companies apply bioremediation technologies very often without well trained staff as a part of their cleanup projects based mainly on physical and physicochemical treatment techniques. It is worth to note that from foreign companies remains in bioremediation business only one French company in the Czech Republic. The following discussion concerns the two Czech companies specialized to bioremediation as their main activity.

Technologies used. The use of on site treatment of excavated soil is applied in various systems with forced aeration, mechanical turning etc. When very low target limits are to be achieved in the final stage a nonionic surfactant is added to enhance bioavailability of pollutant molecules adsorbed to solid particles. This technology was successfully applied to the treatment of soil contaminated with following pollutants: jet fuel, heating oil, diesel oil, phenols, PAHs, butylacetate, perchloroethylene, trichloroethylene, dichloroethylenes and various sludges with high concentration of oil hydrocarbons and raw oil.

Commonly used is bioventing and biosparging in combination with soil vapour extraction and treatment of extracted soil air in biofilters or by adsorption on activated carbon. Many variations of these methods have been employed.

One company uses specially selected bacteria with enhanced production of biosurfactants for *in-situ* treatment of sites contaminated with creosote or coal tar. In this case it was demonstrated that most of the present PAHs is not completely degraded but only biotransformed. Analytical determination showed important decrease of total PAH concentration but following study of toxicity and mutagenicity of treated excavated soil shows an increase in mutagenicity of treated soil in comparison to untreated one [118, 119]. This case shows importance of evaluation of toxicity, ecotoxicity and mutagenicity of bioremediated soil, because chemical analysis alone is not enough for determination of quality of bioremediated soil.

Cooperation between a bioremediation company and a chemical company producing surface active compounds yielded a new composition of nonionic surface active compound which is non toxic for biodegraders in concentration up to 5 % v/v, possesses very low critical micelle concentration and high biodegradability. The new technology for enhancement of bioavailability of sorbed pollutant molecules have been designed and successfully applied. The new surfactant is biodegraded very easily and there is no danger of its accumulation in the environment. Combined soil flushing with biological polishing of freed pollutant molecules and surfactant proved as a powerful mean for treatment of contaminated sites.

Bioremediation was successfully applied for *in-situ* treatment at a site contaminated by various fire extinguishers. These foaming agents were degraded in soil as well as in groundwater by enhancement of activity of autochthonous bacteria and an amendment with nutrients and terminal electron acceptor. The process was laboratory tested before implementation at the site.

Natural attenuation was applied at an industrial site contaminated with chlorinated ethenes. Biological degradation governs natural attenuation at the site. Natural attenuation was used after physical and physicochemical treatment lost their efficiency. After two years of monitoring of natural attenuation it is clear that this process proceeds with rate which ensures protection of recipient surface water [120].

The use of bioremediation in the Czech Republic has a long history. This method has been applied at approx. 50 sites *in-situ*. The estimate of amount of excavated soil treated by bioremediation reached 500,000 tons to 750,000 tons in the Czech Republic [114]. Bioremediation never has been employed as a sole remediation technique. It was combined with physical or physicochemical treatment technologies Even if more recent results of bioremediation were diverse the quality of bioremediation services has improved substantially.

5. Acknowledgement

Presented research was among other grants supported by the grant CLG NATO #SA(EST.GLC.977477)5941, grant No. 104/01/0461 and the grant 526/01/1292 of the Grant Agency of the Czech Republic, by the grant of the Czech Academy of Sciences No. B6127901, David L. Boren Graduate International Fellowships and a grant from the Integrated Petroleum and Environmental Consortium.

6. References

1. Hooper, C.P., Pettigrew and Sayler, G.S. (1990) Ecological fate, effects and prospects for the elimination of environmental polychlorinated biphenyls (PCBs), *Environ. Toxicol. Chem.* **9**, 655-667.
2. Kalantzi, O.I., Alcock, R.E., Johnston, P.A., Santillo, D., Stringer, R.L., Thomas, G.O., and Jones, K.C. (2001) The global distribution of PCBs and organochlorine pesticides in butter, *Environ. Sci. Technol.* **36**, 1013 –1018.
3. Quensen III,J.F., Tiedje,J.M. (1988) Reductive dechlorination of polychlorinated biphenyls by anaerobic microorganisms from sediments, *Science* 242, 752-754.
4. Brown, J.F., Bedard, D.L., Brennan, M.J., Carnahan, J.C., Feng, H., and Wagner, R.E. (1989) Polychlorinated biphenyl dechlorination in aquatic sediments, Science **236**, 709-712.
5. Morris, P.J., Mohn, W.W., Quensen, J.F., Tiedje, J.M. and Boyd, S.A. (1992) Establishment of polychlorinated biphenyl/degrading enrichment culture predominantly meta dechlorination, *Appl. Environm. Microbiol.* **58**, 3088-3094.
6. Niels, L. and Vogel, T.M. (1990) Effects of organic substrates on dechlorination of Aroclor 1242 in anaerobic sediment, *Appl. Environm. Microbiol.* **56**, 2612-2617
7. Dorn E., and Knackmuss, H.J. (1978) Chemical structure and biodegradability of halogenated aromatic compounds. Substituent effects on 1,2-dioxygenation of catechol, *Biochem. J.* **174**, 85-94.
8. Abramowicz, D.A. (1990) Aerobic and anaerobic degradation of PCBs>a review, CRC *Crit Rev. Biotechnol.* **10**, 241-251.
9. Furukawa K., Tonomura, K. and Kamibayashi, A. (1978) Effect of chlorine substitution on biodegradability of polychlorinated biphenyls, *Appl. Environ. Microbiol.* **35**, 223-227.
10. Bedard, D.L., Unterman, R., Bopp, L.H., Brennan, M.J., Haberl, M.L., Johnson, C. (1986) Rapid assay for screening and characterizing microorganisms for the ability to degrade polychlorinated biphenyls, *Appl. Environ. Microbiol.* **51**, 761-768.
11. Furukawa, K. (1994) Molecular genetics and evolutionary relationship of PCB-degrading bacteria, *Biodegradation* **5**, 289-300.
12. Haddock, J.D. and Gibson D. T. (1995) Purification and characterization of the oxygenase component of biphenyl 2,3-dioxygenase from *Pseudomonas* sp. Strain LB400.
13. Seeger, M., Timmis, K.N., Hofer B. Degradation of chlorobiphenyls catalyzed by the bph-encoded biphenyl-2,3 dioxygenase and biphenyl-2,3,-dihydrodiol-2,3-dehydrogenase of *Pseudomonas* sp. LB 400, *FEMS Microbiol. Lett.* **133**, 259-264.
14. Seeger,M., Zielinski, M., Timmis, K.N., and Hofer, B. (1999) Regiospecifity of dioxygenation of di- to pentachlorobiphenyls and their degradation to chlorobenzoates by *bph*-encoded catabolic pathway of *Burkholderia* sp. strain LB 400, *Appl. Environ. Microbiol.* **65**, 3614 – 3621.
15. Massé, R., F. Messier, L. Péloquin, C., Ayotte, M., Sylvestre (1984) Microbial biodegradation of 4-chlorobiphenyl, a model compound of chlorinated biphenyl, *Appl. Environ. Microbiol.* **47**: 947-951.
16. Bokvajová, A., Burkhard, J., Demnerová, K. and Pazlarová, J. (1994) Screening and separation of microorganisms degrading PCBs, *Environ. Health Persp.* **102**, 552-554.

17. Kaštánek, F., Kuncová, G., Demnerová, K., Pazlarová, J., Burkhard, J., Maléterová, Y. (1995) Laboratory and pilot-scale sorption and biodegradation of polychlorinated biphenyls from ground water, *Inter. Biodeterior. Biodegr.*, 287-300.
18. Pazlarová, J., Demnerová, K., Macková, M. and Burkhard J. (1997) Analysis of PCB-degrading bacteria: physiological aspects, *Lett. Appl. Microbiol.* **24,** 334-336.
19. Macková, M., Macek, T., Očenášková, J., Burkhard J., Demnerová, K., Pazlarová, J. (1997) Biodegradation of polychlorinated biphenyls by plant cells, *Int. Biodeterior. Biodegrad.* **39**, 317-325.
20. Kučerová P., Macková M., Chromá L., Burkhard J., Tříska J., Demnerová K. and Macek T. (2000) Metabolism of polychlorinated biphenyls by *Solanum nigrum* hairy root clone SNC-9O and analysis of transformation products, *Plant and Soil* **225**, 109-115.
21. Sondossi, M., Sylvestre M., Ahmad D. (1992) Effects of chlorobenzoate transformation on the *Ps. testosteroni* biphenyl and chlorobiphenyl degradation pathway, *Appl. Environ. Microbiol.* 58, 485-489.
22. Guibeault, B., Sondossi, M., Darakhsan, A., and Sylvestre, M. (1994) Factors affecting the enhancement of PCB degradative ability of soil microbial populations, *Int. Biodeterior. Biodegrad.* **33**, 73-91.
23. Stratford, J., Wright, M.A., Reineke, W., Mokross, H., Havel, J., Knowles, Ch. J., Robinson G. K. (1996) Influence of chlorobenzoates on the utilization of chlorobiphenyls and chlorobenzoate mixtures by chlorobiphenyl/chlorobenzoate - mineralising hybrid bacterial strains, *Arch. Microbiol.* **165**, 213-218.
24. Nováková, H., Pazlarová, J., Demnerová, K., Macková, M. and Burkhard J. (2000) Biphenyl dioxygenase specifity of indigenous bacterial strains, Proceedings of the Fifth International symposium and exhibition on environmental contamination in Central and Eastern Europe p. 102-103 (poster No.145).
25. Amorim, H., V., Dougall D.K., Sharp W.R. (1977) The effect of carbohydrate and nitrogen concentration on phenol synthesi in Paul Scarlet rose cells grown in tissue culture, *Physiol. Plantarum* **39**, 91-95.
26. Donelly P.K., Hedge R.S., Fletcher J.S. (1994) Growth of PCB-degrading bacteria on compounds from photosynthetic plants, *Chemosphere* **28**, 981-988.
27. Gilbert, E.S., Crowley D.E. (1997) Plant compounds that induce PCB biodegradation by *Arthrobacter* sp. strain B1B, *Appl. Environm. Microbiol.* 63, 1933-1938.
28. Pederson, T.F. (1999) The enhancement of PCB degradation by bacterial cultures, *Diploma thesis*, ICTP, Prague 1999.
29. Mc Cullar, M., Brenner, V., Adams, R.H., and D.D. Focht (1994) Construction of a novel polychlorinated biphenyl-degrading bacterium: Utilization of 3,4'-dichlorobiphenyl by the strain *Pseudomonas acidovorans* M3GY. *Appl. Environ. Microbiol.* **60**, 3833-3839.
30. Brenner, V., Arensdorf, J.J., and Focht, D.D. (1994) Genetic construction of PCB degraders, *Biodegradation* **5**, 359-377.
31. Hickey,W.J., D.D.Focht (1990) Degradation of mono-, di-, and trihalogenated benzoic acid by *Pseudomonas aeruginosa* JB2, *Appl. Environ.Microbiol.* **56**, 3842-3850.
32. Adams RH, Huang CM, Higson FK, Brenner V and Focht DD (1992) Construction of a 3-chlorobiphenyl-utilizing recombinant from an intergeneric mating. *Appl Environ Microbiol* **58**, 647-654.
33. Ahmed M and Focht DD (1973) Degradation of polychlorinated biphenyls by two species of *Achromobacter, Can J Microbiol* **19**, 47-52.

34. Arensdorf JJ and Focht DD (1994) Formation of chlorocatechol *meta* cleavage products by a pseudomonad during metabolism of monochlorobiphenyls, *Appl Environ Microbiol* **60,** 2884-2889.

35. Arensdorf JJ and Focht DD (1995) A *meta* cleavage pathway for 4-chlorobenzoate, and intermediate in the metabolism of 4-chlorobiphenyl by *Burkholderia cepacia* P166, *Appl Environ Microbiol* **61**, 443-447.

36. Bartels I, Knackmuss HJ, Reineke W (1984) Suicide inactivation of catechol 2,3-dioxygenase from *Pseudomonas putida* mt-2 by 3-halocatechols, *Appl Environ Microbiol* **47,** 500-505.

37. Bedard DL and Haberl ML (1990) Influence of chlorine substitution pattern on the degradation of polychlorinated biphenyls by eight bacterial strains, *Microbiol Ecol* **20**, 87-102.

38. Blasco R, Mallavarapu M, Wittich RM, Timmis KN, Pieper DH (1997) Evidence that formation of protoanemonin from metabolites of 4-chlorobiphenyl degradation negatively affects the survival of 4-chlorobiphenyl-cometabolizing microorganisms, *Appl Environ Microbiol* **63**, 427-434.

39. Focht DD (1993) Microbial degradation of chlorinated biphenyls, *Soil Biochemistry*, vol 8, Marcel Dekker Inc., New York..

40. Frantz B, Aldrich T and Chakrabarty AM (1987) Microbial degradation of synthetic recalcitrant compounds, *Biotech Adv* **5,** 85-99.

41. Havel J and Reineke W (1991) Total degradation of various chlorobiphenyls by cocultures and in vivo constructed hybrid pseudomonads. *FEMS Microbiol Lett* **78,** 163-170.

42. Hickey WJ, Brenner V and Focht DD (1992) Mineralization of 2-chloro- and 2,5-dichlorobiphenyl by *Pseudomonas* sp. strain UCR2, *FEMS Microbiol Lett* **98,** 175-180.

43. Kaschabek SR, Kasberg T, Müller D, Mars AE, Janssen DB, Reineke W (1998) Degradation of chloroaromatics: purification and characterization of a novel type of chlorocatechol 2,3-dioxygenase of *Pseudomonas putida* GJ31, *J Bacteriol* **180**, 296-302.

44. Keen NT, Tamaki S, Kobayashi D and Trollinger D. (1988) Improved broad-host-range plasmids for DNA cloning in Gram-negative bacteria, *Gene* **70,** 191-197.

45. McCullar MV, Brenner V, Adams RH and Focht DD (1994) Construction of a novel PCB-degrading bacterium: utilization of 3,4'-dichlorobiphenyl by *Pseudomonas acidovorans* M3GY, *Appl Environ Microbiol* **60**, 3833-3839.

46. Mokross H, Schmidt E and Reineke W (1990) Degradation of 3-chlorobiphenyl by in vivo constructed hybrid pseudomonads, *FEMS Microbiol Lett* **71,** 179-186.

47. Pavlů L, Vosáhlová J, Klierová H, Prouza M, Demnerová K, Brenner V (1999) Characterization of chlorobenzoate degraders isolated from polychlorinated biphenyl-contaminated soil and sediment in the Czech Republic, *J Appl Microbiol* **87,** 381-386.

48. Reineke W (1998) Degradation of chloroaromatics: purification and characterization of a novel type of chlorocatechol 2,3-dioxygenase of *Pseudomonas putida* GJ31, *J Bacteriol* **180**, 296-302.

49. Reineke W (1998) Development of hybrid strains for the mineralization of chloroaromatics by patchwork assembly, *Annu Rev Microbiol* **52,** 287-331.

50. Reineke W and Knackmuss HJ (1988) Microbial degradation of haloaromatics, *Annu Rev Microbiol* **42**, 263-287.

51. Sayler GS and Fox R (1990) Environmental Biotechnology and waste treatment, Plenum Press, New York.

52. Sambrook J, Fritsch EF and Maniatis T (1989) Molecular cloning: a laboratory manual, Cold Spring Harbor Laboratory, NY.

53. van der Meer JR, van Neerven ARW, de Vries EJ, de Vos WM and Zehnder AJB (1991) Cloning and characterization of plasmid-encoded genes for the degradation of 1,2-dichloro-, 1,4-dichloro-, and 1,2,4-trichlorobenzoate of *Pseudomonas* sp. strain P51, *J Bacteriol* **173,** 6-15.

54. Focht, D.D. (1995) Strategies for the improvement of aerobic metabolism of polychlorinated biphenyls, *Curr. Opinion Biotechnol* **6**, 341-346.

55. Harkness, M.R., McDermott, J.B., *et al.* (1993) In situ stimulation of aerobic PCB biodegradation in Hudson river sediments, *Science* **259**, 503-507.

56. Flanagan, W.P. and May R.J. (1993) Metabolite detection as evidence for naturally occurring aerobic PCB biodegradation in Hudson River sediments, *Environ. Sci. Technol.* **27**, 2207-2212.

57. Abramowicz, D.A., Brown, J.F.Jr., Harkness, M.R. and O'Donnell, M.K. (1996) *Biotechnology in Industrial waste treatment and bioremediation*, CRC Press, Inc. Lewis Publishers, Boca Raton, Florida.

58. Bedard, D.L. and Quensen, J.F. (1995) *Microbial Transformation and degradation of toxic organic chemicals*, Wiley-Liss. USA.

59. Robinson, G.K., and Lenn, M.J. (1994) The bioremediation of polychlorinated biphenyls (PCBs): problems and perspectives, *Biotechnol. Gen. Engineer. Rev.* **12**, 139-188.

60. Fiebig, R., Schulze, D., Erlemann, P., Slawinski, M., and Dellweg, H. (1993) Microbial degradation of polychlorinated biphenyls in contaminated soil, *Biotechnol. Lett.* **15**, 93-98.

61. Barriault, D., and Sylvestre, M. (1993) Factors affecting PCB biodegradation by implanted bacterial strain in soil microcosms, *Can. J. Microbiol.* **39**, 594-602.

62. Fava, F., Di Gioia, D. and Marchetti, L. (1998) Cyclodextrin effects on the *ex-situ* bioremediation of a chronically polychlorobiphenyl-contaminated soil, *Biotechnol. Bioeng.* **58**, 345-355.

63. Fava, F. and Di Gioia, D. (1998) Effects of Triton X-100 and Quillaya Saponin on the *ex-situ* bioremediation of a chronically polychlorobiphenyl-contaminated soil, *Appl. Microbiol. Biotechnol.* **50**, 623-630.

64. Anid, P.J., Nies, L. and Vogel, T.M. (1991) *On-site Bioreclamation*, Butterworth-Heinemann, MA, USA.

65. Evans, B.S., Dudley, C.A., and Klasson, K.T. (1996) Sequential anaerobic-aerobic biodegradation of PCBs in soil slurry microcosms, *Appl. Biochem. Biotechnol.,* **57/58**, 885-894.

66. Kastanek, F., Demnerova K., Pazlarova J., Burkhard, J., and Maleterova, Y. (1999) Biodegradation of polychlorinated biphenyls and volatile chlorinated hydrocarbons in contaminated soils and ground water in field conditions, *International Biodeterioration and Biodegradation* **44**, 39-47.

67. Robinson, G.K. (1998) (Bio)remediation of polychlorinated biphenyls (PCBs) : problems, perspectives and solutions, *Biochem. Soc. Trans.*, **26**, 686-690.
68. Providenti, M.A., Lee, H., and Trevors, J.T. (1993) Selected factors limiting the microbial degradation of recalcitrant compounds, *J. Ind. Microbiol.* **12**, 379-395.
69. Fava, F. and Bertin, L. (1999) Use of exogenous specialised bacteria in the biological detoxification of a dump-site polychlorobiphenyl-contaminated soil in slurry-phase conditions, *Biotechnol. Bioeng.*, **64**, 240-249.
70. Fava, F., Di Gioia, D., and Marchetti, L. (2000) Role of the reactor configuration in the biological detoxification of a dump site-polychlorobiphenyl-contaminated soil in lab slurry phase conditions, *Appl. Microbiol. Biotechnol.*, **53**, 243-248.
71. Verstraete, W. and Devliegher, W. (1996) Formation of non-bioavailable organic residues in soil: perspectives for site remediation. *Biodegradation* **7**, 471-485
72. Volkering, F., Breure, A.M., and Rulkens, W.H. (1998) Microbiological aspects of surfactant use for biological soil remediation, *Biodegradation* **8**, 401-417.
73. Rouse, J.D., Sabatini, D.A., Suflita, J.M. and Harwell, J.H. (1994) Influence of surfactant on microbial degradation of organic compounds, *Crit. Rev. Environ. Sci. Technol.*, **24**,325-370.
74. Lajoie, C.A., Layton, A.C., Easter J.P., Menn, F.-M. and Sayler, G.S. (1997) Degradation of nonionic surfactants and polychlorinated biphenyls by recombinant field application vectors, *J. Ind. Microbiol. Biotechnol.***19**,252-262.
75. Singer A.C., Gilbert, E.S., Luepromchai, E. and Crowley, D.E. (2000) Bioremediation of polychlorinated biphenyl-contaminated soil using carvone and surfactant-grown bacteria, *Appl. Microbiol. Biotechnol.* **54**, 838-843.
76. Billingsley, K.A., Backus, S.M. and Ward, O.P. (1999) Effect of surfactant solubilization on biodegradation of polychlorinated biphenyl congeners by *Pseudomonas* LB400. *Appl. Microbiol. Biotechnol.* **52**, 255-260.
77. Fiebig, R., Schulze, D., Chung, J.C. and Lee, S.T. (1997) Biodegradation of polychlorinated biphenyls (PCBs) in the presence of a bio-emulsifier produced on sunflower oil, *Biodegradation* **8**, 67-75.
78. Fava, F. and Di Gioia, D. (2001) Soya lecithin effects on the aerobic biodegradation of polychlorinated biphenyls in an artificially-contaminated soil. *Biotechnol. Bioeng.* **72**,177-184.
79. Saenger, W. (1980) Cyclodextrin inclusion compounds in research and industry. *Angew. Chem. Int. Ed. Engl.* **19**, 344-362.
80. Frömming, K.-H. and Szejtli, J. (1994), in Frömming, K.-H. and Szejtli, J. (Eds.), *Cyclodextrins in pharmacy*, Kluwer Academic Publishers, The Netherlands.
81. Fava, F. and Grassi, F. (1996) Cyclodextrins enhance the aerobic degradation and dechlorination of low-chlorinated biphenyls, *Biotechnol. Tech.* **10**, 291-296.
82. Soeder, C.J., Papaderos, A., Kleespies, M., Kneifel, H., Haegel, H., and Weeb, L. (1996) Influence of phytogenic surfactants (quillaya saponin and soya lecithin) on bio-elimination of phenanthrene and fluoranthene by three bacteria. *Appl. Microbiol. Biotechnol.* **44**, 654-659.
83. Fava, F. (1996) The presence of glass beads or Triton X-100 in the medium enhances the aerobic dechlorination of Aroclor 1221 in *Pseudomonas* sp. CPE1 culture. *Chemosphere* **32**, 1469-1475.
84. Viney, I., and Bewley, R.J. (1990) Preliminary studies on the development of a microbial treatment for polychlorinated biphenyls, *Arch. Environ. Contam. Toxicol.* **19**, 789-796.
85. Cookson, J.T. (1995) in Cookson, J.T (Ed.), *Bioremediation Engineering, Design and Application*, McGraw-Hill Inc., NY.

86. Bedard, D.L., R.E. Wagner, M.J. Brennan, M.L. Haberl and J.F. Brown, Jr. (1987) Extensive degradation of aroclors and environmentally transformed polychlorinated biphenyls by *Alcaligenes eutrophus* H850, *Appl. Environ. Microbiol.* 53: 1094-1102.
87. Butler, J.M., Groeger A.W. and Fletcher J.S (1992) Characterization of polychlorinated biphenyl products formed by plant cells, *Bull. Environ. Contam. Toxicol.* **49**: 821-826.
88. Fletcher, J.S., A.W. Groeger and James C. McFarlane (1987) Metabolism of 2-chlorobiphenyl by suspension cultures of Paul's Scarlet rose, *Bull. Environ. Contam. Toxicol.* **39**, 960-965.
89. Fletcher, J.S. and R.S. Hegde. (1995), Release of phenols by perennial plant roots and their potential importance in bioremediation. *Chemosphere* **31**, 3009-3016.
90. Fletcher, J.S., P.K. Donnelly and R.S. Hegde. (1995) Biostimulation of PCB-degrading bacteria by compounds released from plant roots. In Bioremediation of Recalcitrant Organics. Battelle Press, Columbus Ohio. pp.131-136.
91. Groeger, A.G. and J.S. Fletcher (1988) The influence of chlorine content on the accumulation and metabolism of polychlorinated biphenyls (PCBs) by Paul's Scarlet Rose cells, *Plant Cell Rep* **7**, 329-332.
92. Harkness, M.R., J.B. McDermott, D.A. Abramowicz, J.J. Salvo, W.P. Flanagan, M.L. Stephens, F.J. Mondello, R.J. May, J.H. Lobos, K.M. Carroll, M.J. Brennan, A.A. Bracco, K.M. Fish, G.L. Warner, P.R. Wilson, D.K. Dietrich, D.T. Lin, C.B. Morgan and W.L. Gately (1992) In situ stimulation of aerobic PCB biodegradation in Hudson River sediments, *Science* **259**, 503-507.
93. Hegde, R.S and J.S. Fletcher. (1996). Influence of plant growth stage and season on the release of root phenolics by mulberry as related to development of phytoremediation technology, *Chemosphere* **32,** 2471-2479.
94. Kucerova P (2001) Study of the metabolism of selected xenobiotics by plant cells. Dissertation Thesis, ICT Prague.
95. Leigh, M.B. (1997) The release of phenolic compounds into the soil by fine root turnover of mulberry (*Morus rubra* L.). MS Thesis, University of Oklahoma, Norman, OK.
96. Leigh, M.B., J.S. Fletcher, D.P. Nagle, M. Mackova and T. Macek. (2001). *Phytoremediation and Wetlands for Remediation of Contaminated Areas.* Battelle Press., Columbus, Ohio.
97. Lutz, Harold J. and Robert F. Chandler, Jr. (1947) *Forest Soils.* John Wiley and Sons, New York.
98. Macková, M..,T. Macek , L. Chromá, P. Kučerová, J. Burkhard, J. Tříska, N. Vrchotová, H. Nováková and K. Demnerová (2000) Bacterial PCB degradation induced by natural compounds of plant origin, *Chemické Listy* **8**, 725.
99. H. Nováková, J. Pazlarová, K. Demnerová, M. Macková a J. Burkhard (2002). Degradation of PCBs by indigenous bacterial strains, *Acta Biotechnologica* in press.
100. Olson, P.E. and J.S. Fletcher. (1999). Field evaluation of a mulberry tree growing on an industrial waste site with reference to its potential role in phytoremediation. *Bioremed Journal* **3**, 27-33.
101. Olson, P.E. and J.S. Fletcher (2000) Ecological recovery of vegetation at a former PAH-contaminated industrial sludge basin and its implications to phytoremediation and ecological risk assessment, *Environ Sci Poll Res* **7**, 95-204.
102. Olson, P.E. and J.S. Fletcher. 2000. Natural attenuation/phytoremediation in the vadose zone of a former industrial sludge basin. Environmental Science and Pollution Research. http://dx.doi.org/10.1065/espr2000.

103. Alexander, M. (1994) *Bioremediation and Biodegradation,* Academic Press, San Diego, CA.

104. Dasappa, S.M., Loehr, R.C. (1991) Toxicity reduction in contaminated soil remediation processes, *Wt Res* **25**, 1121-1130.

105. Gibson, D.T., Saylor, G.S. (1992) *Scientific Foundation of Bioremediation: Current Status and Future Needs,* American Academy of Microbiology, Washington, D.C.

106. Pavlu, L., Stuchlikova, K., Brenner, V. (2000) Biodegradation capacity of autochtonous microflora at a PAH-contaminated site in O. Halouskova (ed.) *Proceedings of the Symposium Bioremediation IV*, March 8-9, Sec, Vodni Zdroje Ekomonitor, ISBN 80-7080-374-6, pp. 8.

107. Zrotalova, K., Cajthaml, T., Demnerova, K., Sasek, V. (2000) Possibilities for biodegradation of polyaromatic hydrocarbons in biological suspension system in O. Halouskova (ed.) *Proceedings of the Symposium Bioremediation IV*, March 8-9, Sec, Vodni Zdroje Ekomonitor, ISBN 80-7080-374-6, p. 144.

108. Siglova, M.,. Masak, J., Cejkova, A., Jirku, V., Kotrba, D. (2000) Stimulation of adhesion of single cell organisms capable to degrade toxic compounds in O. Halouskova (ed.), *Proceedings of the Symposium Bioremediation IV*, March 8-9, Sec, Vodni Zdroje Ekomonitor, ISBN 80-7080-374-6, p. 137.

109. Sykora, V., Pitter, P. (2000) Biodegradability of ethyleneamine derivatives in O. Halouskova (ed.) *Proceedings of the Symposium Bioremediation IV*, March 8-9, Sec, Vodni Zdroje Ekomonitor, ISBN 80-7080-374-6, p. 144.

110. Jasenska, A., Sedladek, I., Damborsky, J. (2000) Dehalogenation of haloalkanes by *Mycobacterium tuberculosis* H37Rv and other mycobacteria, *Appl Environ Microbiol* **66**, 219-222.

111. Damborsky, J., Kuty, M., Nemec, M., Koca, J. (1997) A molecular modelling study of the catalytic mechanism of haloalkane dehalogenase: I. quantum chemical study of the first reaction step, *J Chem Inf Comp Sci* **37**, 562-568.

112. Damborsky, J., Bohac, M., Prokop, M., Kuty, M., Koca, J. (1998) Computational site-directed mutagenesis of haloalkane dehalogenase in position 172, *Prot.Eng.* **11**, 901-907.

113. Damborsky, J., Jesenska, J., Hynkova, K., Kmunicek, J., Bohac, M., Kuta-Smatanova, I.,Marek, J., Koca, J. (2000) Screening and construction of biocatalysts in O. Halouskova (ed.) *Proceedings of the Symposium Bioremediation IV*, March 8-9, Sec, Vodni Zdroje Ekomonitor, ISBN 80-7080-374-6, p. 14-18.

114. Ruzicka, J. (1993) Decontamination of former Soviet military bases in the Czech Republic, *Planeta* **1**(8), 23-24.

115. Svoma, J. (1993) Decontamination of Soviet military bases, *Planeta*, **1**(9), 25-28.

116. Thacker, B.K., Ford, C.G. (1999): In situ bioremediation technique for sites underlain by silt and clay, *J.Environ.Engin.,* ASCE **125**(12), 1169-1172.

117. Robertiella, A., Lucchese, G., Leo, C. di, Boni, R., Carren, P. (1994) In situ bioremediation of a gasoline and diesel fuel contaminated site with integrated laboratory simulation experiments in Hinchee, R.E., Alleman, B.C., Hoeppel, R.E., Miller, R.N. (eds.), *Hydrocarbon Bioremediation*, Lewis Publishers, Boca Raton, CA, pp.133-139.

118. Malachova, K. (1999) Using short-term mutagenicity tests for the evaluation of genotoxicity of contaminated soils, *J.Soil Contam.* **8**(6), 667-680.

119. Malachova, K., Lednicka, D., Dobias, L. (1998) Bacterial assays of mutagenicity in estimating the effectivness of biological decontamination of soils, *Biologia* (Bratislava) **53**(6), 699-704.

120. Mateju, V., Kyclt, R. (2000): Natural biological attenuation of chlorinated hydrocarbons, Screening and construction of biocatalysts in O. Halouskova (ed.) *Proceedings of the Symposium Bioremediation IV*, March 8-9, Sec, Vodni Zdroje Ekomonitor, ISBN 80-7080-374-6, p. 65-68.